Frontiers in Mathematics

Advisory Editorial Board

Leonid A. Kurdachenko
Javier Otal
Igor Ya. Subbotin

Artinian
Modules
over Group
Rings

Birkhäuser Verlag
Basel · Boston · Berlin

Authors:

Leonid A. Kurdachenko
Department of Algebra
School of Mathematics and Mechanics
National University of Dnepropetrovsk
Vul. Naukova 13
Dnepropetrovsk 49050
Ukraine
e-mail: lkurdachenko@hotmail.com

Igor Ya. Subbotin
Department of Mathematics and Natural Sciences
National University
Los Angeles Campus
5245 Pacific Concourse Drive
Los Angeles, CA 90045
USA
e-mail: isubboti@nu.edu

Javier Otal
Departamento de Matemáticas
Universidad de Zaragoza
Pedro Cerbuna 12
50009 Zaragoza
Spain
e-mail: otal@unizar.es

2000 Mathematical Subject Classification 16P20, 16P70, 20E15, 20E45, 20F14, 20F16, 20F18, 20F19, 20F22, 20F24, 20F50

Library of Congress Control Number: 2006934215

Bibliographic information published by Die Deutsche Bibliothek
Die Deutsche Bibliothek lists this publication in the Deutsche Nationalbibliografie;
detailed bibliographic data is available in the Internet at <http://dnb.ddb.de>.

ISBN 978-3-7643-7764-9 Birkhäuser Verlag, Basel – Boston – Berlin

© 2007 Birkhäuser Verlag, P.O. Box 133, CH-4010 Basel, Switzerland
Part of Springer Science+Business Media
Cover design: Birgit Blohmann, Zürich, Switzerland
Printed on acid-free paper produced from chlorine-free pulp. TCF ∞
Printed in Germany
ISBN-10: 3-7643-7764-X   e-ISBN-10: 3-7643-7765-8
ISBN-13: 978-3-7643-7764-9   e-ISBN-13: 978-3-7643-7765-6

9 8 7 6 5 4 3 2 1   www.birkhauser.ch

To Tamara, Marisol, and Milla

Contents

Preface   ix

1 Modules with chain conditions   1

2 Ranks of groups   13

3 Some generalized nilpotent groups   25

4 Artinian modules and the socle   37

5 Reduction to subgroups of finite index   45

6 Modules over Dedekind domains   53

7 The Kovacs–Newman theorem   63

8 Hartley's classes of modules   81

9 The injectivity of some simple modules   93

10 Direct decompositions in artinian modules   111

11 On the countability of artinian modules over FC-hypercentral groups   131

12 Artinian modules over periodic abelian groups   153

13 Nearly injective modules   161

14 Artinian modules over abelian groups of finite section rank   167

15 The injective envelopes of simple modules over group rings   179

16 Quasifinite modules   189

17 Some applications: splitting over the locally nilpotent residual   211

Bibliography   227

Index   245

Preface

Let G be a group and suppose that G has an abelian normal subgroup A. If $H = G/A$, then H acts on A by $ah = a^g$, where $h = gA \in H$ and $a \in A$, and this action transforms A into a $\mathbb{Z}H$-module (see all details below). If A is periodic, then very often we may replace A by one of its primary p–components. This allows us to assume that A is a p-subgroup, where p is a prime. This way we arrive at a p-module over the ring $\mathbb{Z}H$. In this case, the structure of the lower layer

$$P_1 = \Omega_1(A) = \{a \in A \mid pa = 0\}$$

of A has a significant influence on the structure of A. Since P_1 is an elementary abelian p-subgroup, we may think of P_1 as a module over the ring $\mathbb{F}_p H$, where $\mathbb{F}_p$ is a prime field of order p.

The above approach allows one to employ module and ring-theoretical methods for the characterization of the groups considered. This relatively old idea has shown itself to be very effective in the theory of finite groups. Progress in the study of finite groups naturally led to the implementation of this approach in infinite groups that are closely related to finite groups, specifically, in infinite groups with some finiteness conditions. It is well known that in the theory of rings many significant results are related to finiteness conditions, especially the classical conditions of minimality and maximality. Thus, both artinian and noetherian rings are the main subjects of the largest and the richest branches of the theories of commutative and non-commutative rings. The minimality and the maximality conditions were introduced in groups side by side and played a crucial role in the development of the theory of infinite groups. The study of groups with the maximal condition on all subgroups (the Max condition) led to the fundamental theory of polycyclic groups and applications to other areas (see, [255]). On the other hand, the exploration of groups with the minimal condition on all subgroups (the Min condition) was extremely fruitful and generated among many other the research associated with the well-known problems stated by O.Yu. Schmidt and S.N. Chernikov (see [43, 311]).

After very detailed investigations, many group theorists conjectured that groups with the Max condition should be polycyclic-by-finite and the groups with the Min condition should be Chernikov. Refuting this, A.Yu. Ol'shanskii surprisingly developed a series of his famous *monsters* ([208, Chapter 28]), emphasizing the inexhaustible wealth of Infinite Group Theory.

These examples also solved other famous problems in Group Theory. Other infinite, finitely generated, residually finite, periodic groups having many unusual properties have been constructed by R. I. Grigorchuk (see, for example the survey [93]). Observe that for his first example [92], R.I. Grigorchuk produced a clear and relatively simple construction which compares favorably with examples that existed before. These brilliant creations of A.Yu. Olshanskii and R.I. Grigorchuk show that methods and approaches that are traditional for the theory of generalized soluble groups do not work effectively beyond this theory. They developed a clear understanding that the theory of generalized soluble groups is just a proper specific part of the general group theory. Along with other branches such as finite groups, abelian groups, and linear groups, the theory of generalized soluble groups has its own specific subject of research, rich history and sophisticated methodology.

Together with the ordinary maximal and minimal conditions on all subgroups, the maximal and minimal conditions on normal subgroups (Max-n and Min-n) began to be studied. It turns out that in locally nilpotent groups these latter conditions coincided with the respective ordinary conditions. However, even soluble groups with Max-n and Min-n required new approaches. The classical papers of P. Hall [95, 96, 98] played a major role in the implementation of both module and ring-theoretical methods in the study of soluble groups. The investigation of abelian-by-nilpotent groups that satisfy the maximal condition on normal subgroups led P. Hall to the consideration of noetherian modules over a ring of the form $\mathbb{Z}H$, where H is a finitely generated nilpotent group. The basis connection here between groups, rings and modules is established by the remarkable theorem due to P. Hall: If R is a noetherian ring and G is a polycyclic-by-finite group, then the group ring RG is likewise noetherian. This result stimulated further development of the theory of group rings of polycyclic- by-finite groups as well as the theory of modules over polycyclic-by-finite groups (see, for example, C.J. Brookes [28], K.A. Brown [30], S. Donkin [57, 58], D.R. Farkas [68], K.W. Gruenberg [94], A.V. Jategaonkar [118, 120] , I.N. Musson [195, 196], D. Passman [216, 217, 218], J.E. Roseblade [248, 249, 250, 251, 252], R.L. Snider [267, 268], and many others).

Exploration of the dual condition — that is, investigation of groups with the minimal condition on normal subgroups — began significantly later. The first investigations showed that there exist metabelian groups satisfying the minimal condition on normal subgroups that are not Chernikov. Investigation of such groups determined the necessity of the study of artinian modules over rings of type $\mathbb{Z}H$, where H is an abelian Chernikov group. The excellent paper by B. Hartley and D. McDougall [110] contains the description of such modules and groups. This paper was also a starting point for the investigation of artinian modules over group rings. It is worth noting here that the situation with artinian modules is rather different. Actually, the group ring of a group that is not polycyclic-by-finite loses the valuable property of being noetherian. Therefore, we have no well-developed deep theory of the respective group rings, and we need to build one. At first glance, the criteria of complementability of modules, which arise from the classical theorems of Maschke and Fitting, seem as likely candidates for this. A result of L.G. Kovacs and M.F. Newman [137] is one of the first important generalizations of Maschke's Theorem on infinite groups. Since then, the following steps of the theory were correlated with the criteria of semisimplicity of artinian modules, which are related to conditions of injectivity of simple modules. Investigation of the questions mentioned above developed approaches to the description of some artinian modules over a ring of the form FG, where F is a field. Note that the transition from a scalar field to a ring significantly complicates the problem because there are not too many situations in which one can get a good description of artinian modules over group rings. Sometimes it is only possible to obtain the description of their injective envelopes, and just in the cases in which they are also artinian modules. This way we arrive at the following important problem: to find conditions under which an injective envelope of an artinian module is likewise artinian.

At present the theory of modules over group rings is a very well developed algebraic theory that is rich in many important results and has its own goals and themes of a different nature. Many famous algebraists have made their contributions to this theory. The main aim of this book is to highlight some important results within the framework of the described circle of issues outlined above. Because of the voluntarily limited scope of this book, we were unable to include all valuable accomplishments. We focused our study on artinian modules because noetherian modules are presented well enough elsewhere. The last chapter is dedicated to some group theoretical results about the splitting of a group over its locally nilpotent residual. Such theorems about splitting of a group over its abelian generalized nilpotent radical are very useful in many different investigations. In particular, we found them to be very effective in the study of just non-$\mathfrak{X}$-groups (see L. A. Kurdachenko, J. Otal and I.Ya. Subbotin [157]). Among these results we chose a general theorem proven by D.J.S. Robinson [246]. In [246] the author's proof is based on homological methods. In the current book we develop a new proof having applied only results pertaining groups and artinian modules. This aptly illustrates the effectiveness of the artinian condition.

We want to note that originally many important results have been obtained for integral group rings. However, quite often in the applications, one needs to deal with other group rings RG. For example, the cases in which $R = F\langle x \rangle$ is the group ring of an infinite cyclic group over a (finite) field F, or $R = \mathbb{Z}_{p^\infty}$ is the ring of integer p-adics, or $R = F[[X]]$ is the ring of the power series over a (finite) field F frequently occur. Therefore, in the majority of situations, we consider artinian modules over a group ring DG, where the ring D of scalars is a Dedekind domain. This requires the insertion of some valuable results of the theory of modules over Dedekind domains.

Of course, the choice of content has also been determined by the interests and tastes of the authors. Our selections have been influenced by the work of many people, and the authors especially owe their gratitude to B. Hartley and D.I. Zaitsev. The contribution of B. Hartley to the development of the theory of artinian modules over group rings is difficult to overestimate. His well-known papers in this and other areas, as well as the work of his numerous students and collaborators around the world, have been incredibly influential. Many of his important contributions are mentioned in this book, but many others are not. For example, we have omitted the detailed construction of the fundamental important counterexamples of uncountable artinian modules over certain nilpotent and soluble groups. D.I. Zaitsev's interest in the implementation of ring and module-theoretical results to groups, and his outstanding achievements in this area, inspired a series of works dedicated to the development of his productive methods (see the survey L.S. Kazarin and L.A. Kurdachenko [132]). In many cases, his influence determined the content of this book. For the first author, D.I. Zaitsev was a mentor and a good friend. Unfortunately, the Chernobyl Nuclear Catastrophe undermined his health, and D.I. Zaitsev passed away in 1990 at the age of forty eight. This catastrophe was also an original cause of death of our friends and colleagues V.E. Goretsky, S.S. Levishenko and V.V. Pylaev, who were at the top of their careers when passing away due to different medical complications brought on by this disaster. The stress experienced during the Chernobyl Catastrophe was the main reason for the heart attack that took away the life of one of the main founders of Infinite Group Theory and the head of the Kiev Group Theory School, S.N. Chernikov. This great mathematician was a teacher for many Ukranian algebraists, including the authors of this book. His influence on the development of Infinite Group Theory could not be overestimated.

We, with genuine gratitude, remember another brilliant algebraist, whose extremely various scientific interests and personality made a great impact on the forming of the first and the third authors, namely Z.I. Borevich. He was a professor of St Petersburg University, but he was born in Ukraine and was interested in the development of Ukrainian algebra and greatly supported many Ukrainian mathematicians, who have became leading researches nowadays.

We would like to express our great appreciation to the official research committees of Spain (CICYT) and Aragón (CONSI+D) for the financing of this project. We also are extremly grateful to the National University (California, USA), and to the University of Zaragoza (Spain) for their thorough support of the authors' work.

Chapter 1

Modules with chain conditions

Finiteness conditions in Algebra play a key role in the evolution from finite to infinite objects. Historically, these conditions have been introduced in rings and modules where they have shown their effectiveness. At the beginning, the finiteness conditions have been most fruitfully used as the minimal and maximal conditions. They shaped the development of modern Algebra, and still have a significant value today. The maximal condition under the form of *the ascending chain condition* was initially used by R. Dedekind [51], who introduced this subject in his study of ideals in algebraic numbers fields. On the other hand, *the descending chain condition* was introduced by J.M. Wedderburn [279], and explicitly defined by W. Krull, E. Noether and E. Artin. Their applications have been decisive in studying maximal and minimal conditions in groups, modules, Lie algebras, and other algebraic structures. The investigation of different algebraic systems satisfying maximal and minimal conditions still remains very active.

An ordered set $\mathfrak{M} = (\mathfrak{M}, \leq)$ is said to satisfy *the maximal condition* if every non-empty subset of $\mathfrak{M}$ has a maximal element. It is rather easy to see that $\mathfrak{M}$ satisfies the maximal condition if and only if $\mathfrak{M}$ satisfies *the ascending chain condition*, that is, given an ascending chain

$$a_1 \leq a_2 \leq \cdots \leq a_n \leq \cdots$$

of elements of $\mathfrak{M}$, there is some $k \in \mathbb{N}$ such that $a_k = a_{k+n}$ for any $n \in \mathbb{N}$.

Dually, $\mathfrak{M}$ satisfies *the minimal condition* if every non-empty subset of $\mathfrak{M}$ has a minimal element, which is equivalent to $\mathfrak{M}$ satisfies *the descending chain condition*, that is, given a descending chain

$$a_1 \geq a_2 \geq \cdots \geq a_n \geq \cdots$$

of elements of $\mathfrak{M}$, there is some $k \in \mathbb{N}$ such that $a_k = a_{k+n}$ for any $n \in \mathbb{N}$.

Throughout, unless specified otherwise, module means *right module*. We recall that any assertion about right modules has a similar assertion for left modules.

For a ring R and an R-module A, we consider the lattice $\mathcal{L}_R(A)$ consisting of all R-submodules of A ordered by inclusion. We say that A is *a noetherian R-module* or *it satisfies the maximal condition on its submodules* if $\mathcal{L}_R(A)$ satisfies the maximal condition, and similarly A is *an artinian R-module* or A satisfies *the minimal condition on its submodules* if $\mathcal{L}_R(A)$ satisfies the minimal condition. As we mentioned above, this is equivalent to the members of $\mathcal{L}_R(A)$ satisfying the ascending chain condition (and the descending chain condition respectively).

A ring R is called *right noetherian* (in short, *R satisfies Max-r*) if the right R-module R is noetherian, and R is called *right artinian* (*R satisfies Min-r*) if the right R-module R is artinian (note, that a ring means a ring with identity). Similarly, we define *left noetherian* and *left artinian* modules and the conditions Max-l or Min-l. Since in this book we consider modules rather than rings, it is worth noting that we shall use the terms noetherian and artinian rings to denote right noetherian and right artinian rings. Modules with chain conditions are standard topics of many books, such as F. Anderson and K. Fuller [2], M. Atiyah and I.G. MacDonald [3], N. Bourbaki [15, 16, 19], A.W. Chatters and C.R. Hajarmavis [41], C. Faith [66, 67], J.S. Golan and T. Head [85], B. Hartley and T.O. Hawkes [109], T. Hungerford [114], A. Kertesz [136], A.G. Kurosh [177], T.Y. Lam [178], S. Lang [179], J.C. MacConnel and J.C. Robson [190], D.G. Northcott [207], D.S. Passman [219], and R. Sharp [263], for example. For the reader's convenience, we will review some elementary results on them, which we will freely use in what follows.

Every field and the ring $\mathbb{Z}$ of all integers are examples of noetherian rings. By Hilbert's Basis Theorem the ring of polynomials $R[X]$ over a commutative noetherian ring is likewise noetherian. As a consequence we obtain that a commutative ring R such that R is finitely generated over its noetherian subring S with the same identity is noetherian. Moreover, the ring of formal power series $R[[X]]$ over a commutative noetherian ring is likewise noetherian. There are distinct generalizations of Hilbert's Basis Theorem on non-commutative noetherian rings. For example: *let R be a ring and S be a subring of R with the same identity. If $U(R)$ (the group of all invertible elements of R) includes a polycyclic-by-finite subgroup G such that $S = S^G$ and G generates R over S, then R is noetherian* [217, Theorem 10.2.7]. The famous Hall's theorem follows from this: *the group ring RG over a noetherian ring R is also noetherian* [95]. By Hopkins theorem (see, for example, [136, Corollary 59.3]) every artinian ring is noetherian. It follows that every artinian ring has a finite composition series of (right) ideals. In particular, a direct sum of finitely many division rings is artinian; a matrix ring over a division ring is artinian.

Let R be a ring, and let A be an R-module. If B is a subset of A, by definition *the R-annihilator of B* is the left ideal of R

$$\operatorname{Ann}_R(B) = \{x \in R \mid ax = 0 \text{ for all } a \in B\}.$$

In particular, the annihilator of A is the two-sided ideal

$$\operatorname{Ann}_R(A) = \bigcap_{a \in A} \operatorname{Ann}_R(a).$$

Lemma 1.1. *Let R be a ring, and A be an R-module.*

(1) *If A is noetherian and $B \leq A$, then B and A/B are noetherian.*

(2) *If B is a submodule of A such that B and A/B are noetherian, then A is noetherian.*

(3) *If A has a finite series of submodules*

$$\langle 0 \rangle = A_0 \leq A_1 \leq \cdots \leq A_n = A,$$

every factor of which is noetherian, then A is noetherian.

(4) *If $A = A_1 + \cdots + A_n$ and $A_1, \ldots, A_n$ are noetherian, then A is noetherian.*

(5) *A is noetherian if and only if every submodule of A is finitely generated.*

(6) *If R is a noetherian ring, then any finitely generated R-module is noetherian.*

Proof. (1) is obvious and (3) is an immediate consequence of (2).

(2) Let

$$A_1 \leq A_2 \leq \cdots \leq A_n \leq \cdots$$

be an ascending chain of submodules of A. Since B and A/B are noetherian, there is some $m \in \mathbb{N}$ such that $A_m \cap B = A_{m+n} \cap B$ and $A_m + B = A_{m+n} + B$ for all $n \in \mathbb{N}$. Therefore

$$A_{m+n} = A_{m+n} \cap (A_{m+n} + B) = A_{m+n} \cap (A_m + B)$$
$$= A_m + (A_{m+n} \cap B) = A_m + (A_m \cap B) = A_m$$

for all $n \in \mathbb{N}$, and then the result follows.

(4) We proceed by induction on n. Assume that $B = A_1 + \cdots + A_{n-1}$ is noetherian. Then by (1) $A/B = (A_n + B)/B \cong A_n/(A_n \cap B)$ is also noetherian. By (2), A is noetherian.

(5) Suppose that A is noetherian, and let B be a non-zero submodule of A. Pick $0 \neq b_1 \in B$, and put $B_1 = b_1 R$. If $B_1 \neq B$, then pick $0 \neq b_2 \in B \setminus B_1$, and put $B_2 = b_1 R + b_2 R$. Since $b_1 \notin B_1$ we have that $B_2 \neq B_1$. Proceeding in this way, we construct an ascending chain of finitely generated submodules of A,

$$B_1 \leq B_2 \leq \cdots \leq B_n \leq \cdots,$$

such that every B_n is finitely generated. By hypothesis, there is some $k \in \mathbb{N}$ such that $B_k = B_{k+m}$ for all $m \in \mathbb{N}$. It follows that $B = B_k$, and thus B is finitely generated.

Conversely, let

$$A_1 \le A_2 \le \cdots \le A_n \le \cdots$$

be an ascending chain of submodules. If $B = \bigcup_{n \in \mathbb{N}} A_n$, then there exist $b_1, \ldots, b_m \in A$ such that $B = b_1 R + \cdots + b_m R$. Choose $r \in \mathbb{N}$ such that $b_1, \ldots, b_m \in A_r$. We have that $B = A_r$, and then $A_r = A_n$ for all $n \ge r$ so that A is noetherian.

(6) Let $A = a_1 R + \cdots + a_t R$. Given $1 \le i \le n$, since $a_i R \cong R/\mathrm{Ann}_R(a_i)$, each $a_i R$ is noetherian by (1). By (4), A is also noetherian. $\qquad\square$

The corresponding result for artinian modules is similar.

Lemma 1.2. *Let R be a ring and A an R-module.*

(1) *If A is artinian and B is a submodule of A, then B and A/B are artinian.*

(2) *If B is a submodule of A such that B and A/B are artinian, then so is A.*

(3) *If A has a finite series of submodules*

$$\langle 0 \rangle = A_0 \le A_1 \le \cdots \le A_n = A,$$

every factor of which is artinian, then A is likewise artinian.

(4) *If $A = A_1 + \cdots + A_n$, where A_i is artinian for $1 \le i \le n$, then A is artinian.*

Now we focus our study on two topics. The first one is the establishment of the existence and the uniqueness of direct decompositions of certain modules. To do so, we consider endomorphisms of modules with chain conditions.

Lemma 1.3. *Let R be a ring and A an R-module. If α is an endomorphism of A and $n \ge 1$, we put $E_n = \mathrm{Im}\,\alpha^n$ and $R_n = \mathrm{Ker}\,\alpha^n$. Then:*

(1) *If $E_1 = E_2$, then $R_1 + E_1 = A$.*

(2) *If $R_1 = R_2$, then $R_1 \cap E_1 = \langle 0 \rangle$.*

(3) *If A is artinian, then there is some $n \in \mathbb{N}$ such that $A = R_n + E_n$.*

Proof. (1) If $E_1 = E_2$, then for every $a \in A$ there is some $b \in A$ such that $a\alpha = b\alpha^2$. It follows that $a - b\alpha \in \mathrm{Ker}\,\alpha = R_1$, and so $a \in E_1 + R_1$. Consequently, $A = E_1 + R_1$.

(2) Let $R_1 = R_2$. If $a \in R_1 \cap E_1$, then $a\alpha = 0$, and there is some $b \in A$ such that $a = b\alpha$. It follows that $b \in \mathrm{Ker}\,\alpha^2 = R_2 = R_1$, and hence $a = b\alpha = 0$.

(3) If A is artinian, then there is some $m \ge 1$ such that $E_m = E_{2m}$, and therefore $A = E_n + R_n$ by (1). $\qquad\square$

Corollary 1.4. *Let R be a ring and A an R-module, and let α be an endomorphism of A.*

(1) *If A is artinian, then α is an automorphism if and only if $\mathrm{Ker}\,\alpha = \langle 0 \rangle$.*

(2) *If A is noetherian, then α is an automorphism if and only if $\mathrm{Im}\,\alpha = A$.*

Proof. Suppose that A is artinian and $\operatorname{Ker} \alpha = \langle 0 \rangle$. Then α^m is a monomorphism for every $m \in \mathbb{N}$, that is, $\operatorname{Ker} \alpha^m = \langle 0 \rangle$ for $m \in \mathbb{N}$. By Lemma 1.3, there is some $n \in \mathbb{N}$ such that $A = \operatorname{Im} \alpha^n$. Then $A = \operatorname{Im} \alpha$, i.e. α is an automorphism.

Suppose that A is noetherian, and $A = \operatorname{Im} \alpha$. Then $A = \operatorname{Im} \alpha^m$ for every $m \in \mathbb{N}$. By the finiteness of ascending chains, there is some $n \in \mathbb{N}$ such that $\operatorname{Ker} \alpha^n = \operatorname{Ker} \alpha^{n+1}$. Taking reciprocal images, we obtain that

$$(\operatorname{Ker} \alpha)(\alpha^n)^{-1} = \operatorname{Ker} \alpha^{n+1} = \operatorname{Ker} \alpha^n = \{0\}(\alpha^n)^{-1}.$$

Since α^n is an epimorphism,

$$\operatorname{Ker} \alpha = ((\operatorname{Ker} \alpha)(\alpha^n)^{-1})\alpha^n = (\langle 0 \rangle \, (\alpha^n)^{-1})\alpha^n = \langle 0 \rangle \, ,$$

and so α is an automorphism. $\qquad\square$

Proposition 1.5 (Fitting lemma). *Let A be an artinian and noetherian module over a ring R. If α is an endomorphism of A, then $A = B \oplus C$, and the following conditions hold:*

(1) *$B\alpha \leq B$ and $C\alpha \leq C$.*

(2) *The restriction of α on B is an automorphism of B.*

(3) *The restriction of α on C is nilpotent.*

Proof. Since the chains

$$A \geq A\alpha \geq \cdots \geq A\alpha^k \geq \cdots$$

and

$$\langle 0 \rangle \leq \operatorname{Ker} \alpha \leq \cdots \leq \operatorname{Ker} \alpha^k \leq \cdots$$

have to be finite, there exists some $r \geq 1$ such that $A\alpha^r = A\alpha^m$, and $\operatorname{Ker} \alpha^r = \operatorname{Ker} \alpha^m$ for all $m \geq r$. Put $B = A\alpha^r$, and $C = \operatorname{Ker} \alpha^r$. Then

$$B\alpha = (A\alpha^r)\alpha = A\alpha^{r+1} = A\alpha^r = B,$$

$$C\alpha = (\operatorname{Ker} \alpha^r)\alpha = (\operatorname{Ker} \alpha^{r+1})\alpha \leq \operatorname{Ker} \alpha^r = C.$$

Let α_1 (respectively α_2) be the restriction of α on B (respectively on C). By Corollary 1.4, α_1 is an automorphism of B since it is clearly surjective. Moreover

$$C\alpha^r = (\langle 0 \rangle \, (\alpha^r)^{-1})\alpha^r = \langle 0 \rangle \, ,$$

and then α_2 is nilpotent.

If $b \in B \cap C$, then $b\alpha^r = 0$. Since $b\alpha^r = b(\alpha_1)^r$ and α_1 is an automorphism of B, $(\alpha_1)^r$ is also an automorphism. It follows that $b = 0$; that is, $B \cap C = \langle 0 \rangle$. Finally, applying Lemma 1.3 we obtain that

$$A = \operatorname{Im} \alpha^r + \operatorname{Ker} \alpha^r = B + C.$$

This finishes the proof. $\qquad\square$

An R-module A is said to be *indecomposable* if it has no non-trivial direct decompositions; that is, if $A = B \oplus C$, then either $B = \langle 0 \rangle$ or $C = \langle 0 \rangle$. Otherwise A is said to be *decomposable*. As a consequence of Proposition 1.5, we have

Corollary 1.6. *Let A be an indecomposable R-module over a ring R, and let α be an endomorphism of A. If A is noetherian and artinian, then either α is an automorphism of A or α is nilpotent.*

A non-zero R-module A is said to be *simple* if A has no non-zero proper submodules. For an arbitrary R-module A, the intersection of all non-zero submodules of A is called the R-monolith of A, and denoted by $\mu_R(A)$. If $\mu_R(A) \neq \langle 0 \rangle$, then A is said to be *monolithic*. In this case, it is clear that $\mu_R(A)$ is the unique simple R-submodule of A.

Theorem 1.7. *Let A be an artinian R-module, where R is a ring. If*

$$A = A_1 \oplus \cdots \oplus A_n = B_1 \oplus \cdots \oplus B_r$$

are two direct decompositions of A, and all the submodules A_i and B_j are R-monolithic, then $n = r$, and there are a permutation $\sigma \in S_n$, and an automorphism ϕ of A such that $(B_j)\phi = A_{j\sigma}$ for $1 \leq j \leq n$.

Proof. For every $1 \leq i \leq n$ and $1 \leq j \leq r$, let $\pi_i : A \longrightarrow A_i$ and $\rho_j : A \longrightarrow B_j$ be the canonical projections. It will suffice to prove the following assertion:

There are an injective mapping

$$\sigma(t) : \{1, \ldots, t\} \longrightarrow \{1, \ldots, r\},$$

and an automorphism ϕ_t of A such that, if $1 \leq j \leq t$, then $(B_j)\phi_t = A_{j\sigma(t)}$.

We proceed by induction on t. If $t = 1$, then it suffices to take $\phi_1 = \varepsilon_A$, the identity automorphism of A. Suppose that $t > 1$, and we are given an injective mapping $\sigma(t-1)$, and an automorphism ϕ_{t-1}, which satisfy the above conditions. Reordering if necessary, and replacing B_j by an isomorphic image, we may assume that $B_1 = A_1, \ldots, B_{t-1} = A_{t-1}$. Given $a \in A$, we have that $a = a\pi_1 + \cdots + a\pi_n$. It follows that

$$a\rho_t = (a\pi_1)\rho_t + \cdots + (a_n\pi_n)\rho_t$$

and so

$$\rho_t = \pi_1\rho_t + \cdots + \pi_n\rho_t.$$

We note that the restriction of ρ_t on B_t is the identity automorphism of B_t. On the other hand, if the kernel of an endomorphism of a monolithic module is non-zero, this kernel must include the monolith. It follows that the sum of finitely many endomorphisms with non-zero kernels has a non-zero kernel.

Since B_t is monolithic, there is some $1 \leq m \leq n$, such that the kernel of the restriction of $\pi_m\rho_t$ on B_t is zero. By Corollary 1.4, this restriction is in fact an automorphism of B_t. If $1 \leq j \leq t-1$, then $\pi_j = \rho_j$ so that $\pi_j \rho_t - \rho_j \rho_t = 0$. This means that $m \geq t$.

Define

$$j\sigma(t) = \begin{cases} j, & \text{if } 1 \leq j \leq t-1, \\ m, & \text{if } j = t, \end{cases}$$

and

$$\phi_t = e_A - \rho_t + \rho_t \pi_m.$$

If $a \in \text{Ker } \phi_t$, then

$$0 = (a\phi_t)\rho_t = a\rho_t - (a\rho_t)\rho_t + (a\rho_t\pi_m)\rho_t$$
$$= a\rho_t - a\rho_t + (a\rho_t)\pi_m\rho_t = (a\rho_t)\pi_m\rho_t.$$

Since $\text{Ker } \pi_m \rho_t = \langle 0 \rangle$, it follows that $a\rho_t = 0$. Thus

$$a = a\phi_t + a\rho_t - (a\rho_t)\pi_m = a\phi_t = 0$$

and therefore $\text{Ker } \phi_t = \langle 0 \rangle$. Then, by Corollary 1.4, ϕ_t is actually an automorphism. Since

$$A = (B_t)\phi_t \oplus (\bigoplus_{j \neq t} B_j)\phi_t \text{ and } (B_t)\phi_t \leq A_m,$$

we conclude that

$$A_m = (B_t)\phi_t \oplus (A_m \bigcap (\bigoplus_{j \neq t} B_j)\phi_t).$$

But A_m is monolithic, so that, in particular, A_m is indecomposable. Hence $A_m = (B_t)\phi_t$. Taking $t = n$, we obtain the required result. $\square$

Let A be a group with a set of (multiplicative) operators Σ. This means that with each $\sigma \in \Sigma$ there is associated a certain endomorphism of A so that to each $a \in A$ there corresponds $a\sigma \in A$ such that $(ab)\sigma = a\sigma \cdot b\sigma$ for all $a, b \in A$. We say that A *is a Σ-operator group* or A *is a Σ-group*. Note that different operators of Σ can correspond to the same endomorphism, that is, it can be $\sigma \neq \pi$ and $a\sigma = a\pi$ for every $a \in A$. This is the great advantage of considering groups with operators instead of groups with a fixed set of endomorphisms. A subgroup B of A is called Σ-*invariant* if for every $b \in B$ and $\sigma \in \Sigma$, we have $b\sigma \in B$. It follows that we can consider Σ-invariant subgroups of a group A as groups with the same set Σ of operators.

Suppose that A is a normal subgroup of a group G. If $\Sigma = G$, we may consider A as a Σ-operator group defining an action of G on A by $(a, g) \longrightarrow g^{-1}ag = a^g$, $a \in A, g \in G$. In this case, the Σ-invariant subgroups are called G-*invariant*.

If A is an abelian Σ-group and $\Sigma = R$ is a ring, then a typical example here is a module over the ring R. In this case Σ-invariant subgroups are exactly R-submodules.

A Σ-operator group A is said to satisfy *the minimal condition on its Σ-invariant subgroups* (in short Min-Σ) if the ordered-by-inclusion set of its Σ-invariant subgroups satisfies the minimal condition. Dually, A is said to satisfy *the maximal condition on its Σ-invariant subgroups* (in short Max-Σ) if the ordered-by-inclusion set of its Σ-invariant subgroups satisfies the maximal condition.

Suppose that A is a normal subgroup of a group G, $\Sigma = G$ and G acts on A by conjugation. If A satisfies Min-Σ (respectively Max-Σ), then we say that A *satisfies the condition Min-G (respectively Max-G)*. In particular, if $A = G$, then we obtain the minimal condition for normal subgroups (Min-n) and the maximal condition for normal subgroups (Max-n).

The case when $A = G$ and $\Sigma = \{\varepsilon_G\}$ is opposite. If A satisfies Min-Σ (respectively Max-Σ), then in this case G satisfies the minimal condition for all subgroups or the condition Min (respectively, the maximal condition for all subgroups or the condition Max). We also say that a group G is *artinian* (respectively *noetherian*), if G satisfies Min (respectively Max). We will denote by $\mathfrak{M}^{\triangledown}$ the class of all artinian groups and by $\mathfrak{M}^{\triangle}$ the class of all noetherian groups.

As we already noted above, the minimal and the maximal conditions have been introduced by German algebraists. The study of groups satisfying these conditions was initiated later than the corresponding investigations in rings and modules. It is not hard to see that a locally (soluble-by-finite) groups satisfying Max is polycyclic-by-finite. Distinct characteristics of noetherian groups have been obtained by R. Baer [6]. In connection with this R. Baer formulated the following problem:

Is every noetherian group polycyclic-by-finite?

A prominent Soviet algebraist S.N. Chernikov was a pioneer in investigation of artinian groups. He was a student of the famous founders of the Soviet Algebra School A.G. Kurosh and O.Yu. Schmidt. It is worth noting, that after WW I, fruitful and cross influential connections between German and Soviet algebraists were established. Some Soviet researchers visited Germany at that time and many prominent German algebraists, including E. Noether, delivered lectures at Moscow State University. Unfortunately, in a few years the Soviet government prohibited these and any other contacts with foreign scientists.

Let p be a prime and

$$C_{p^\infty} = < a_n \mid a_1^p = 1, a_{n+1}^p = a_n, n \in N > .$$

The group C_{p^∞} is called *a Prüfer p-group*. A group G is called *a Chernikov group* if G includes a normal subgroup of finite index which is a direct product of finitely many Prüfer p-groups. It is not hard to see that every Chernikov group satisfies Min. We will denote by $\mathfrak{C}$ the class of all Chernikov groups and by $\mathfrak{P}$ the class of all polycyclic-by-finite groups. These groups are among the oldest classical subjects in the theory of infinite groups. The properties of Chernikov and polycyclic-by-finite groups have been considered in detail in many books (for example, S.N. Chernikov [43], M.I. Kargapolov and Yu.I. Merzlyakov [129], O.H. Kegel and B.A.F. Wehrfritz [134], A.G. Kurosh [176], D.J.S. Robinson [230, 234, 235, 242], and D. Segal [255]). The following description of locally soluble artinian groups was one of the first results of S.N. Chernikov:

A locally soluble group satisfies Min *if and only if it is Chernikov*

(see, for example [43, Theorem 1.1]). V.P. Shunkov and O.H. Kegel and B.A.F. Wehrfritz independently obtained the following theorem which one can consider as the latest positive result on artinian groups:

Every locally finite group satisfying Min *is Chernikov*

(see, for example [134, Theorem 5.8]). S.N. Chernikov formulated the following problem:

Is every artinian group Chernikov?

Negative answers both to Baer's and Chernikov's questions have been obtained by A.Yu. Olshanskii who constructed exotic examples of torsion-free groups whose proper subgroups are cyclic and periodic groups whose proper subgroups are finite (such groups are called *quasifinite*); in particular, an example of an infinite p-group whose proper subgroups have order p (the *Tarsky Monster*). Furthermore, there exist uncountable artinian groups (see A.Yu. Olshanskii [208, chapters 28, 35, 38]). In connection with this we observe that R.I. Grigorchuk [92, 93] constructed very interesting examples of finitely generated p-groups with highly diverse properties, which are some kind of dual to the quasifinite groups. These groups are *just infinite*; that is, they have only finite proper factor-groups.

Let A be an abelian group. The generalization of the concept of the layer of an abelian p-group leads us to the following definition. If α is an endomorphism of A, for each $n \in \mathbb{N}$, we define

$$\Omega_{\alpha,n}(A) = \{a \in A \mid a\alpha^n = 0\}.$$

Note that $\Omega_{\alpha,n}(A) = \operatorname{Ker} \alpha^n$ is a subgroup of A, and $\Omega_{\alpha,n}(A) \leq \Omega_{\alpha,n+1}(A)$ for any $n \in \mathbb{N}$. In terms of these subgroups, D.I. Zaitsev [306] proved the following criterion, which is very useful in the study of artinian modules.

Proposition 1.8. *Let A be an abelian Γ-group, where Γ is a semigroup of distributive endomorphisms of A. Suppose that we are given an element $\alpha \in \Gamma$ such that*

(i) $\alpha\gamma = \gamma\alpha$ *for every $\gamma \in \Gamma$; and*

(ii) $A = \bigcup_{n \in \mathbb{N}} \Omega_{\alpha,n}(A)$.

Then the group A satisfies Min-Γ if and only if $\Omega_{\alpha,1}(A)$ satisfies Min-Γ.

Proof. Since α satisfies (i), each $\Omega_{\alpha,n}(A)$ is a Γ-invariant subgroup of A. Therefore, if A satisfies Min-Γ, then $\Omega_{\alpha,1}(A)$ satisfies Min-Γ.

Conversely, suppose that $B = \Omega_{\alpha,1}(A)$ satisfies Min-Γ. Let

$$A_1 \geq A_2 \geq \cdots \geq A_n \geq \cdots$$

be a descending chain of Γ-invariant subgroups of A. If $k, m \in \mathbb{N}$, we put

$$B_{k,m} = (A_k \cap \Omega_{\alpha,m+1}(A))\alpha^m.$$

Obviously each $B_{k,m}$ is Γ-invariant, $B_{k,m} \geq B_{k+1,m}$, and $B_{k,m} \geq B_{k,m+1}$ for any pair $k, m \in \mathbb{N}$. It follows that $B_{k,m} \geq B_{s,t}$ whenever $s \geq k$, and $t \geq m$. Since B satisfies Min-Γ, there are some $q, r \in \mathbb{N}$ such that $B_{k,m} = B_{q,r}$ whenever $k \geq q$ and $m \geq r$. The mapping $\phi : a \mapsto a\alpha^t$, $a \in \Omega_{\alpha,t+1}(A)$ is a Γ-endomorphism, and so Im ϕ, and Ker $\phi = \Omega_{\alpha,t}(A)$ are Γ-invariant. Therefore

$$\Omega_{\alpha,t+1}(A)/\Omega_{\alpha,t}(A) \cong \text{Im } \phi \leq \Omega_{\alpha,1}(A).$$

It follows that $\Omega_{\alpha,t+1}(A)/\Omega_{\alpha,t}(A)$ satisfies Min-Γ, and therefore so does $\Omega_{\alpha,n}(A)$, for any $n \in \mathbb{N}$. Therefore, there is some $q(1) \in \mathbb{N}$ such that

$$A_k \cap \Omega_{\alpha,r+1}(A) = A_{q(1)} \cap \Omega_{\alpha,r+1}(A)$$

for every $k \geq q(1)$. Let $q(2)$ be the maximum of $\{q, q(1)\}$. If $k \geq q(2)$ and $n \geq r+1$, then we claim that the following expression holds:

$$(E): \qquad A_k \cap \Omega_{\alpha,n}(A) = A_{q(2)} \cap \Omega_{\alpha,n}(A).$$

To see this, we proceed by induction on n. Suppose that (E) has been shown for some $n \geq r + 1$. Pick $a \in A_{q(2)} \cap \Omega_{\alpha,n+1}(A)$. Then $a\alpha^n \in B_{q(2),n} = B_{k,n}$ so that $a\alpha^n = c\alpha^n$ for some $c \in A_k \cap \Omega_{\alpha,n+1}(A)$. It follows that $(a - c)\alpha^n = 0$; that is,

$$a - c \in A_{q(2)} \cap \Omega_{\alpha,n}(A) = A_k \cap \Omega_{\alpha,n}(A).$$

In particular, $a \in A_k$, that is $a \in A_k \cap \Omega_{\alpha,n+1}(A)$. Consequently,

$$A_k \cap \Omega_{\alpha,n+1}(A) = A_{q(2)} \cap \Omega_{\alpha,n+1}(A),$$

and then our claim follows. Then we have

$$A_k = A_k \cap A = A_k \cap \left(\bigcup_{n \geq r+1} \Omega_{\alpha,n}(A) \right) = \bigcup_{n \geq r+1} (A_k \cap \Omega_{\alpha,n}(A))$$

$$= \bigcup_{n \geq r+1} (A_{q(2)} \cap \Omega_{\alpha,n}(A)) = A_{q(2)} \cap \left(\bigcup_{n \geq r+1} \Omega_{\alpha,n}(A) \right) = A_{q(2)},$$

and so A satisfies Min-Γ. $\qquad\qquad\qquad\qquad\qquad\qquad\qquad\qquad\qquad\square$

Let P be an abelian p-group, where p is a prime. If $n \in \mathbb{N}$, by definition, *the nth-layer of the p-group P* is the characteristic subgroup

$$\Omega_n(P) = \{x \in P \mid |x| \leq p^n\}.$$

We clearly have that

$$\Omega_1(P) \leq \Omega_2(P) \leq \cdots \leq \Omega_n(P) \leq \cdots$$

and $P = \bigcup_{n \in \mathbb{N}} \Omega_n(P)$. The first layer $\Omega_1(P)$ concentrates a lot of information about P, and it is simply known as *the lower layer of the p-group P*. Let ρ be the endomorphism of P given by $\rho : a \to pa$, $a \in P$. Applying Proposition 1.8, we deduce the following elegant characterization due to B. Hartley and D. McDougall [110].

Corollary 1.9. *Let A be an abelian p-group, where p is a prime, and suppose that A is a Γ-group, where Γ is a semigroup of distributive endomorphisms of A. Then A satisfies* Min-Γ *if and only if $\Omega_1(A)$ satisfies* Min-Γ.

Chapter 2

Ranks of groups

In this and successive chapters we consider some extensions of the class of finite groups whose members are groups G related to certain numerical invariants, which are generically known as *ranks of the group* G. The concept of rank has appeared in the theory of modules over arbitrary rings as a natural generalization of the idea of dimension of a vector space over a field. Recall that an abelian group is exactly a module over the ring $\mathbb{Z}$ of integers. Since the theory of abelian groups is the initial part of the theory of groups, it is not surprising that the inherited notion of rank of a module has been successfully employed in many areas of Group Theory. In fact, these ideas have become the foundation for new important extensions of finite groups, most of which were created by A.I. Maltsev [181, 182] and D.J.S. Robinson [230]. Some of such extensions are our next goals. However, we are not going to consider in full the properties of groups of finite rank, which has been done already in a wide variety of different papers. Note that one can find some fundamental results about groups of different finite ranks in the books of M.R. Dixon [52], M.I. Kargapolov and Yu.I. Merzlyakov [129], A.G. Kurosh [176], B.I. Plotkin [221], D.J.S. Robinson [230, 234, 235, 242], and others. We are going to focus on some important definitions and results that are suitable for our purposes.

A group G is said to have *finite 0-rank* $r_0(G) = r$ if G has a finite subnormal series with exactly r infinite cyclic factors, the others being periodic. We note that every refinement of one of these series has only r factors that are infinite cyclic. Since any two finite subnormal series have isomorphic refinements, we conclude that 0-rank is independent of the chosen series. This numerical invariant is also known as *the torsion-free rank of* G. In this general form the concept of the 0-rank was introduced by D.I. Zaitsev [303]. Polycyclic-by-finite groups are first examples of groups having finite 0-rank. For these groups 0-rank is exactly its *Hirsch number*.

Lemma 2.1. *Let G be a group of finite 0-rank. Suppose that H is a subgroup of G and L is a normal subgroup of G. Then*
(1) *H has finite 0-rank; concretely, $r_0(H) \leq r_0(G)$; and*
(2) *G/L has finite 0-rank; concretely, $r_0(G) = r_0(L) + r_0(G/L)$.*

Proof. Suppose that $r_0(G) = r$. Then G has a finite subnormal series

$$\langle 1 \rangle = G_0 \trianglelefteq G_1 \trianglelefteq \cdots \trianglelefteq G_n = G$$

such that $r \leq n$, r factors are infinite cyclic and the remaining $n - r$ factors are periodic.

(1) If $H \leq G$, then

$$\langle 1 \rangle = G_0 \cap H \trianglelefteq \cdots \trianglelefteq G_n \cap H = H$$

is a subnormal series of H. Since the factors of this series are isomorphic to subgroups of factors of the given series, we deduce immediately that $r_0(H) \leq r_0(G)$, as claimed.

(2) Given $L \trianglelefteq G$, we have the subnormal series

$$\langle 1 \rangle = G_0 \cap L \trianglelefteq \cdots \trianglelefteq G_n \cap L = L = LG_0 \trianglelefteq LG_1 \trianglelefteq \cdots \trianglelefteq LG_n = G.$$

Since $\{G_j \cap L \mid 0 \leq j \leq n\}$ is a subnormal series of L , and $\{G_j L/L \mid 0 \leq j \leq n\}$ is a subnormal series of G/L with exactly r infinite cyclic factors, the others being periodic, we find out that $r_0(G) = r_0(L) + r_0(G/L)$, as required. $\qquad\square$

If A is an abelian group, then

$$r_0(A) = \dim_{\mathbb{Q}}(A \otimes_{\mathbb{Z}} \mathbb{Q}).$$

In fact, an abelian group A has finite 0-rank r if and only if $A/t(A)$ is isomorphic to a subgroup of the additive group

$$\underbrace{\mathbb{Q} \oplus \cdots \oplus \mathbb{Q}}_{r}$$

(see L. Fuchs [78, Lemma 24.3]), where $t(A)$ is the periodic part of A (that is, the unique maximal periodic subgroup of A). Note also that $r_0(A)$ is exactly the $\mathbb{Z}$-rank of the $\mathbb{Z}$-module $A/t(A)$. Related to abelian groups of finite 0-rank are the soluble groups of finite 0-rank. This class of groups is exactly the class of *soluble A_1-groups*, which was introduced by A.I. Maltsev [182]. The linear group $\mathbf{UT}_n(Q)$ (even $\mathbf{UT}_n(F)$ where F is a finite field extension of $\mathbb{Q}$) is a natural example of a non-abelian group of finite 0-rank.

Let A be an abelian group. If p is a prime, then *the p-rank $r_p(A)$ of A is* defined as follows. Let P be the Sylow p-subgroup of A , and consider $\Omega_1(P)$ as a vector space over the prime field $F_p = \mathbb{Z}/p\mathbb{Z}$. By definition,

$$r_p(A) = \dim_{F_p} \Omega_1(P).$$

We note that $r_p(A)$ is finite if and only if the Sylow p-subgroup P of A is Chernikov (see L. Fuchs [78, Theorem 25.1]).

Following D.J.S. Robinson [230, 6.1], we define the classes $\mathfrak{A}_0$ and $\mathfrak{G}_0$ as follows:

- An abelian group A belongs to the class $\mathfrak{A}_0$ (or A *is an* $\mathfrak{A}_0$-*group*) if and only if $r_0(A)$ is finite and $r_p(A)$ is finite for all primes p.

- A soluble group G belongs to the class $\mathfrak{S}_0$ (or G is *an* $\mathfrak{S}_0$-*group*) if and only if G has a finite subnormal series, every factor of which is an abelian $\mathfrak{A}_0$-group.

A direct product $U \times \mathbf{Dr}_{p \in P} C_p$, where U is a direct product of finitely many copies of $\mathbb{Q}$ and C_p is a direct product of finitely many copies of a Prüfer p-groups for all $p \in \mathbb{P}$ (here $\mathbb{P}$ is the set of all primes), is a natural example of an abelian group from the class $\mathfrak{A}_0$. Moreover, every group from the class $\mathfrak{A}_0$ is isomorphic to some subgroup of a direct product of this kind. We can find some natural examples of groups from class $\mathfrak{S}_0$ among linear groups. If F is a field and p is a prime such that $p \neq \operatorname{char} F$, then every p-subgroup of $\mathbf{GL}_n(F)$ is Chernikov (see, for example [283, 9.1]). It follows that every periodic soluble subgroup S of $\mathbf{GL}_n(F)$ such that char $F \notin \mathbf{\Pi}(S)$ belongs to $\mathfrak{S}_0$.

These classes $\mathfrak{A}_0$ and $\mathfrak{S}_0$ admit the following almost obvious characterizations.

Proposition 2.2. (1) *An abelian group A is an $\mathfrak{A}_0$-group if and only if every elementary abelian p-factor of A is finite, for each prime p.*

(2) *A soluble group G is an $\mathfrak{S}_0$-group if and only if every elementary abelian p-section of G is finite, for each prime p.*

The above result shows that $\mathfrak{S}_0$-groups can be called *groups of finite (abelian) section rank*. In fact Proposition 2.2 provides us the possibility of extending the concept of p-rank from abelian groups to arbitrary groups in the following way. We say that *a group G has finite p-rank $r_p(G) = r$* if every abelian p-section of G is finite of order p^r, and there is an abelian p-section U/V such that $|U/V| = p^r$.

Let A be an abelian group. Following D.J.S. Robinson [230, 6.5], we define *the reduced rank of A* by

$$r_{red}(A) = r_0(A) + \max\{r_p(A) \mid p \in \Pi(A)\}.$$

The abelian group A belongs to the class $\hat{\mathfrak{U}}$ (or A is a $\hat{\mathfrak{U}}$-group) if and only if $r_{red}(A)$ is finite. In other words, there is some $m \in \mathbb{N}$ such that $r_0(A) \leq m$ and $r_p(A) \leq m$, for each $p \in \Pi(A)$. A soluble group G belongs to the class $\hat{\mathfrak{S}}$ (or G is an $\hat{\mathfrak{S}}$-group) if and only if G has a finite subnormal series, every factor of which is an abelian $\hat{\mathfrak{U}}$-group.

A group G is said to have *finite special rank $r(G) = r$* if every finitely generated subgroup of G can be generated by r elements and r is the least positive integer with this property. This notion is due to A.I. Maltsev [181]. The special rank of a group is called also *the Prüfer–Maltsev rank*

To characterize these classes in a simpler way, we need the next almost obvious result.

Lemma 2.3. *Let G be a group. If H is a subgroup of G and L is a normal subgroup of G, then:*

(1) *If G has finite special rank, then H and G/L have finite special rank, and $r(H) \leq r(G)$, $r(G/L) \leq r(G)$.*

(2) *If L and G/L have finite special ranks, then so do have G.*

Corollary 2.4. (1) *An abelian group A is an $\mathfrak{A}^\wedge$-group if and only if A has finite special rank.*

(2) *A soluble group G is an $\mathfrak{S}^\wedge$-group if and only if G has finite special rank.*

This result shows that the class $\mathfrak{S}^\wedge$ is exactly the class of soluble A_2-groups, formerly introduced by A.I. Maltsev [182].

Let A be an abelian group. Following D.J.S. Robinson [230, 6.2], we define *the total rank of A* as

$$r_{tot}(A) = r_0(A) + \sum_{p \in \Pi(A)} r_p(A).$$

- An abelian group A belongs to the class $\mathfrak{A}_1$ (or A is an $\mathfrak{A}_1$-group) if and only if $r_{tot}(A)$ is finite. In particular, this means that $\Pi(A)$ is finite. In other words, $A \in \mathfrak{A}_1$ if and only if A has finite 0-rank , and the periodic part of A is a Chernikov group.

- A soluble group G belongs to $\mathfrak{S}_1$ (or G is an $\mathfrak{S}_1$-group) if and only if G has a finite subnormal series, every factor of which is an abelian $\mathfrak{A}_1$-group.

The class $\mathfrak{S}_1$ is exactly the class of soluble A_3-groups, which was introduced by A.I. Maltsev [182]. Note that Maltsev also considered the subclass of soluble A_4-groups, which consists of the groups having a finite subnormal series, every factor of which is an abelian $\mathfrak{A}_1$-group with a finite periodic part.

A group G is called *minimax* if it has a finite subnormal series, all factors of which belong to $\mathfrak{M}^\triangledown$ or $\mathfrak{M}^\triangle$. The class of soluble minimax groups is denoted by $\mathfrak{S}_2$ (D.J.S. Robinson [230]). In particular, $\mathfrak{S}_2\mathfrak{F}$ is exactly the class of groups having finite subnormal series, every factor of which is Chernikov or polycyclic-by-finite. The following combinatorial characteristic of groups relates to the minimax groups. Following D.I. Zaitsev [295] we say that a group G has *finite minimax rank* $r_{mmx}(G) = m$, if for each finite chain of subgroups

$$\langle 1 \rangle = G_0 \leq G_1 \leq \cdots \leq G_n = G$$

such that all indexes $|G_{j+1} : G_j|$ are infinite, we have $n \leq m$ and there exists a chain of this kind for which $n = m$. If such a number does not exist, then *the minimax rank of the group is infinite*; if G is finite, then $r_{mmx}(G) = 0$. Observe, that in the paper [295], D.I. Zaitsev used another term *the index of minimality*.

This term happened to be not suitable. Therefore Zaitsev stopped using it and introduced instead the *minimax rank*. As it is proved by Zaitsev [291] *a locally (soluble-by-finite) group has a finite minimax rank if and only if it is minimax.*

The theory of soluble-by-finite minimax groups is well developed now. These groups have been studied by many authors from different points of view. General properties of soluble minimax groups have been considered in the paper [229] (see, also [235, 10.3, 10.4]). Minimax groups appear in the study of distinct finiteness conditions (see, for example, the following papers: R. Baer [7], G. Cutolo and L.A. Kurdachenko [50], M. Karbe [126], M. Karbe and L.A. Kurdachenko [127], L.S. Kazarin and L.A. Kurdachenko [132], L.S. Kazarin, L.A. Kurdachenko and I.Ya. Subbotin [133], P.H. Kropholler [138], L.A. Kurdachenko [141, 142, 144, 145, 146, 147, 150], L.A. Kurdachenko and V.E. Goretsky [153], L.A. Kurdachenko and H. Smith [165, 166, 167, 168], L.A. Kurdachenko, A.V. Tushev and D.I. Zaitsev [175], M.L. Newell [202, 203], H. Smith [266], D.I. Zaitsev [289, 290, 291, 292], D.I. Zaitsev, L.A. Kurdachenko and A.V. Tushev [312]). D.J.S. Robinson showed that every finitely generated hyperabelian group of finite special rank (respectively finite section rank) is minimax [228, 233, 238, 243] (see also D.I. Zaitsev [293]). Properties of subnormal and permutable subgroups in minimax groups have been considered by D.C. Brewster and J.C. Lennox [24], D.J. McCaughan [189], D. McDougall [191] and H.H.G. Seneviratne [262]. S. Franciosi S. and F. de Giovanni [72] considered maximal subgroups in soluble minimax groups. In the papers by L.A. Kurdachenko [149] and L.A. Kurdachenko and J. Otal [156] the groups with minimax conjugacy classes have been studied; J. S. Wilson [286, 287] investigated a soluble product of minimax groups; in the articles by M. Karbe and L.A. Kurdachenko [127], D.I. Zaitsev, L.A. Kurdachenko and A.V. Tushev [312], D. Segal [256, 257], D.I. Zaitsev, L.A. Kurdachenko and A.V.Tushev [312] the authors explored modules over minimax groups.

Now we consider relations among different ranks of groups. As usual, if G is a group, by $\Pi(G)$ we denote the set of all primes that occur as divisors of the orders of periodic elements of G.

Lemma 2.5. *Let G be a group and let H be a periodic subgroup of G. If H is ascendant in G, then $H \leq O_\pi(G)$, where $\pi = \Pi(H)$.*

Proof. Let

$$H = H_0 \vartriangleleft H_1 \vartriangleleft \cdots H_\alpha \vartriangleleft H_{\alpha+1} \vartriangleleft \cdots H_\gamma = G$$

be an ascending series. We will prove by transfinite induction on α that $H \leq O_\pi(H_\alpha)$. If $\alpha = 1$, this inclusion is obvious. Suppose that we have already proved that $H \leq O_\pi(H_\beta)$ for all $\beta < \alpha$. If α is a limit ordinal, then $H_\alpha = \bigcup_{\alpha < \beta} H_\beta$ and hence $H \leq \bigcup_{\alpha < \beta} O_\pi(H_\beta) = O_\pi(H_\alpha)$. Suppose now that $\alpha - 1$ exists. Since $H_{\alpha-1}$ is normal in H_α and $O_\pi(H_{\alpha-1})$ is a characteristic subgroup of $H_{\alpha-1}$, $O_\pi(H_{\alpha-1})$ is normal in H_α and therefore $O_\pi(H_{\alpha-1}) \leq O_\pi(H_\alpha)$. Hence $H \leq O_\pi(H_{\alpha-1}) \leq O_\pi(H_\alpha)$. $\square$

If G is a group, then by $t(G)$ we will denote *the largest normal periodic subgroup of* G.

Corollary 2.6. *Let* G *be a group and suppose that* G *includes a non-identity ascending periodic subgroup. Then* $t(G) \neq \langle 1 \rangle$.

Let A be an abelian torsion-free group of finite 0-rank and let T be a periodic automorphism group of A. Then T is finite (see, for example, [280, Theorem 9.33]). Moreover, if $r_0(A) = r$, then there is a function $\mathbf{f}_1 : N \longrightarrow N$ such that $|T| \leq \mathbf{f}_1(r)$.

Lemma 2.7 ([55]). *Let* G *be a group and suppose that* $t(G) = \langle 1 \rangle$. *If* G *has an ascending series* $H = H_0 \lhd H_1 \lhd \cdots \lhd H_\alpha \lhd H_{\alpha+1} \lhd \cdots \lhd H_\gamma = G$ *such that* H *is an abelian torsion-free subgroup of finite 0-rank and the rest of the factors are locally finite, then* G *includes a characteristic abelian torsion-free subgroup* A *such that* $r_0(A) = r_0(H)$, $H \leq A$ *and* $|G/A| \leq \mathbf{f}_1(r_0(H))$.

Proof. We proceed by transfinite induction on α. Let $\alpha = 1$ and put $K = H_1$, $L = C_K(H)$. By the above remark $|K/L| \leq \mathbf{f}_1(r_0(H))$. The subgroup L is a central extension of H by a locally finite group. By the generalized Schur theorem (see, for example, [234, Corollary to Theorem 4.12], the subgroup $[L, L]$ is locally finite. It follows that $t(L)$ is locally finite, and $L/t(L)$ is abelian and torsion-free. Moreover, $r_0(L/t(L)) = r_0(H)$. Since $t(L)$ is a characteristic subgroup of L, $t(L)$ is normal in K. If we suppose that $t(L) \neq \langle 1 \rangle$, then Corollary 2.6 yields that $t(G) \neq \langle 1 \rangle$. This show that L is an abelian and torsion-free subgroup of 0-rank $r_0(H)$. In this case let A be the locally nilpotent radical of K. Then clearly A is an abelian torsion-free subgroup of 0-rank $r_0(H)$, $H \leq A$ and $|K/A| \leq \mathbf{f}_1(r_0(H))$.

Assume that we have already proved that H_α includes a normal subgroup $L\alpha$ such that $|H_\beta/L_\beta| \leq \mathbf{f}_1(r_0(H))$, and L_β is a torsion-free abelian subgroup of 0-rank $r_0(H)$ for all $\beta < \alpha$. Assume first that α is not a limit ordinal. Put $H_\alpha = U, H_{\alpha-1} = V, L_{\alpha-1} = W$. Since $|V/W| = s \leq \mathbf{f}_1(r_0(H))$, $V^s \leq W$. In other words, U includes a normal torsion-free abelian subgroup V^s such that $r_0(V^s) = r_0(H)$ and U/V^s is locally finite. By the above argument the locally nilpotent radical D of U is a torsion-free abelian subgroup of 0-rank $r_0(H)$ such that $|U/D| \leq \mathbf{f}_1(r_0(H))$. Since W is a subnormal abelian subgroup of U, $W \leq D$.

Suppose now that α is a limit ordinal. The above proof shows that we can assume that $L_\beta \leq L_{\beta+1}$ for all $\beta < \alpha$. Put $A = \bigcup_{\beta < \alpha} L_\beta$. Clearly, A is a normal abelian torsion-free subgroup of 0-rank $r_0(H)$ such that H_α/L_α is locally finite. But this case has been considered above. So the result is proved. $\square$

Corollary 2.8 ([55]). *Let* G *be a group and suppose that* $t(G) = \langle 1 \rangle$. *If* G *has an ascending series*

$$H = H_0 \lhd H_1 \lhd \cdots H_\alpha \lhd H_{\alpha+1} \lhd \cdots H_\gamma = G$$

such that r *factors of this series are infinite cyclic and the rest of the factors are locally finite, then* G *has a finite series*

$$\langle 1 \rangle = K_0 \leq L_1 \leq K_1 \leq L_2 \leq K_2 \leq \cdots \leq L_n \leq K_n = G$$

of normal subgroups such that L_{j+1}/K_j are abelian torsion-free subgroups of finite 0-rank, $0 \leq j \leq n - 1, n \leq r$, and the factors K_j/L_j are finite of order at most $f_1(r)$.

Proof. We proceed by induction on r. Since $t(G) = \langle 1 \rangle$, Lemma 2.5 yields that H_1 is an infinite cyclic subgroup. Since G includes an ascendant abelian subgroup, its locally nilpotent radical L is non-identity. Moreover, $H_1 \leq L$. Note, that the set of all elements of finite order of a locally nilpotent group is a characteristic subgroup. Together with the equation $t(G) = \langle 1 \rangle$ it implies that L is torsion-free. By Maltsev's theorem [182] a subgroup L is nilpotent, in particular, $Z = \zeta(L) \neq \langle 1 \rangle$.

Put $K_1/Z = t(G/Z)$; then K_1/Z is locally finite. By Lemma 2.7, K_1 includes a characteristic abelian torsion-free subgroup L_1 of finite 0-rank such that K_1/L_1 is finite and $|K_1/L_1| \leq f_1(r)$. By the choice of K_1 we have $t(G/K_1) = \langle 1 \rangle$, and G/K_1 has an ascending series, in which the $r - 1$ factors are infinite cyclic and the rest factors are locally finite. Hence we may apply to G/K_1 the inductive hypothesis. □

Let $\mathfrak{M}_n$ be the set of all non-isomorphic finite groups of order at most n and let $\mathfrak{A}_n = \{Aut(G) \mid G \in \mathfrak{M}_n\}$. Then the set $\mathfrak{A}_n$ is finite and each member of $\mathfrak{A}_n$ is finite. Hence there is $D \in \mathfrak{A}_n$ such that $|D|$ is maximal. Put $\mathbf{a}(n) = |D|$. Clearly, $\mathbf{a}(n) \leq n!$.

Corollary 2.9 ([55]). *Let G be a group and suppose that $t(G) = \langle 1 \rangle$. If G has an ascending series*

$$H = H_0 \lhd H_1 \lhd \cdots H_\alpha \lhd H_{\alpha+1} \lhd \cdots H_\gamma = G$$

such that r factors of this series are infinite cyclic and the rest of the factors are locally finite, then G includes a normal soluble subgroup S of finite index. Moreover, $|G/S| \leq a(f_1(r))^r$.

Proof. By Corollary 2.8, G has a finite series

$$\langle 1 \rangle = K_0 \leq L_1 \leq K_1 \leq L_2 \leq K_2 \leq \cdots \leq L_n \leq K_n = G$$

of normal subgroups such that L_{j+1}/K_j are abelian torsion-free subgroups of finite 0-rank, $0 \leq j \leq n - 1$, and the factors K_j/L_j are finite of order at most $f_1(r)$, $1 \leq j \leq n, n \leq r$. Put $C_j = C_G(K_j/L_j)$, $1 \leq j \leq n$. Then G/C_j is finite, moreover, $G_j/C_j \leq \mathbf{a}(f_1(r))$.

Put $S = \bigcap_{1 \leq j \leq n} C_j$. By Remak's theorem $G/S \hookrightarrow \mathbf{Dr}_{1 \leq j \leq n} G/C_j$, in particular, G/S is finite and $|G/S| \leq |G/S_1| \cdots |G/S_n| \leq \mathbf{a}(f_1(r))^n \leq \mathbf{a}(f_1(r))^r$. By its choice, S is soluble. □

If G is a soluble group, then denote by $s(G)$ the class of solubility of G. Let G be a finite soluble group and $|G| = n = p_1^{k_1} \cdots p_m^{k_m}$. Put $d(n) = k_1 + \cdots + k_m$.

Then clearly $s(G) \le d(|G|)$. We have

$$d(n) = k_1 + \cdots + k_m$$
$$= \log_{p_1}(p_1^{k_1}) + \cdots + \log_{p_m}(p_m^{k_m})$$
$$\le \log_2(p_1^{k_1}) + \cdots + \log_2(p_m^{k_m}) = \log_2(p_1^{k_1} \cdots p_m^{k_m}) = \log_2 n.$$

Thus $d(n) \le \log_2 n$.

Let A be an abelian torsion-free group and G be an automorphism group of A. We say that A *is rationally G -irreducible* if for every non-identity G-invariant subgroup B of A the factor-group A/B is periodic.

Lemma 2.10. *Let G be a group and suppose that G includes a finite normal subgroup T such that G/T is an abelian torsion-free group of finite 0-rank r. Then G includes a characteristic abelian torsion-free subgroup A such that G/A is finite. Moreover, if $|T| = t$, then $|G/A| \le a(t)t^{2r+1}$.*

Proof. Put $C = C_G(A)$; then G/C is finite and $|G/C| \le d = \mathbf{a}(t)$. If $g \in C$, then clearly $g^t \le \zeta(C)$, so that $|C/\zeta(C)| \le t^r$. A subgroup $\zeta(C)$ is abelian and has a finite periodic part, so that $\zeta(C) = K \times Z$ for some torsion-free subgroup Z where $K = T \cap \zeta(C)$ (see, for example, [78, Theorem 27.5]). It follows that $A = (\zeta(C))^t \le Z$, in particular, A is torsion-free. Clearly, A is a characteristic subgroup of G. Finally, $|G/A| \le \mathbf{a}(t)t^r tt^r = \mathbf{a}(t)t^{2r+1}$. □

Lemma 2.11 ([55]). *Let G be a group and suppose that $t(G) = \langle 1 \rangle$. Assume also that G has an ascending series*

$$H = H_0 \triangleleft H_1 \triangleleft \cdots H_\alpha \triangleleft H_{\alpha+1} \triangleleft \cdots H_\gamma = G$$

such that r factors of this series are infinite cyclic and the rest of the factors are locally finite. Then G has normal subgroups $L \le K \le S \le G$ such that L is nilpotent and torsion-free, K/L is abelian torsion-free and finitely generated, G/K is finite and S/K is soluble. Moreover, there are functions $f_2, f_3 : N \longrightarrow N$ such that $|G/K| \le f_2(r)$ and $s(S) \le f_3(r)$.

Proof. By Corollary 2.9, G includes a maximal normal soluble subgroup S of finite index such that $|G/S| \le \mathbf{a}(f_1(r))^r$. Let L be the locally nilpotent radical of G. Since $t(G) = \langle 1 \rangle$ and the set of all elements having finite order of a locally nilpotent group is a characteristic subgroup, L is torsion-free. By Maltsev's theorem [182] the subgroup L is nilpotent. It is not hard to prove that L has a finite series

$$\langle 1 \rangle = A_0 \le A_1 \le \cdots \le A_n = L$$

of G-invariant subgroups which are L-central and all the factors A_{j+1}/A_j are abelian torsion-free and rationally G-irreducible, $0 \le j \le n - 1$.

Put $C_j = C_G(A_{j+1}/A_j), 0 \le j \le n - 1$. Then we can consider $G_j = G/C_j$ as an irreducible soluble-by-finite subgroup of $GL_k(\mathbb{Q})$ where $k = r_0(A_{j+1}/A_j)$. Let

S_j be a normal soluble subgroup of G_j having finite index. Then we can consider S_j as a completely irreducible soluble subgroup of $GL_k(\mathbb{Q})$. By a Maltsev theorem [182] S_j is abelian-by-finite. By a result due to V.S. Charin [36], S_j includes a normal free abelian subgroup of finite index. Hence G_j includes a normal free abelian subgroup U_j such that G_j/U_j is finite. Put $D_j = C_{G_j}(U_j)$, then G_j/D_j is finite, moreover $|G_j/D_j| \leq \mathbf{f_1}(r)$. A subgroup D_j is central-by-finite, therefore by a Schur theorem (see, for example, [234, Theorem 4.12]) the set T_j of all its elements having finite order is a finite characteristic subgroup. By the choice of G_j we have $|T_j| \leq \mathbf{f_1}(r)$. By Lemma 1.7, D_j includes a characteristic abelian torsion-free subgroup K_j such that D_j/K_j is finite. Moreover, $|D_j/K_j| \leq \mathbf{a}(t)t^{2r+1}$, where $t = \mathbf{f_1}(r)$. Hence G_j/K_j is finite and $|G_j/K_j| \leq \mathbf{a}(t)t^{2r+2}$.

Put $C = \bigcap_{0 \leq j \leq n-1} C_j$. By Remak's theorem $G/L \leq \mathbf{Dr}_{0 \leq j \leq n-1} G/C_j$, in particular, G/C includes a finitely generated torsion-free abelian subgroup K/C such that G/K is finite. Moreover,

$$|G/K| \leq |G/K_j|^n \leq (\mathbf{a}(t)t^{2r+2})^n \leq (\mathbf{a}(t)t^{2r+2})^r$$
$$= (\mathbf{a}(\mathbf{f_1}(r))^r (\mathbf{f_1}(r))^{(2(r+1)r} = \mathbf{f_2}(r).$$

It is not hard to prove that $C \leq L$. Finally, clearly L is nilpotent of class nilpotency at most $r - 1$. We have now

$$\mathbf{s}(S) \leq \mathbf{s}(K) + \mathbf{s}(S/K) = r - 1 + 1 + \mathbf{s}(S/K) = r + \mathbf{s}(S/K).$$

The factor-group S/K is finite and soluble, hence $\mathbf{s}(S/K) \leq \mathbf{d}(|S/K|)$. From the choice of S it follows that $\mathbf{s}(S/K) \leq \mathbf{d}(\mathbf{a}(\mathbf{f_1}(r))(\mathbf{f_1}(r))^{2(r+1)})$. Now we may put $\mathbf{f_3}(r) = r + \mathbf{d}(\mathbf{a}(\mathbf{f_1}(r))(\mathbf{f_1}(r))^{2(r+1)}) \leq r + \log_2(\mathbf{a}(\mathbf{f_1}(r))(\mathbf{f_1}(r))^{2(r+1)})$. □

Another way of calculating the function $\mathbf{f_3}(r)$ follows from the proof above. By a Maltsev theorem [182] an irreducible soluble subgroup of $GL_r(\mathbb{Q})$ includes a normal abelian subgroup of finite index at most $\mu(r)$ for some function $\mu : N \longrightarrow N$. It follows that $\mathbf{s}(S) \leq r + \mathbf{d}(\mu(r))$. Note that $\mu(r) \leq r!(r^2\mathbf{a}(r^2))^r$ (see, for example, [280, pp. 45]), so that we may put $\mathbf{f_4}(r) = r + \mathbf{d}(r!(r^2\mathbf{a}(r^2))^r) \leq r + \log_2((r!(r^2\mathbf{a}(r^2))^r)$. Thus $\mathbf{s}(S) \leq \mathbf{f_4}(r)$.

Theorem 2.12 ([55]). *Let G be a group and suppose that G has an ascending series*

$$H = H_0 \triangleleft H_1 \triangleleft \cdots H_\alpha \triangleleft H_{\alpha+1} \triangleleft \cdots H_\gamma = G$$

such that r factors of this series are infinite cyclic and the other factors are locally finite. Then G has normal subgroups $T \leq L \leq K \leq S \leq G$ such that T is locally finite, L/T is nilpotent and torsion-free, K/L is abelian torsion-free and finitely generated, G/K is finite and S/K is soluble. Moreover, there are functions $f_2, f_3 : N \longrightarrow N$ such that $|G/K| \leq f_2(r)$ and $s(S/T) \leq f_3(r)$.

Proof. Put $T = t(G)$ and apply Lemma 2.11 to the factor-group G/T. □

A group G is said to be *generalized radical*, if G has an ascending series whose factors are locally nilpotent or locally finite. It is not hard to see that a generalized radical group G has an ascending series of normal subgroups, whose factors are locally nilpotent or locally finite. Note that a periodic generalized radical group is locally finite. Therefore a periodic locally generalized radical group is locally finite too. Thus we have

Corollary 2.13 ([55]). *Let G be a locally generalized radical group. If G has finite 0-rank r, then G has normal subgroups $T \leq L \leq K \leq S \leq G$ such that T is locally finite, L/T is nilpotent and torsion-free, K/L is abelian torsion-free and finitely generated, G/K is finite and S/K is soluble. Moreover, $|G/K| \leq f_2(r)$ and $s(S/T) \leq f_3(r)$.*

Corollary 2.14 ([55]). *Let G be a locally generalized radical group and T be the maximal normal periodic subgroup of G. If G has a finite 0-rank r, then G/T has finite special rank. Moreover, there is a function $f_5 : N \longrightarrow N$ such that $r(G) \leq f_5(r)$.*

Indeed we can put $\mathbf{f}_5(r) = r + \mathbf{f}_2(r)$.

Corollary 2.15 ([55]). *Let G be a locally (soluble-by-finite) group. If G has a finite 0rank r, then G has normal subgroups $T \leq L \leq K \leq S \leq G$ such that T is locally finite, L/T is nilpotent and torsion-free, K/L is abelian torsion-free and finitely generated, G/K is finite and S/K is soluble. Moreover, $|G/K| \leq f_2(r)$ and $s(S/T) \leq f_3(r)$.*

This assertion is an essential specification [73, Lemma 2.12].

Corollary 2.16 (D.J.S. Robinson [245]). *Let G be a locally (soluble-by-finite) group of finite 0-rank. Then $r_0(G) \leq r(G)$.*

Proof. Put $r_0(G) = r$. By Corollary 2.15, G has normal subgroups $T \leq L \leq K \leq S \leq G$ such that T is locally finite, L/T is nilpotent and torsion-free, K/L is abelian torsion-free and finitely generated, G/K is finite and S/K is soluble. By [293, Theorem 1], $r_0(K/T) = r(K/T)$. It follows that $r_0(G/T) \leq r(G/T)$, and because $r_0(G/T) = r_0(G)$ we have $r_0(G) \leq r(G)$. $\square$

We consider now the relationship between the 0-rank, and the p-rank of a group. If A is an abelian group of finite p-rank, then, obviously, A has finite 0-rank, and moreover $r_0(A) \leq r_p(A)$. Now we show that the situation is the same for soluble groups.

Lemma 2.17. *Let G be a nilpotent torsion-free group of finite p-rank for some prime p. Then G has finite 0-rank. Moreover $r_0(G) \leq \frac{1}{2}(r_p(G) + 1)r_p(G)$.*

Proof. The group G has a central series

$$\langle 1 \rangle = Z_0 \leq Z_1 \leq \cdots \leq Z_n = G$$

every factor of which is torsion-free (see D.J.S. Robinson [234, Theorem 2.25]). Since every free abelian group of 0-rank s has an elementary abelian factor-group of order p^s, every factor of this series has finite 0-rank. Therefore, G has finite 0-rank. Put $r_p(G) = k$. Let A be a maximal abelian normal subgroup of G. Then $C_G(A) = A$, and A is a pure subgroup. Since G is nilpotent, $G/C_G(A)$ can be embedded in the unitriangular group $UT_m(\mathbb{Q})$, where $m = r_0(A)$. It follows that

$$r_0(G/C_G(A)) \leq \frac{1}{2}(m-1)m$$

and so

$$r_0(G) \leq m + \frac{1}{2}(m-1)m = \frac{1}{2}(m+1)m \leq \frac{1}{2}(k+1)k,$$

as required. $\qquad\square$

Theorem 2.18. *Let G be a soluble group of finite section p-rank for some prime p. Then G has finite 0-rank. Moreover $r_0(G) \leq \frac{1}{2}(r_p(G)+3)r_p(G)$.*

Proof. If U/V is a torsion-free abelian section of G, then clearly $r_0(U/V) \leq r_p(G)$. It follows that G has finite 0-rank. Then G has a series of normal subgroups $T \leq L \leq G$ satisfying the following conditions: T is periodic, L/T has a finite subnormal series, every factor of which is torsion-free abelian locally cyclic, and G/L is finite. Without loss of generality, we may assume that $T = \langle 1 \rangle$. By V.S. Charin [36, Theorem 5], G has a series of normal subgroups $L \leq S \leq G$ such that L is a torsion-free nilpotent subgroup, S/L is a free abelian group of finite 0-rank, and G/S is finite. By Lemma 2.9,

$$r_0(L) \leq \frac{1}{2}(r_p(G)+1)r_p(G),$$

and hence

$$r_0(G) \leq r_0(L) + r_0(S/L) \leq \frac{1}{2}(r_p(G)+1)r_p(G) + r_p(G) \leq$$
$$\leq \frac{1}{2}(r_p(G)+3)r_p(G).$$

$\qquad\square$

Chapter 3

Some generalized nilpotent groups

In this book we will consider modules over group rings when the groups have certain features, for example, are near to nilpotency. In particular, we will consider generalized nilpotent and generalized hypercentral groups, whose specific properties are the goals of this chapter. Nilpotent and hypercentral groups have arisen from abelian groups with the help of the upper central series. We use a similar construction for the groups that seem to be natural extensions of abelian groups.

We begin with the following generalization of an abelian group. Let G be a group, and let S be a G-invariant subset of G. This means that for every $a \in S$ and $g \in G$, we have $a^g \in S$. It follows that $C_G(S)$ is a normal subgroup of G. The corresponding factor-group $\mathrm{Coc}_G(S) = G/C_G(S)$ is called *the cocentralizer of the G-invariant set S in the group G* (L.A. Kurdachenko [149]). Note that the group $G/\mathrm{Coc}_G(S)$ is isomorphic to some subgroup of $\mathrm{Aut}(\langle S \rangle)$. The influence of the cocentralizers of many objects related to a group on the structure of the group itself is a subject of study in many branches of Group Theory. For instance, in the theory of finite groups, many examples have been developed with the consideration of the cocentralizers of chief factors, which play there a significant role. For example, we can mention Formation Theory since local formations are defined by cocentralizers of the chief factors of the groups involved. In the theory of infinite groups, many types of groups appear in studying cocentralizers of their conjugacy classes. This is our point of origin.

Let $\mathfrak{X}$ be a class of groups. A group G is said to have $\mathfrak{X}$-*conjugacy classes* (or G *is an* $\mathfrak{X}C$-*group*) if $\mathrm{Coc}_G(\langle g \rangle^G) \in \mathfrak{X}$ for each $g \in G$. Thus, if $\mathfrak{X} = \mathfrak{I}$ is the class of all identity groups, then the class of $\mathfrak{I}C$-groups is exactly the class $\mathfrak{A}$ of all abelian groups. Therefore, the class of $\mathfrak{X}C$-groups is a generalization of the class of abelian groups associated to the given class $\mathfrak{X}$. If $\mathfrak{X} = \mathfrak{F}$ is the class of all finite groups, then the $\mathfrak{F}C$-groups are *the groups with finite conjugacy classes* or FC-*groups*. This class is a suitable extension both of $\mathfrak{F}$ and $\mathfrak{A}$, so that it inherits many properties of these classes. Currently, the theory of FC-groups is one of the best-developed branches of the theory of infinite groups, and many authors have made significant contributions to it (see the books Yu.M. Gorchakov [90], S.N. Chernikov [43],

D.J.S. Robinson [234, 242], M.J. Tomkinson [272], and the updated survey paper
M.J. Tomkinson [274]). Also the influence of FC-groups on the structure of some
rings was a widely popular topic (see G. H. Cliff and S.K. Sehgal [44], I.N. Herstein
[113], D.S. Passman [216, 217, 218], C. Polcino-Milies [222, 223], C. Polcino-Milies
and S.K. Sehgal [224], S.K. Sehgal [258, 259], S.K. Sehgal and H.J. Zassenhaus
[260, 261]). If $\mathfrak{X} = \mathfrak{C}$ is the class of all Chernikov groups, then $\mathfrak{C}C$-groups are
the groups with Chernikov conjugacy classes or CC-*groups*. Ya.D. Polovicky [225]
introduced this class and obtained some initial results. Although CC-groups have
not been investigated as far as FC-groups, they are subjects of many recent papers
(see J. Alcázar and J. Otal [1], S. Franciosi, F. de Giovanni and M.J. Tomkinson
[77], M. González and J. Otal [86, 87, 88], M. González, J. Otal and J.M. Peña
[89], J. Otal and J.M. Peña [209, 210, 211, 212, 213], J. Otal, J.M. Peña and M.J.
Tomkinson [214]). The study of $\mathfrak{X}C$-groups for other important classes $\mathfrak{X}$ has only
recently begun (see S. Franciosi, F. de Giovanni and L.A. Kurdachenko [73], S.
Franciosi, F. de Giovanni and M.J. Tomkinson [76], L.A. Kurdachenko [149, 151],
L.A. Kurdachenko, V.N. Polyakov and I.Ya. Subbotin [161], L.A. Kurdachenko
and J. Otal [156], and L.A. Kurdachenko and I.Ya. Subbotin [169]).

The nature of FC-groups allows one to construct distinct types of such groups
using finite and abelian groups. Note that a direct product of FC-groups is also
an FC-group. In particular, any direct product of finite groups and abelian groups
is an FC-group. Moreover, let $G = \mathbf{Cr}_{\lambda \in \Lambda} G_\lambda$ be the Cartesian product. Put

$$\mathbf{Cdr}_{\lambda \in \Lambda} G_\lambda = \{(g_\lambda)_{\lambda \in \Lambda} \mid g_\lambda \in \zeta(G_\lambda) \text{ for all } \lambda \in \Lambda \backslash \Lambda_g \text{ where } \Lambda_g \text{ is finite}\}.$$

The group $\mathbf{Cdr}_{\lambda \in \Lambda} G_\lambda$ is called the *central direct product* of the groups G_λ, $\lambda \in \Lambda$.
Clearly, a central direct product of finite and abelian groups is an FC-group.
Central-by-finite groups and groups with finite derived subgroups (finite-by-abel-
ian groups) deliver other examples of FC-groups. However, not every abelian-by-
finite group is an FC-group (but it is FC-nilpotent). Other classes of $\mathfrak{X}C$-groups
are more diverse. For example, a central direct product of Chernikov groups is a
CC-group, and a central-by-Chernikov group is a CC-group. However, not every
group with a Chernikov derived subgroup is a CC-group. This is true only for
periodic groups. This fact follows from the assertion that a periodic automorphism
group of a Chernikov group is Chernikov (see, for example [43, Corollary 1.18]).

Let G be a group. If $\mathfrak{X}$ is a class of groups, then we put

$$\mathfrak{X}C(G) = \{x \in G \mid \mathrm{Coc}_G(\langle x \rangle^G) \in \mathfrak{X}\}.$$

In general $\mathfrak{X}C(G)$ is not a subgroup of G, although, depending on $\mathfrak{X}$, the set $\mathfrak{X}C(G)$
enjoys richer properties. We select the most common classes of groups to study.
A class of groups $\mathfrak{X}$ is said to be *a formation of groups* if the following conditions
are satisfied:

(F1) If $G \in \mathfrak{X}$, and H is a normal subgroup of G, then $G/H \in \mathfrak{X}$; and

(F2) If H, L are normal subgroups of a group G such that $G/H, G/L \in \mathfrak{X}$, then $G/(H \cap L) \in \mathfrak{X}$.

Actually, we have

Lemma 3.1. *Let $\mathfrak{X}$ be a formation of groups. If G is a group, then $\mathfrak{X}C(G)$ is a characteristic subgroup of G.*

Proof. Since

$$C_G(\langle x^{-1} \rangle^G) = C_G(\langle x \rangle^G)$$

and

$$C_G(\langle xy \rangle^G) \geq C_G(\langle x \rangle^G) \cap C_G(\langle y \rangle^G),$$

for every $x, y \in G$, and $\mathfrak{X}$ is a formation of groups, it readily follows that $\mathfrak{X}C(G)$ is a subgroup of G. If $\phi \in \mathrm{Aut}\, G$, then

$$C_G(\langle x \rangle^G)\phi = C_G(\langle x\phi \rangle^G)$$

and thus

$$G/C_G(\langle x \rangle^G) \cong G/C_G(\langle x\phi \rangle^G)$$

for each $x \in G$. Thus $\mathfrak{X}C(G)$ is a characteristic subgroup of G. $\qquad\square$

If $\mathfrak{X}$ is a formation of groups, then the subgroup $\mathfrak{X}C(G)$ is said to be *the $\mathfrak{X}C$-center of the group G*. The reason for this term appears to be very clear. If $\mathfrak{X} = \mathfrak{I}$, $\mathfrak{F}$ and $\mathfrak{C}$, we obtain the ordinary center $\zeta(G)$, the FC-center $FC(G)$, and the CC-center $CC(G)$ of the group G, respectively. Of course, it is worth noting that the group G is an $\mathfrak{X}C$-group if and only if $G = \mathfrak{X}C(G)$.

Starting from the $\mathfrak{X}C$-center of a group G, we construct *the upper $\mathfrak{X}C$-central series of the group G*,

$$\langle 1 \rangle = C_0 \leq C_1 \leq \cdots \leq C_\alpha \leq C_{\alpha+1} \leq \cdots \leq C_\gamma$$

where $C_1 = \mathfrak{X}C(G)$, $C_{\alpha+1}/C_\alpha = \mathfrak{X}C(G/C_\alpha)$ if $\alpha < \gamma$, and $\mathfrak{X}C(G/C_\gamma) = \langle 1 \rangle$. The term C_α of the series is said to be *the αth-$\mathfrak{X}C$-hypercenter of G*, whereas the last term C_γ is called *the upper $\mathfrak{X}C$-hypercenter of G*, and denoted by $\mathfrak{X}C_\infty(G)$. If $\mathfrak{X}C_\infty(G) = G$, then G is said to be *$\mathfrak{X}C$-hypercentral*. If moreover, γ is finite, then G is called *$\mathfrak{X}C$-nilpotent*. Once again, if $\mathfrak{X} = \mathfrak{I}$, $\mathfrak{X} = \mathfrak{F}$, and $\mathfrak{X} = \mathfrak{C}$, then $\mathfrak{X}C_\infty(G) = \zeta_\infty(G)$, $\mathfrak{X}C_\infty C = FC_\infty(G)$, and $\mathfrak{X}C_\infty(G) = CC_\infty(G)$, which are called *the upper hypercenter, the upper FC-hypercenter*, and *the upper CC-hypercenter* of the group G, respectively.

Lemma 3.2. *Let $\mathfrak{X}$ be a formation of groups closed under taking subgroups (that is, $\mathfrak{X} = \mathbf{S}\mathfrak{X}$). If G is a group, $g_1, \ldots, g_s \in \mathfrak{X}C(G)$, and $H = \langle g_1 \rangle^G \cdots \langle g_s \rangle^G$, then H is a central-by-$\mathfrak{X}$-group.*

Proof. Clearly,

$$C = C_G(\langle g_1 \rangle^G) \cap \cdots \cap C_G(\langle g_s \rangle^G) \cap H \leq \zeta(H).$$

Moreover

$$H/C \hookrightarrow H/(H \cap C_G(\langle g_1 \rangle^G)) \times \cdots \times H/(H \cap C_G(\langle g_s \rangle^G)).$$

Since

$$H/(H \cap C_G(\langle g_j \rangle^G)) \cong HC_G(\langle g_j \rangle^G))/C_G(\langle g_j \rangle^G) \leq G/C_G(\langle g_j \rangle^G)),$$

$1 \leq j \leq s$, we obtain that $H/\zeta(H) \in \mathfrak{X}$, as required. $\qquad\square$

Corollary 3.3. *Let $\mathfrak{X}$ be a formation of locally finite groups closed under taking subgroups, and let G be an $\mathfrak{X}C$-group. Then*

(1) *The commutator subgroup $[G, G]$ is locally finite.*

(2) *If G is torsion-free, then G is abelian.*

(3) *If H is a torsion-free normal subgroup, then $H \leq \zeta(G)$.*

(4) *If $g \in G$ and $L = \langle g \rangle^G$, then either L is locally finite or L includes a G-invariant locally finite subgroup T such that H/T is infinite cyclic.*

Proof. (1) Let $g_1, \ldots, g_s \in G$, and put $H = \langle g_1 \rangle^G \cdots \langle g_s \rangle^G$. By Lemma 3.2, H is a central-by-$\mathfrak{X}$-group. As a consequence of a result due to Schur (see D. J. S. Robinson [234, Corollary of Theorem 4.12]), $[H, H]$ is locally finite. Therefore $[G, G]$ is locally finite.

(2) and (3) are obvious.

(4) By (1), $[L, L]$ is locally finite. It is immediate that the elements of finite order of the abelian group $L/[L, L]$ form a characteristic subgroup $T/[L, L]$. It follows that T is a G-invariant subgroup of L, and either L/T is a non-identity torsion-free abelian or $T = L$. In the first case, we note that $L/T \leq \zeta(G/T)$ by (3). Note that $L/T = \langle gT \rangle^{G/T}$ so that $L/T = \langle gT \rangle$. $\qquad\square$

Corollary 3.4. *Let G be a group, $x_1, \ldots, x_s \in FC(G)$, and $H = \langle g_1 \rangle^G \cdots \langle g_s \rangle^G$. Then:*

(1) *H is a finitely generated subgroup.*

(2) *H is central-by-finite.*

(3) *Either H is finite or H includes a G-invariant finite subgroup T such that H/T is a finitely generated free abelian group.*

Proof. Since each $x_j \in FC(G)$, $G/C_G((x_j)^G)$ is finite, in particular, $|G : C_G(x_j)|$ is finite. It follows that each x_j has finitely many conjugates in G, and then H is finitely generated, which gives (1).

(2) follows from Lemma 3.2.

(3) By a Theorem of Schur (see D. J .S. Robinson [234, Corollary of Theorem 4.12]), the commutator subgroup $[H, H]$ is finite. The group $H/[H, H]$ is finitely generated and abelian, hence its elements of finite order form a characteristic finite subgroup $T/[H, H]$. Then H/T is a torsion-free finitely generated abelian group. □

Our next goal is to obtain some elementary properties of the class of $\mathfrak{X}C$-groups, when $\mathfrak{X}$ is a formation of groups, which is a slight extension of the class $\mathfrak{F}$ of finite groups.

Lemma 3.5. *Let G be a group, and suppose that G/C is an abelian Chernikov group, where $C = \zeta(G)$. Assume that H is a subgroup of G such that $H \geq C$, $H/C = \langle h_1 C \rangle \cdots \langle h_n C \rangle$ and $|h_j C| = p^{s_j}$, where p is a prime and $s_j \in \mathbb{N}$ for $1 \leq j \leq n$. Then $[G, H]$ is a finite p-subgroup of C, and $r_p([G, H]) \leq n r_p(G/C)$.*

Proof. If $1 \leq j \leq n$, we consider the mapping $\phi_j : G \longrightarrow C$ given by $g\phi_j = [g, h_j]$, $g \in G$. We have

$$(g_1 g_2)\phi_j = [g_1 g_2, h_j] = (g_2)^{-1}[g_1, h_j]g_2[g_2, h_j]$$
$$= [g_1, h_j][g_2, h_j] = g_1\phi_j g_2\phi_j$$

because $[g_1, h_j] \in \zeta(G)$. Hence ϕ_j is a homomorphism so that $\mathrm{Im}\ \phi_j = [G, h_j]$ and $\mathrm{Ker}\ \phi_j = C_G(h_j)$ are normal subgroups of G. Furthermore,

$$[g, h_j^2] = [g, h_j](h_j)^{-1}[g, h_j]h_j = [g, h_j]^2$$

and it follows that $[g, h_j^t] = [g, h_j]^t$ for every $t \in \mathbb{N}$. Since $h_j^{p^{s_j}} \in C$, we have $1 = [g, h_j^{p^{s_j}}] = [g, h_j]^{p^{s_j}}$. By the isomorphism

$$[G, h_j] = \mathrm{Im}\ \phi_j \cong G/\mathrm{Ker}\ \phi_j = G/C_G(h_j)$$

and the inclusion $C \leq C_G(h_j)$, we obtain that $[G, h_j]$ is a finite p-subgroup, and $r_p([G, h_j]) \leq r_p(G/C)$.

It is easy to see that

$$[G, H] = [G, h_1] \cdots [G, h_n].$$

It follows that $[G, H]$ is a finite p-subgroup, and $r_p([G, H]) \leq n r_p(G/C)$. □

Corollary 3.6. *Let G be a group, and suppose that G/C is an abelian Chernikov group, where $C = \zeta(G)$. If P/C is the Sylow p-subgroup of G/C, where p is a prime, then $[G, P]$ is a Chernikov p-subgroup of C, and $r_p([G, P]) \leq r_p(G/C)^2$.*

Proof. We have

$$P/C = U_1/C \times \cdots \times U_n/C$$

where $U_1/C, \ldots, U_n/C$ either are Prüfer p-subgroups or cyclic p-subgroups. Therefore, P has an ascending chain of subgroups

$$G = P_0 \le P_1 \le \cdots \le P_s \le \cdots$$

such that P_s/C is a direct product of n cyclic subgroups , and $P = \bigcup_{s \in \mathbb{N}} P_s$. It follows that $[G, P]$ has an ascending chain of subgroups

$$[G, P_1] \le [G, P_2] \le \cdots \le [G, P_s] \le \cdots,$$

such that $[G, P] = \bigcup_{s \in \mathbb{N}} [G, P_s]$. By Lemma 3.5, $[G, P_s]$ is a finite p-subgroup of C, and $r_p([G, P_s]) \le n^2$, for each $s \in \mathbb{N}$. Thus $r_p([G, P]) \le n^2$. This implies that [G,P] is Chernikov (see L. Fuchs [78, Theorem 25.1]). □

Corollary 3.7. *Let G be a group, and suppose that G/C is an abelian Chernikov group, where $C = \zeta(G)$. Then the commutator subgroup $[G, G]$ is Chernikov.*

Proof. Suppose that

$$G/C = \mathbf{Dr}_{p \in \Pi(G/C)} S_p/C$$

where S_p/C is the Sylow p-subgroup of G/C. By Corollary 3.6, each $[G, S_p]$ is a Chernikov p-subgroup. It is easy to see that

$$[G, G] = \mathbf{Dr}_{p \in \Pi(G/C)} [G, S_p].$$

Since $\Pi(G/C)$ is finite, $[G, G]$ is Chernikov. □

Lemma 3.8. *Let G be an abelian-by-finite group, and put $C = \zeta(G)$. If G/C is Chernikov, then $[G, G]$ is Chernikov.*

Proof. Let A be a normal abelian subgroup of finite index, and suppose that $\{x_1, \ldots, x_s\}$ is a transversal to H in G. For every $1 \le j \le s$, we consider the mapping $\phi_j : A \longrightarrow A$ given by $a\phi_j = [g_j, a]$, $a \in A$. Since

$$(a_1 a_2)\phi_j = [g_j, a_1 a_2] = [g_j, a_2] a_2^{-1} [g_j, a_1] a_2$$
$$= [g_j, a_1][g_j, a_2] = a_1 \phi_j a_2 \phi_j,$$

ϕ_j is an endomorphism of A. Therefore

$$[g_j, A] = \mathrm{Im} \, \phi_j \cong A/\mathrm{Ker} \, \phi_j = A/C_A(g_j).$$

Since $C \cap A \le C_A(g_j)$, $A/C_A(g_j)$ is Chernikov , and it follows that for any $1 \le j \le s$ $[g_j, A]$ is Chernikov too. Put

$$B = [g_1, A] \cdots [g_s, A]$$

so that B is likewise Chernikov. If $g \in G$, then there is some $1 \le j \le s$ such that $g \in g_j A$, and so $g = g_j u$, for some $u \in A$. Then

$$[g, a] = [g_j u, a] = u^{-1}[g_j, a] u [u, a] = [g_j, a] \in [g_j, A].$$

This means that $[G, A] = B$. Thus $A/B \leq \zeta(G/B)$, and, in particular, G/B is central-by-finite. By a classical result due to Schur (see D.J.S. Robinson [234, Theorem 4.12]), the commutator subgroup $K/B = [G/B, G/B]$ is finite. Hence K is Chernikov, and G/K is abelian. $\qquad\square$

Theorem 3.9. *Let G be a group. If $G/\zeta(G)$ is a Chernikov group, then the commutator subgroup $[G, G]$ is likewise Chernikov.*

Proof. Put $C = \zeta(G)$, and let D/C be the divisible part of G/C. By Corollary 3.7, $[D, D]$ is Chernikov. Since $G/[D, D]$ is abelian-by-finite, it suffices to apply Lemma 3.8. $\qquad\square$

The above result is a generalization of the already mentioned theorem of Schur. Now we are making use of it to describe the normal closures of elements of the CC-groups.

Lemma 3.10 (L.A. Kurdachenko [144]). *Let G be a hypercentral group. Suppose that A is an abelian normal p-subgroup of G, where p is a prime, such that $G/C_G(A)$ has no subgroups of index p. Then $A \leq \zeta(G)$.*

Proof. Suppose the contrary, that is $\zeta(G)$ contains no A. Then $A\zeta(G)/\zeta(G)$ is a non-identity normal subgroup of the hypercentral group $G/\zeta(G)$, and hence there exists some $a \in (A \cap \zeta_2(G)) \setminus \zeta(G)$. Since A is a p-subgroup, we may assume that $a^p \in \zeta(G)$. Then the mapping $\phi : g \mapsto [g, a]$, $g \in G$ is an endomorphism of G such that $[G, a] = \operatorname{Im} \phi \neq \langle 1 \rangle$ since $a \notin \zeta(G)$. Actually, since $a^p \in \zeta(G)$, $[G, a]^p = \langle 1 \rangle$, and so $\operatorname{Im} \phi$ is an elementary abelian p-subgroup. Since

$$[G, a] = \operatorname{Im} \phi \simeq G/\operatorname{Ker} \phi = G/C_G(a),$$

we deduce that $G/C_G(a)$ is an elementary abelian p-group as well. Since $C_G(A) \leq C_G(a)$, and $G/C_G(A)$ has no subgroups of index p, this is a contradiction. $\qquad\square$

Lemma 3.11. *Let G be a CC-group, and suppose that A is a normal abelian p-subgroup of G, where p is a prime. Then $G/C_G(A)$ has no abelian divisible subgroups.*

Proof. Suppose the contrary, and let $D/C_G(A)$ be a non-identity abelian divisible subgroup. Pick $g \in D \setminus C_G(A)$. Then the mapping $\phi_g : a \mapsto [g, a]$, $a \in A$ is an endomorphism of A. For each $n \in \mathbb{N}$, we put $A_n = \Omega_{p,n}(A)$. Since G is a CC-group, $G/C_G(\langle g \rangle^G)$ is a Chernikov group, and, in particular, $A_n/(A_n \cap C_G(\langle g \rangle^G))$ is Chernikov. This means that $A_n/(A_n \cap C_G(g))$ is likewise Chernikov. Since the orders of the elements of $A_n/(A_n \cap C_G(g))$ are bounded, this factor-group has to be finite. As $D/C_G(A)$ is abelian, ϕ_g is in fact a $\mathbb{Z}D$-endomorphism. Now $D/C_D(A_n/(A_n \cap C_G(g)))$ is finite, and $C_D(A_n/(A_n \cap C_G(g))) \geq C_G(A)$, so, we deduce $D = C_D(A_n/(A_n \cap C_G(g)))$. In other words, $[D, A_n] \leq A_n \cap C_G(g)$. As the above argument holds for any $n \in \mathbb{N}$, we conclude that $[D, A] \leq C_A(g)$. This means that

$$[D, A] \leq \bigcap_{g \in D} C_A(g) = C_A(D),$$

so that $[D, [D, A]] = \langle 1 \rangle$. By Lemma 3.8, $A \leq \zeta(D)$, a contradiction. $\qquad\square$

Corollary 3.12. *Let G be a CC-group, and suppose that A is a periodic abelian normal subgroup of G. Then $G/C_G(A)$ has no abelian divisible subgroups. In particular, if $G/C_G(A)$ is Chernikov, then it is finite.*

Proof. Suppose the contrary, and let $D/C_G(A)$ be a non-identity divisible abelian subgroup. If $p \in \Pi(A)$, we decompose $A = A_p \times E_p$, where A_p is the Sylow p-subgroup of A, and E_p is the Sylow p'-subgroup of A. By Lemma 3.11, $[D, A] \leq E_p$, and then

$$[D, A] \leq \bigcap_{p \in \Pi(A)} Ep = \langle 1 \rangle .$$

Therefore $D \leq C_G(A)$, a contradiction. $\qquad\square$

Theorem 3.13 (Ya.D. Polovicky [225]). *Let G be a CC-group. If $g \in G$ and $H = \,< g >^G$, then either H is a Chernikov subgroup or H has a G-invariant Chernikov subgroup T such that H/T is infinite cyclic.*

Proof. By Lemma 3.2, H is central-by-Chernikov, and, by Theorem 3.9, $[H, H]$ is Chernikov. Thus, we may assume that H is abelian. Let T be the periodic part of H (definition below). By Corollary 3.5, H/T is an infinite cyclic group. If H is periodic, then $G/C_G(H)$ is finite by Corollary 3.12. It follows that $g \in FC(G)$. Thus H is finite by Corollary 3.4.

Suppose now that $H \neq T$. Then $H = \langle g, T \rangle$, and the order of g has to be infinite. Let $D/C_G(H)$ be the divisible part of $G/C_G(H)$. By Corollary 3.12, $D \leq C_G(T)$. Also $H/T \leq \zeta(G/T)$ by Corollary 3.3. Consider the mapping $\phi_g : d \mapsto [g, d]$, $d \in D$. If $d_1, d_2 \in D$, then

$$\begin{aligned}(d_1 d_2)\phi_g &= [g, d_1 d_2] = [g, d_2]d_2^{-1}[g, d_1]d_2 \\ &= [g, d_1][g, d_2] = d_1\phi_g \cdot d_2\phi_g\end{aligned}$$

and therefore ϕ_g is an endomorphism of D. Thus, Im $\phi_g = [g, D]$, and Ker $\phi_g = C_D(g)$ are D-invariant. Since G is a CC-group, $G/C_G(\langle g \rangle^G)$ is a Chernikov group, so that $D/(D \cap C_G(g))$ is Chernikov too. As

$$[g, D] = \text{Im } \phi_g \cong D/\text{Ker } \phi_g = D/C_D(g),$$

$[g, D]$ is Chernikov. Let $x \in G$. Since $H/T \leq \zeta(G/T)$, there exists some $v \in T$ such that $g^x = gv$. If $d \in D$, then

$$\begin{aligned}[g, d]^x &= [g^x, d^x] = [gv, d^x] = [g, d^x]^v[v, d^x] \\ &= [g, d^x] \in [g, D],\end{aligned}$$

since D is normal. This means that $[g, D]$ is a normal subgroup of G. Since $D \leq C_G(T)$, and $H = \langle g, T \rangle$, $D/[g, D] \leq C_{G/[g,D]}(H/[g, D])$. In other words, $(G/[g, D])/C_{G/[g,D]}(H/[g, D])$ is finite. By Corollary 3.4, $T/[g, D]$ is finite, and then T is Chernikov. $\qquad\square$

Lemma 3.14. *Let G be a CC-group. If $\zeta(G) = \langle 1 \rangle$, then G is periodic.*

Proof. Since $\zeta(G) = \langle 1 \rangle$,

$$\bigcap_{g \in G} C_G(\langle g \rangle^G) = \langle 1 \rangle.$$

Moreover, every factor-group $G/C_G(\langle g \rangle^G)$ is Chernikov since G is a CC-group. Then, if $S \subset G$ a finite subset of non-identity elements of G, there is a normal subgroup U of G such that $S \cap U = \emptyset$, and G/U is a Chernikov group. Let $1 \neq x \in G$, and put $X = \langle x \rangle^G$. By Theorem 3.13, either X is Chernikov or X has a G-invariant Chernikov subgroup Y such that $X/Y = \langle xY \rangle$. Let $C = \mathrm{Soc}_G(X)$. Since $C \leq Y$, C is finite. Therefore, there exists a normal subgroup H such that $H \cap C = \langle 1 \rangle$, and G/H is Chernikov. Put $B = H \cap X$, and $K = [G, B]$. By Corollary 3.3, the commutator subgroup $[G, G]$ is locally finite , and then $K \leq Y$. If $K \neq \langle 1 \rangle$, then $K \cap C \neq \langle 1 \rangle$. But $B \leq H$, and H is a normal subgroup of G, hence $K \leq H$, and it follows that $K \cap C = \langle 1 \rangle$. This means that $[G, B] = \langle 1 \rangle$, that is $B \leq \zeta(G) = \langle 1 \rangle$, and $H \cap X = \langle 1 \rangle$. Since

$$X \cong X/(X \cap H) \cong XH/H \leq G/H,$$

X is Chernikov. In particular, the order of x has to be finite. $\qquad \square$

By Theorem 3.13, every CC-group is FC-hypercentral. To finish this chapter, we deduce some elementary properties of FC-hypercentral groups.

Lemma 3.15. *Let $\mathfrak{X}$ be a formation of groups. Given a group G, we put $P = \mathfrak{X}C(G)$, and $Q = \mathfrak{X}C_\infty(G)$. If H is a non-identity G-invariant subgroup of Q, then $H \cap P \neq \langle 1 \rangle$.*

Proof. Let

$$\langle 1 \rangle = C_0 \leq C_1 \leq \cdots \leq C_\alpha \leq C_{\alpha+1} \leq \cdots \leq C_\gamma = Q$$

be the upper $\mathfrak{X}C$-central series of G. There exists an ordinal α such that $H \cap C_\alpha \neq \langle 1 \rangle$. Let β be the least ordinal with this property. Clearly β cannot be a limit, so that $H \cap C_{\beta-1} = \langle 1 \rangle$. Pick $1 \neq x \in H \cap C_\beta$, and put $X = \langle x \rangle^G$. Since $xC_{\beta-1} \in C_\beta/C_{\beta-1}$, $G/C_G(XC_{\beta-1}/C_{\beta-1}) \in \mathfrak{X}$. Let $g \in C_G(XC_{\beta-1}/C_{\beta-1})$. Then $[g, X] \leq C_{\beta-1}$. On the other hand $[g, X] \leq H$ since H is a normal subgroup of G. Thus $[g, X] \leq H \cap C_{\beta-1} = \langle 1 \rangle$, which means that $C_G(XC_{\beta-1}/C_{\beta-1}) \leq C_G(X)$. In particular, $G/C_G(X) \in \mathfrak{X}$ since $\mathfrak{X}$ is a formation. Hence $X \leq P$, and this gives the required result. $\qquad \square$

Corollary 3.16. *Let $\mathfrak{X}$ be a formation of groups. If A is a non-identity normal subgroup of an $\mathfrak{X}C$-hypercentral group G, then $A \cap \mathfrak{X}C(G) \neq \langle 1 \rangle$.*

Lemma 3.17. *Let $\mathfrak{X}$ be a formation of finite groups closed under taking subgroups, and let G be a finitely generated $\mathfrak{X}C$-nilpotent group. Then G is a nilpotent-by-$\mathfrak{X}$-group. In particular, G is polycyclic-by-finite.*

Proof. Let

$$\langle 1 \rangle = F_0 \leq F_1 \leq \cdots \leq F_n = G$$

be the upper $\mathfrak{X}C$-central series of G. It suffices to show that $G/C_G(F_{j+1}/F_j) \in \mathfrak{X}$, for every $0 \leq j \leq n-1$. We carry this out by induction on n.

If $n = 1$, then G is an $\mathfrak{X}C$-group , and it suffices to apply Lemma 3.2. Let now $n > 1$, and assume that we have already proved that $G/C_G(F_{j+1}/F_j) \in \mathfrak{X}$ for every $1 \leq j < n$. Put

$$H = C_G(F_2/F_1) \cap \cdots \cap C_G(F_n/F_{n-1}).$$

Then $G/H \in \mathfrak{X}$, and H/F_1 is nilpotent (see O. H. Kegel and B.A.F. Wehrfritz [134, Theorem 1.C.1]). In particular, G/F_1 is finitely presented (see D.J.S. Robinson [234, Corollary 1.43]), and it follows that

$$F_1 = \langle g_1 \rangle^G \cdots \langle g_s \rangle^G$$

where $g_1, \ldots, g_s \in \mathfrak{X}C(G)$ (D.J.S. Robinson [234, Corollary 1.43]). Put

$$U = C_G(\langle g_1 \rangle^G) \cap \cdots \cap C_G(\langle g_s \rangle^G).$$

Then $G/U \in \mathfrak{X}$, and $U = C_G(F_1)$. If $C = H \cap U$, then $G/C \in \mathfrak{X}$, and C is nilpotent (see O.H. Kegel and B.A.F. Wehrfritz [134, Theorem 1.C.1]). $\square$

Corollary 3.18. *A finitely generated FC-nilpotent group G is nilpotent-by-finite. In particular, G is polycyclic-by-finite.*

Proposition 3.19. *Let $\mathfrak{X}$ be a formation of finite groups closed under taking subgroups, and let G be an $\mathfrak{X}C$-hypercentral group. If Y is a finitely generated subgroup of G, then Y is a nilpotent-by-$\mathfrak{X}$-group.*

Proof. Let

$$\langle 1 \rangle = G_0 \leq G_1 \leq \cdots \leq G_\alpha \leq \cdots \leq G_\gamma = G$$

be the upper $\mathfrak{X}C$-central series of G. We consider the following set of ordinals,

$$\Xi = \{\alpha \mid Y/(Y \cap G_\alpha) \text{ is a nilpotent-by-}\mathfrak{X}\text{-group}\}.$$

Since

$$Y = \bigcup_{\alpha \leq \gamma} (Y \cap G_\alpha),$$

$\Xi \neq \emptyset$. Let β be the least element of Ξ. If $\beta = 0$, then Y is a nilpotent-by-$\mathfrak{X}$-group, as claimed. Suppose that $\beta > 0$. If β is not a limit, then

$$(Y \cap G_\beta)/(Y \cap G_{\beta-1}) = (Y \cap G_\beta)/(G_\beta \cap Y \cap G_{\beta-1})$$
$$\cong (Y \cap G_\beta)G_{\beta-1}/G_{\beta-1},$$

and, in particular,

$$(Y \cap G_\beta)/(Y \cap G_{\beta-1}) \leq \mathfrak{X}C(Y/(Y \cap G_{\beta-1})).$$

Since $Y/(Y \cap G_\beta)$ is $\mathfrak{X}C$-nilpotent, by Lemma 3.17, $Y/(Y \cap G_{\beta-1})$ is $\mathfrak{X}C$-nilpotent too. This fact contradicts the choice of β.

Hence β has to be a limit ordinal whenever $\beta > 0$. Since $Y/(Y \cap G_\beta)$ is finitely presented,

$$Y \cap G_\beta = \langle y_1, \ldots, y_s \rangle^G,$$

for some $y_1, \ldots, y_s \in Y \cap G_\beta$ (see D.J.S. Robinson [234, Corollary 1.43] again). Then there exists an ordinal $\delta < \beta$ such that $y_1, \ldots, y_s \in G_\delta$, so that $Y \cap G_\beta = Y \cap G_\delta$, and we get a new contradiction, which shows that the case $\beta > 0$ cannot occur. □

Corollary 3.20. *Let G be an FC-hypercentral group. If X is a finitely generated subgroup of G, then X is nilpotent-by-finite.*

Chapter 4

Artinian modules and the socle

In studying further properties of artinian modules, the concept of the *socle of the module* has appeared to be very functional. We develop here a slight extension of this concept, which in its original form is due to R. Remak [226].

Let R be a ring, and let A be an R-module. The submodule $\mathrm{Soc}_R(A)$ generated by all minimal R-submodules of A is said to be *the R-socle of A*. If A has no such minimal submodules, we define $\mathrm{Soc}_R(A) = \langle 0 \rangle$.

Lemma 4.1. *Let R be a ring, and let A be an R-module. If $\mathfrak{M}$ is a family of minimal R-submodules of A and B is the submodule generated by all members of $\mathfrak{M}$, then B is a direct sum of some elements of $\mathfrak{M}$.*

Proof. If $\emptyset \neq \mathfrak{S} \subseteq \mathfrak{M}$, then $\mathfrak{S}$ is said to be *independent* if the submodule S generated by all members of $\mathfrak{S}$ is the direct sum of them, that is, $S = \bigoplus_{X \in \mathfrak{S}} X$. Clearly, every atomic subfamily is independent. Let $\mathcal{L}$ be the set of all independent subfamilies of $\mathfrak{M}$. If $\mathcal{S}$ is a linearly ordered (by inclusion) subset of $\mathcal{L}$, we put

$$\mathfrak{C} = \bigcup_{\mathfrak{H} \in \mathcal{S}} \mathfrak{H}$$

and denote by C the submodule generated by all members of $\mathfrak{C}$. Given a finite subset $\mathfrak{Z} = \{X_1, \ldots, X_r\}$ of $\mathfrak{C}$, pick some $c \in X_j \cap \langle X_i \mid l \neq j \rangle$. Since $\mathcal{S}$ is linearly ordered, there is some $\mathfrak{D} \in \mathcal{S}$ such that $\mathfrak{Z} \subseteq \mathfrak{D}$. As the submodule generated by all members of $\mathfrak{D}$ is the direct sum $\bigoplus_{X \in \mathfrak{D}} X$, it follows that $c = 0$, so that $C = \bigoplus_{X \in \mathfrak{C}} X$, and therefore $\mathfrak{C}$ is also independent. By Zorn's Lemma, $\mathfrak{M}$ contains a maximal independent subfamily $\mathfrak{R}$.

Put $E = \sum_{X \in \mathfrak{R}} X = \bigoplus_{X \in \mathfrak{R}} X$. If we assume that $E \neq B$, then there is a minimal R-submodule $U \in \mathfrak{M}$ such that E contains no U. Then $U \cap E = \langle 0 \rangle$, and so $E + U = E \oplus U$. This means that $\mathfrak{R} \cup \{U\}$ is an independent family, which contradicts the choice of $\mathfrak{R}$. This contradiction shows that $E = B$, as required. $\square$

Corollary 4.2. *Let A be a module over a ring R. Suppose that $A = \mathrm{Soc}_R(A)$. If B is an R-submodule of A, then:*

(1) $A/B = Soc_R(A/B)$.

(2) $A = B \oplus M$ whenever M is a maximal R-submodule of A under $M \cap B = \langle 1 \rangle$.

(3) If $\mathfrak{M}$ denotes the family of all minimal R-submodules of B, then B is generated by all members of $\mathfrak{M}$.

Proof. (1) is obvious.

(2) Suppose that $A \neq B \oplus M$. Then $A/M \neq BM/M$. Since A/M is generated by minimal R-submodules, there is a minimal R-submodule U/M such that BM/M contains no U/M. Hence $U/M \cap BM/M = \langle 0 \rangle$, and then $U \cap BM = M$. Thus $M = U \cap BM = M(U \cap B)$; that is, $U \cap B \leq M$. It follows that

$$U \cap B = M \cap (U \cap B) = (M \cap U) \cap B = M \cap B = \langle 0 \rangle,$$

which contradicts the choice of M. Therefore $A = B \oplus M$.

(3) is an immediate consequence of (1) and (2). $\square$

Corollary 4.3. *Let R be a ring, and let A be an R-module. Then the following statements are equivalent.*

(1) *A is a sum of simple R-submodules.*

(2) *A is a direct sum of simple R-submodules.*

(3) *For every R-submodule B, there is an R-submodule C such that $A = B \oplus C$.*

Proof. (1) $\Rightarrow$ (2) has been proved in Lemma 4.1.

(2) $\Rightarrow$ (3). Let B be a non-zero R-submodule of A. By Corollary 4.2, $B = \bigoplus_{\lambda \in \Lambda} M_\lambda$, where M_λ is a simple R-submodule for each $\lambda \in \Lambda$. This means that the family $\{M_\lambda \mid \lambda \in \Lambda\}$ is independent. Then there is a maximal independent family $\mathfrak{M}$, such that $\{M_\lambda \mid \lambda \in \Lambda\} \subseteq \mathfrak{M}$. We have already seen that $A = \bigoplus_{M \in \mathfrak{M}} M$. If $\mathfrak{L} = \mathfrak{M} \setminus \{M_\lambda \mid \lambda \in \Lambda\}$, and $C = \bigoplus_{M \in \mathfrak{L}} M$, then $A = B \oplus C$, as required.

(3) $\Rightarrow$ (1). Let $0 \neq a \in A$, and put $E = aR$. Obviously a does not lie in any proper R-submodule of E. Therefore the union of any ascending chain of proper submodules of E cannot reach E itself. By Zorn's Lemma, E has a maximal proper R-submodule V. In particular, E/V is a simple R-module. By (3), there is a submodule U such that $A = V \oplus U$. Since $V \leq E$, the modular law gives that $E = V \oplus (E \cap U)$. Since $E/V \cong E \cap U$, $E \cap U$ is a simple R-submodule. In particular, $S = Soc_R(A) \neq \langle 0 \rangle$. If $S \neq A$, then $A = S \oplus W$ for some R-submodule W. Again we choose an element $0 \neq w \in W$, and consider the submodule wR. We have already proved that wR has a simple R-submodule M. It follows that $M \leq S$. But $M \leq wR \leq W$ and $W \cap S = \langle 0 \rangle$. This contradiction shows that $A = Soc_R(A)$. $\square$

A non-zero module A over a ring R is said to be *a semisimple R-module* if $A = Soc_R(A)$. By Lemma 4.1, A is a semisimple R-module if and only if A is a (direct) sum of simple R-submodules so that *semisimplicity* is an equivalent condition to the corresponding condition of Corollary 4.3. Moreover, it is immediate to obtain

Corollary 4.4. *Let A be a semisimple R-module. Then:*

(1) *If B is a an R-submodule of A, then B and A/B are likewise semisimple R-modules.*

(2) *If $A = \bigoplus_{\lambda \in \Lambda} A_\lambda$, where every A_λ is a simple R-submodule, and C is a simple R-submodule of A, then there is an index $\nu \in \Lambda$ such that $C \cong A_\nu$.*

Proposition 4.5. *Let R be a commutative ring and A an R-module. Suppose that $A = \bigoplus_{\lambda \in \Lambda} A_\lambda$, where the A_λ are simple R-submodules isomorphic each to other. If $0 \neq a \in A$, then aR is a simple R-submodule.*

Proof. We have
$$a = a_1 + \cdots + a_k,$$

where $a_i \in A_{\lambda(j)}$, and $\lambda(j) \in \Lambda, 1 \leq j \leq k$. Suppose that $a_1 \neq 0$. Put $B = aR$, and let ϕ be the restriction to B of the canonical projection from A to $A_{\lambda(1)}$. If $ax \in \operatorname{Ker} \phi$, then
$$0 = (ax)\phi = (a\phi)x = a_1 x.$$

Let $\psi : A_{\lambda(1)} \longrightarrow A_{\lambda(1)}$ be the mapping given by $b\psi = bx$ if $b \in A_{\lambda(1)}$. Then $\operatorname{Ker} \psi \neq \langle 0 \rangle$ because $a_1 x = 0$. Since R is commutative, $\operatorname{Ker} \psi$ is an R-submodule of $A_{\lambda(1)}$. It follows that $\operatorname{Ker} \psi = A_{\lambda(1)}$; that is, $A_{\lambda(1)} x = \langle 0 \rangle$. Since $A_{\lambda(1)} \cong_R A_\lambda$ for every $\lambda \in \Lambda$, $A_\lambda x = \langle 0 \rangle$, and so

$$ax = a_1 x + \cdots + a_k x = 0.$$

Thus $\operatorname{Ker} \psi = \langle 0 \rangle$, and hence ψ is an isomorphism from B to $A_{\lambda(1)}$. ⊔

With the aid of the socle we obtain the next characterization of artinian modules. The group-theoretical analogue of this result was obtained by R. Baer [5]. Recall that an R-module A has *finite composition length* if there is a finite series of submodules
$$\langle 0 \rangle = A_0 \leq A_1 \leq \cdots \leq A_n = A$$

whose factors A_{i+1}/A_i are all simple (*a finite composition series*). It is known that a module has finite composition length if and only if it is noetherian and artinian. However we have

Theorem 4.6. *An R-module A is artinian if and only if the socle of every non-zero factor-module of A is non-zero, and has finite composition length.*

Proof. If A is artinian and B is a submodule of A, by Lemma 1.2, A/B is artinian. It follows that $\operatorname{Soc}_R(A/B) \neq \langle 0 \rangle$ and, by Corollary 4.3, the socle clearly has finite composition length.

Conversely, let

$$S_0 = \langle 0 \rangle, \quad S_1 = \operatorname{Soc}_R(A), \quad S_{\alpha+1}/S_\alpha = \operatorname{Soc}_R(A/S_\alpha),$$

and $S_\alpha = \bigcup_{\beta < \alpha} S_\beta$ for every limit ordinal α. Let B be a non-zero submodule of A. Then there is a least ordinal β such that $C = B \cap S_\beta \neq \langle 0 \rangle$. Clearly, β cannot be a limit ordinal. In particular, $S_{\beta-1} \cap B = \langle 0 \rangle$, and then

$$C \cong C/(C \cap S_{\beta-1}) \cong (C + S_{\beta-1})/S_{\beta-1} \leq S_\beta/S_{\beta-1}.$$

By Corollary 4.4, C is a semisimple submodule. In particular, C contains a non-zero simple submodule, and then $B \cap S_1 \neq \langle 0 \rangle$. Hence every non-zero submodule of A has non-zero intersection with the R-socle.

Suppose that A has an infinite strictly descending chain of submodules

$$A_1 > A_2 > \cdots > A_n > \cdots .$$

Replacing A by $A/(\bigcap_{n \in \mathbb{N}} A_n)$, we may assume that $\bigcap_{n \in \mathbb{N}} A_n = \langle 0 \rangle$. Since S_1 has finite composition length, and $A_n \cap S_1 \neq \langle 0 \rangle$ for each $n \in \mathbb{N}$, there is some $m \in \mathbb{N}$ such that $A_j \cap S_1 = A_m \cap S_1$ for every $j \geq m$. However, in this case

$$\bigcap_{n \in \mathbb{N}} A_n \geq A_m \cap S_1.$$

This contradiction proves that A is an artinian module. $\qquad\square$

Let A be an R-module. Starting from the socle we define *the upper socular series* or *the ascending Loewy series* of A as

$$\langle 0 \rangle = S_0 \leq S_1 \leq \cdots \leq S_\alpha \leq \cdots \leq S_\rho,$$

where $S_1 = \mathrm{Soc}_R(A)$, and $S_{\alpha+1}/S_\alpha = \mathrm{Soc}_R(A/S_\alpha)$ for a given ordinal α. Note that $S_\mu = \bigcup_{\beta < \mu} S_\beta$ for any limit ordinal μ. The least ordinal ρ such that $S_\rho = S_{\rho+1}$ is called *the socular height* of A.

Lemma 4.7. *Let R be a ring and A an R-module. Suppose that*

$$\langle 0 \rangle = S_0 \leq S_1 \leq \cdots \leq S_\alpha \leq \cdots \leq S_\rho$$

is the upper socular series of the R-module A. If B is a submodule of A having an ascending series of submodules

$$\langle 0 \rangle = B_0 \leq B_1 \leq \cdots \leq B_\alpha \leq \cdots \leq B_\lambda = B$$

all of whose factors are semisimple, then $B_\lambda \leq S_\lambda$.

Proof. We proceed by transfinite induction. Clearly, $B_1 \leq \mathrm{Soc}_R(A) = S_1$. Let $\alpha > 1$, and suppose that $B_\beta \leq S_\beta$ for all ordinals $\beta < \alpha$. If α is a limit ordinal, then

$$B_\alpha = \bigcup_{\beta < \alpha} B_\beta \leq \bigcup_{\beta < \alpha} S_\beta = S_\alpha.$$

Suppose now that $\alpha - 1$ exists. We have

$$(B_\alpha + S_{\alpha-1})/S_{\alpha-1} \cong B_\alpha/(B_\alpha \cap S_{\alpha-1})$$

and

$$B_\alpha \cap S_{\alpha-1} \geq B_{\alpha-1}.$$

By Corollary 4.4, $(B_\alpha + S_{\alpha-1})/S_{\alpha-1}$ is semisimple, and hence

$$(B_\alpha + S_{\alpha-1})/S_{\alpha-1} \leq \mathrm{Soc}_R(A/S_{\alpha-1}) = S_\alpha/S_{\alpha-1}.$$

Therefore $B_\alpha \leq S_\alpha$ in this case. Applying induction, $B = B_\lambda \leq S_\lambda$. □

Corollary 4.8. *Let B be a submodule of an R-module A. If*

$$\langle 0 \rangle = B_0 \leq B_1 \leq \cdots \leq B_\alpha \leq \cdots$$

and

$$\langle 0 \rangle = S_0 \leq S_1 \leq \cdots \leq S_\alpha \leq \cdots$$

are the upper socular series of B, and A respectively, then $B_\alpha = S_\alpha \cap B$ for any α.

Proof. Indeed $B_\alpha \leq S_\alpha$ by Lemma 4.7. Since $S_\alpha \cap B$ has an ascending series of length $\leq \alpha$ with semisimple factors, Lemma 4.7 yields that $S_\alpha \cap B \leq B_\alpha$, and then we obtain the equality $S_\alpha \cap B = B_\alpha$. □

Concerning the height of the socular series of A and related questions, we will show some interesting results due to B. Hartley [105].

Lemma 4.9. *The socular height of an R-module A is the supremum of the socular heights of the cyclic submodules of A.*

Proof. Let β be the supremum of the socle heights of all cyclic submodules of A. Let

$$\langle 0 \rangle = S_0 \leq S_1 \leq \cdots \leq S_\alpha \leq \cdots \leq S_\rho$$

be the upper socular series of A. Given an ordinal $\lambda < \beta$, then there exists a cyclic submodule B having an upper socular series

$$\langle 0 \rangle = B_0 \leq B_1 \leq \cdots \leq B_\delta$$

such that $\delta > \lambda$. In particular, $B_\lambda \neq B_{\lambda+1}$. By Corollary 4.8,

$$B_\lambda = B \cap S_\lambda < B_{\lambda+1} = B \cap S_{\lambda+1}, \text{ so that } S_\lambda \neq S_{\lambda+1}.$$

On the other hand, assume that $S_\beta \neq S_{\beta+1}$. Then $S_{\beta+1}$ has a cyclic submodule D such that $D \nleq S_\beta$. Let

$$\langle 0 \rangle = D_0 \leq D_1 \leq \cdots \leq D_\sigma$$

be the upper socular series of D. By Corollary 4.8, $D_\beta = S_\beta \cap D \neq S_{\beta+1} \cap D = D_{\beta+1}$, which contradicts the choice of β. Consequently, $\rho \leq \beta$, and the result has just been shown. □

Corollary 4.10 ([105]). *Let R be a ring with cardinality $\mathfrak{m}$, and let λ be the least ordinal of cardinal greater than $\mathfrak{m}$. Then the socular height of any R-module is at most λ.*

Proof. Let A be an R-module, and pick a cyclic submodule C of A. Then $C \cong R/\mathrm{Ann}_R(C)$. It follows that $|C| \leq \mathfrak{m}$, and then the socular height of C is an ordinal of cardinal at most $\mathfrak{m}$. Since λ is an upper bound for all these ordinals, the result follows from Lemma 4.9. $\hfill\square$

Corollary 4.11 ([105]). *Let R be an infinite ring with cardinality $\mathfrak{m}$, and let $\mathfrak{m}^+$ be the least cardinal greater than $\mathfrak{m}$. Then every artinian R-module has cardinal at most $\mathfrak{m}^+$.*

Proof. Let A be an artinian R-module, and let

$$\langle 0 \rangle = S_0 \leq S_1 \leq \cdots S_\rho$$

be the upper socular series of A. Since A is an artinian module, $S_\rho = A$ and $S_{\alpha+1}/S_\alpha$ has finite composition length for every $\alpha < \rho$. By Corollary 4.10, $\rho \leq \lambda$, where λ is the least ordinal of cardinal $\mathfrak{m}^+$. Every factor $S_{\lambda+1}/S_\lambda$ is a direct sum of simple R-submodules, and finitely many of them can appear. Since a simple R-module is a homomorphic image of the R-module R, its cardinal is at most $\mathfrak{m}$. Therefore $|S_{\alpha+1}/S_\alpha| \leq \mathfrak{m}$ for each $\alpha < \lambda$. Since the number of these factors is at most $\mathfrak{m}^+$, we conclude $|A| \leq \mathfrak{m}\mathfrak{m}^+ = \mathfrak{m}^+$ as required. $\hfill\square$

We are now in a position to deduce a result due to P.H. Neumann (see B. Hartley [105]).

Theorem 4.12. *An artinian module A over a commutative ring R has socular height at most ω, the first infinite ordinal. In particular, if R is countable so is A.*

Proof. Let $a \in A$. Then $aR \cong R/\mathrm{Ann}_R(a)$. Since R is commutative, $J = \mathrm{Ann}_R(a)$ is a two-sided ideal of R, and then R/J becomes an artinian ring. It follows that R/J is also noetherian (see R. Sharp [263, Theorem 8.44]). Hence aR is simultaneously artinian and noetherian, and therefore it has a finite composition series. In particular, the upper socular series of aR is finite. It follows from Lemma 4.9 that the socle height of A is at most ω.

Since A is artinian, A itself is the last term of the upper socle series. Hence A has an ascending series of submodules

$$\langle 0 \rangle = A_0 \leq A_1 \leq \cdots \leq A_{\mathfrak{m}} = A$$

whose factors are simple. In particular, A_{n+1}/A_n is R-isomorphic to some factor module of the R-module R.

Suppose that R is countable. Then every factor A_{n+1}/A_n is countable. Hence A is likewise countable. $\hfill\square$

Even in the case of artinian modules over countable group rings, we cannot guarantee that an artinian module is countable, as the following result shows. Recall that a module A is called *uniserial* if the set of all submodules of A is well ordered by inclusion. The ordinal corresponding to the order type of the set of the proper submodules of a uniserial module is called *the length* of this module. Clearly every uniserial module is artinian.

Proposition 4.13 ([105]). *Let F be an arbitrary field, and let G be the free group freely generated by a countably infinite subset X. Then there exists a uniserial module A over the group ring FG of length Ω, where Ω denotes the first uncountable ordinal.*

Proof. Suppose that $X = \{x_n \mid n \in \mathbb{N}\}$, and let A be a vector space over F with basis $\{a_\alpha \mid \alpha < \Omega\}$. For each $0 < \alpha < \Omega$, the set of ordinals $\beta < \alpha$ is countable, and so we may choose a bijective mapping f_α from $\mathbb{N}$ onto that set. For each $n \in \mathbb{N}$, we define a linear transformation ξ_n of A by

$$a_\alpha \xi_n = a_\alpha + a_{f_\alpha(n)} \text{ if } 0 < \alpha < \Omega, \text{ and } a_0 \xi_n = a_0.$$

Since $f_\alpha(n) < \alpha$, every a_α is annihilated by some power of $\xi_n - 1$, and so the power series for $(1 + (\xi_n - 1))^{-1}$ represents a well-defined linear transformation of A that will be the inverse of ξ_n. Therefore ξ_n is an F-automorphism of A, and we can make A into an FG-module via the mapping $x_n \to \xi_n$, $n \in \mathbb{N}$.

For each $\alpha < \Omega$, let $A_\alpha = \sum_{\beta < \alpha} Fa_\beta$. Clearly every A_α is an FG-submodule of A, and the proof of this proposition will be completed by showing that the set of all submodules of A is exactly $\{A_\alpha \mid \alpha < \Omega\}$. To do this, let $a = \sum_{\beta < \alpha} \lambda_\beta a_\beta$ be an arbitrary non-zero element of A, where all but finitely many of the coefficients λ_β are zero, $\lambda_\beta \in F$, $\beta < \Omega$. We put

$$\operatorname{Supp} a = \{\beta \mid \lambda_\beta \neq 0\}.$$

We define *the height of a* to be the largest ordinal β such that $\lambda_\beta \neq 0$, and prove by transfinite induction that any element of height β generates $A_{\beta+1}$ as FG-module. This clearly holds with $\beta = 0$. Let $0 < \beta < \Omega$, and suppose that the claim holds for all $\gamma < \beta$. Let

$$b = \sum_{\gamma < \beta} \mu_\gamma a_\gamma$$

be an element such that $\operatorname{Supp} b = \{\beta_1, \ldots, \beta_k, \beta\}$. We may assume that

$$\beta_1 < \cdots < \beta_k < \beta.$$

Since f_β is surjective, $f_\beta(t) = \beta_k$ for some $t \in \mathbb{N}$. Then $b(\xi_t - 1) = c + \mu_\beta a_\nu$, where $\nu = \beta_k$ and c is a linear combination of certain a_γ with $\gamma < \beta_k$. The FG-submodule B generated by b contains an element $b(\xi_t - 1)$, which has height β_k, and hence this submodule contains $A_{\nu+1}$ by induction hypothesis. But then B also contains

$$\mu_\beta a_\beta = b - \sum_{\gamma \in \Gamma} \mu_\gamma a_\gamma,$$

where $\Gamma = \{\beta_1, \ldots, \beta_k\}$, and hence a_β itself. Finally B contains $a_\beta(\xi_n - 1) = a_\sigma$ where $\sigma = f_\beta(n)$ for all $n \in \mathbb{N}$. Since f_β is surjective, $A_{\beta+1} \le B$. Equality is clear.

It now follows that if T is an arbitrary submodule of A and β is the largest ordinal such that $A_\beta \le T$, then every non-zero element of T has height less than β; that is, $T \le A_\beta$. Hence $T = A_\beta$, and the proof is complete. $\qquad\square$

B. Hartley [105] has also constructed examples of uncountable artinian modules over group rings. For the reader's convenience, we mention here some of his important results without proof.

Let $p \ne q$ be primes, and let Q be a Prüfer q-group. There exists a simple $\mathbb{F}_p Q$-module A such that $C_Q(A) = \langle 0 \rangle$ (see L.A. Kurdachenko, J. Otal and I.Ya. Subbotin [157, Corollary 2.4]). Consider the natural semidirect product $G = A \leftthreetimes Q$. This group is called *the Charin group* (it was constructed by V.S. Charin [33]).

Theorem 4.14 ([105]). *Let G be a Charin group, and let r be a prime such that $r \notin \Pi(G)$. Then there exists a uniserial $\mathbb{F}_r G$-module of length Ω.*

Let $D = \langle d \rangle \leftthreetimes \langle c \rangle$ be the dihedral group of order 8, such that $|d| = 4$, $|c| = 2$, and $d^c = d^{-1}$. For every $n \in \mathbb{N}$ denote by $D_n = \langle d_n \rangle \leftthreetimes \langle c_n \rangle$ be an isomorphic copy of D. Put $R = \mathrm{Dr}_{n \in \mathbb{N}} D_n$, $E = \langle d_n^2 d_{n+1}^{-2} \mid n \in \mathbb{N} \rangle$, and $U = R/E$.

Theorem 4.15 ([105]). *Let F be a field of characteristic other than 2. Then there exists a uniserial FU-module of length ω.*

Chapter 5

Reduction to subgroups of finite index

Let R be a ring, G be a group, and A be an RG-module. If H is a subgroup of G, then A is trivially an RH-module, and the following questions naturally arise:

> *What properties of the RG-module A does the RH-module A inherit?*
> *What can be said of A as an RH-module?*

For example, if A is noetherian or artinian as an RG-module, it is rather easy to see that these properties are no longer true when A is viewed as an RH-module. Despite this, a famous result due to J.S. Wilson [284] shows that the properties of *being artinian* and *being noetherian* are inherited when H has finite index in G. This result is the key to significant applications and further results in the study of simplicity, the existence of important decompositions, and many other topics. Also we consider here another main result, the classical Maschke's theorem, which we will discuss in the most general form.

We will give here a module version of Wilson's theorem.

Let R be a ring, G a group, and A an RG-module. If B is an R-submodule of A and T is a subset of G, then we define BT as the RT-submodule generated by B; that is,

$$BT = \sum_{t \in T} Bt.$$

Lemma 5.1. *Let R be a ring, G a group and A an RG-module. Suppose that H is a subgroup of G and T is a subset of G such that $G = HT$. If B is an RH-submodule of A, then $\{Bx \mid x \in G\} = \{Bt \mid t \in T\}$ and the subsets BT and $\bigcap_{t \in T} Bt$ are RG-submodules.*

The proof is immediate.

Theorem 5.2 (J.S. Wilson [284]). *Let R be a ring, G a group and A an RG-module. Suppose that H is a normal subgroup of G having finite index. If A is an artinian RG-module, then A is an artinian RH-module.*

Proof. Suppose the contrary: that is, A does not satisfy Min-RH. Then the family $\mathfrak{M}$ of all RG-submodules B of A that does not satisfy Min-RH is not empty. Since A is an artinian RG-module, $\mathfrak{M}$ has a minimal element M. Let $\mathcal{S}$ be the set of all non-empty finite subsets X of G with the following property: *If $\{C_n \mid n \in N\}$ is a strictly infinite descending chain of RH-submodules of M, then $M = C_n X$ for all $n \in \mathbb{N}$.*

Let T be a transversal to H in G. Then $G = HT$. By Lemma 5.1, $C_n T$ is an RG-submodule of M. We claim that $C_n T = M$. Otherwise, by the choice of M, $C_n T$ is an artinian RH-module. It follows that there exists some $m \geq n$ such that $C_{m+j} = C_m$ for every $j \in \mathbb{N}$, contradicting the choice of the chain $\{C_n \mid n \in \mathbb{N}\}$. Thus our claim has just been proven and so $C_n T = M$ for all $n \in \mathbb{N}$. Therefore $T \in \mathcal{S}$ and then $\mathcal{S} \neq \emptyset$.

We now choose a minimal element X of $\mathcal{S}$. If $x \in X$, then $Xx^{-1} \in \mathcal{S}$. In particular, Xx^{-1} is a minimal element of $\mathcal{S}$ containing 1. Replacing X by Xx^{-1}, we may assume that $1 \in X$. Clearly $X \neq \langle 1 \rangle$, so that $Y = X \backslash \langle 1 \rangle$ is a non-empty finite subset of G. By the minimality of X, $Y \notin \mathcal{S}$. Given a strictly infinite descending chain of RH-submodules $\{C_n \mid n \in \mathbb{N}\}$ of M, we put $E_n = C_n \cap C_n Y$ for every $n \in \mathbb{N}$. Clearly E_n is an RH-submodule and $E_n \geq E_{n+1}$ for every $n \in \mathbb{N}$. Suppose that there exists some $k \in \mathbb{N}$ such that $E_{k+1} = E_k$. Since $X \in \mathcal{S}$, $C_{k+1} X = M$ and

$$C_k = C_k \cap M = C_k \cap C_{k+1} X = C_k \cap (C_{k+1} + C_{k+1} Y)$$
$$= C_{k+1} + (C_k \cap C_{k+1} Y) \leq C_{k+1} + (C_k \cap C_k Y)$$
$$= C_{k+1} + E_k = C_{k+1} + E_{k+1} = C_{k+1}.$$

Thus $C_k = C_{k+1}$, contradicting the choice of $\{C_n \mid n \in \mathbb{N}\}$. This contradiction shows that $E_n \neq E_{n+1}$ for every $n \in \mathbb{N}$. In other words, $\{E_n \mid n \in \mathbb{N}\}$ is a strictly infinite descending chain of RH-submodules of M. Consequently, $E_n X = M$ for all $n \in \mathbb{N}$ and

$$C_n = C_n \cap M = C_n \cap E_n X = C_n \cap (E_n + E_n Y)$$
$$= E_n + (C_n \cap E_n Y) \leq E_n + (C_n \cap C_n Y) = E_n.$$

We have now

$$C_n X = C_n + C_n Y = E_n + C_n Y = (C_n \cap C_n Y) + C_n Y = C_n Y$$

for every $n \in \mathbb{N}$. This means that $Y \in \mathcal{S}$, which contradicts the choice of X. Hence, A is an artinian RH-module, as required. $\square$

Theorem 5.3 ([284]). *Let R be a ring, G a group, and A an RG-module. Suppose that H is a normal subgroup of G having finite index. If A is a noetherian RG-module, then A is a noetherian RH-module.*

Proof. Suppose the contrary: that is, A does not satisfy Max-RH. Then the family $\mathfrak{M}$ of all RG-submodules B of A such that A/B does not satisfy Max-RH is not empty. Since A is a noetherian RG-module, $\mathfrak{M}$ has a maximal element M. Let $\mathcal{S}$ be the set of all non-empty finite subsets X of G with the following property: *If* $\{C_n \mid n \in \mathbb{N}\}$ *is a strictly infinite ascending chain of RH-submodules containing* M, *then*

$$M = \bigcap_{x \in X} C_n x = \operatorname{Core}_X C_n$$

for all $n \in \mathbb{N}$.

Let T be a transversal to H in G. Then $G = HT$. Let $n \in \mathbb{N}$. By Lemma 5.1, $\operatorname{Core}_T C_n$ is an RG-submodule containing M. We claim that $\operatorname{Core}_T C_n = M$. Otherwise, by the choice of M, $A/\operatorname{Core}_T C_n$ is a noetherian RH-module. It follows that there exists some $m \geq n$ such that $C_{m+j} = C_m$ for every $j \in \mathbb{N}$, contradicting the choice of the chain $\{C_n \mid n \in \mathbb{N}\}$. Thus our claim has just been proven and so $\operatorname{Core}_T C_n = M$ for all $n \in \mathbb{N}$. Therefore $T \in \mathcal{S}$ and then $\mathcal{S} \neq \emptyset$.

We now choose a minimal element X of $\mathcal{S}$. As in the proof of Theorem 5.2, we may assume that $1 \in X$. Clearly $X \neq \langle 1 \rangle$, so that $Y = X \setminus \langle 1 \rangle$ is a non-empty finite subset of G. By the minimality of X, $Y \notin \mathcal{S}$. Given a strictly infinite ascending chain of RH-submodules $\{C_n \mid n \in \mathbb{N}\}$ including M, we put $E_n = C_n + \operatorname{Core}_Y C_n$ for every $n \in \mathbb{N}$. Clearly E_n is an RH-submodule and $E_n \leq E_{n+1}$ for every $n \in \mathbb{N}$. Suppose that there exists some $k \in \mathbb{N}$ such that $E_{k+1} = E_k$. Since $X \in \mathcal{S}$, $\operatorname{Core}_X C_{k+1} = M$ and

$$\begin{aligned}
C_k &= C_k + \operatorname{Core}_X C_{k+1} = C_k + (C_{k+1} \cap \operatorname{Core}_Y C_{k+1}) \\
&= C_{k+1} \cap (C_k + \operatorname{Core}_Y C_{k+1}) \geq C_{k+1} \cap E_k = C_{k+1} \cap E_{k+1} \\
&= C_{k+1} \cap (C_{k+1} + \operatorname{Core}_Y C_{k+1}) \\
&= C_{k+1} + (C_{k+1} \cap \operatorname{Core}_Y C_{k+1}) = C_{k+1}.
\end{aligned}$$

Thus $C_k = C_{k+1}$, which contradicts the choice of $\{C_n \mid n \in \mathbb{N}\}$. This contradiction shows that $E_n \neq E_{n+1}$ for every $n \in \mathbb{N}$. In other words, $\{E_n \mid n \in \mathbb{N}\}$ is a strictly infinite ascending chain of RH-submodules including M. Consequently, $\operatorname{Core}_X E_{k+1} = M$ for all $n \in \mathbb{N}$. Hence,

$$\begin{aligned}
C_n &= C_n + \operatorname{Core}_X E_n = C_n + (E_n \cap \operatorname{Core}_Y E_n) \\
&= E_n \cap (C_n + \operatorname{Core}_Y E_n) \geq E_n
\end{aligned}$$

and so $C_n = E_n$. Thus

$$\begin{aligned}
M &= \operatorname{Core}_X C_n = C_n \cap \operatorname{Core}_Y C_n = E_n \cap \operatorname{Core}_Y C_n \\
&= (C_n + \operatorname{Core}_Y \cap C_n) \cap \operatorname{Core}_Y C_n = \operatorname{Core}_Y C_n
\end{aligned}$$

for every $n \in \mathbb{N}$. This means that $Y \in \mathcal{S}$. This contradicts the choice of X. Hence, A is a noetherian RH-module, as required. $\qquad\square$

We will study the behaviour of the semisimplicity.

Lemma 5.4. *Let R be a ring, G be a group, and A be an RG-module. Suppose that H is a normal subgroup of G and let B be an RH-submodule of A.*

(1) *If $g \in G$, then Bg is likewise an RH-submodule. Moreover:*

   (a) *If B is noetherian, then so is Bg.*

   (b) *If B is artinian, then so is Bg.*

   (c) *If B is simple, then so is Bg.*

(2) *If A is a simple RG-module and B is a simple RH-submodule, then $A = \bigoplus_{g \in S} Bg$, for some $S \subseteq G$. In particular, A is a semisimple RH-module.*

Proof. (1) Since H is normal in G, $gRH = RHg$, for each $g \in G$. Therefore

$$(Bg)RH = B(gRH) = B(RHg) = (BRH)g = Bg,$$

and, then Bg is stable under RH.

   (a) Let $\{C_n \mid n \in \mathbb{N}\}$ be an ascending chain of RH-submodules of Bg. Then $\{C_n g^{-1} \mid n \in \mathbb{N}\}$ is an ascending chain of RH-submodules of B. Since B is noetherian, there is some $m \in \mathbb{N}$ such that $C_m g^{-1} = C_{m+n} g^{-1}$, for all $n \in \mathbb{N}$. It follows that $C_m = (C_{m+n} g^{-1})g = C_{m+n}$, for all $n \in \mathbb{N}$, which shows the required result.

   (b) This proof is similar.

   (c) Let C be a non-zero RH-submodule of Bg. If $Cg^{-1} = \langle 0 \rangle$, then $C = \langle 0 \rangle$, a contradiction. Thus $Cg^{-1} = B$ and so $C = (Cg^{-1})g = Bg$. Hence, Bg is simple.

   (2) Put $C = \sum_{g \in G} Bg$ so that C is an RG-submodule of A. Since $C \neq \langle 0 \rangle$, we have that $C = A$. Then it suffices to apply Corollary 4.3. $\square$

The following result appears in the papers of B. Hartley [102] and D.I. Zaitsev [296].

Theorem 5.5. *Let R be a ring, G be a group, and A be a simple RG-module. If H is a normal subgroup of G and $|G : H| = n$ is finite, then A has a simple RH-submodule B and $A = \bigoplus_{g \in S} Bg$, for some finite subset S such that $|S| \leq n$.*

Proof. Obviously A is an artinian RG-module. By Theorem 5.2, A is an artinian RH-module. It follows that the family of all non-zero RH-submodules of A has a minimal element B. Then B is clearly simple. Let X be a transversal to H in G so that $|X| = n$. By Lemma 5.4, $A = \sum_{g \in X} Bg$. It suffices to apply Corollary 4.3. $\square$

Corollary 5.6. *Let R be a ring, G a group, and H a normal subgroup of G of finite index. If A is a semisimple RG-module, then A is a semisimple RH-module.*

This extends the celebrated theorem of Clifford (H. Clifford [45]) to infinite groups.

The paper of D.I. Zaitsev [297] contains the following extension of Theorem 5.5.

Theorem 5.7. *Let F be a field, G an abelian group, and H a periodic subgroup of G. If A is a simple FG-module, then A is a semisimple FH-module.*

Proof. Let $\mathcal{L}$ be a local system of finite subgroups of H and choose $K \in \mathcal{L}$. If $0 \neq b \in A$, then $\dim_F bFK$ is finite and so bFK has a simple FK-submodule B. By Lemma 5.4, $A = \bigoplus_{x \in M} Bx$, for some subset $M \subset G$. Since FG is commutative, B and Bx are isomorphic for every $x \in M$. By Proposition 4.5, $aFK \cong B$, for each $0 \neq a \in A$. In particular, aFK is a simple FK-submodule. Since $H = \bigcup_{K \in \mathcal{L}} K$ and $aFH = \bigcup_{K \in \mathcal{L}} aFK$, we deduce that aFH is further a simple FH-submodule. Then, it suffices to apply Lemma 5.4. $\qquad\square$

If A is a simple $\mathbb{Z}G$-module, then the underlying additive group of A either is a divisible torsion-free abelian group or is a p-elementary abelian group, for some prime p. In the first case we may think of A as a $\mathbb{Q}G$-module whereas in the second one we think of A as an $\mathbb{F}_pG$-module. Thus, in any case, we deduce from Theorem 5.7 the following result.

Corollary 5.8. *Let G be an abelian group and H a periodic subgroup of G. If A is a simple $\mathbb{Z}G$-module, then A is a semisimple $\mathbb{Z}G$-module.*

The next results describe other criteria of semisimplicity obtained with the aid of the well-known theorem of Maschke. We are considering here one of the most general versions of this result, which was obtained in the paper of S. Franciosi, F. de Giovanni and L.A. Kurdachenko [74], and from it we will deduce several consequences.

Theorem 5.9. *Let R be a ring, G a group and H a normal subgroup of G such that G/H has finite order n. If A is an RG-module and B is an RG-submodule admitting an R-complement, then there exists an RG-submodule E such that $nA \leq B + E$ and $n(B \cap E) = \langle 0 \rangle$.*

Proof. By hypothesis, there is an R-submodule C such that $A = B \oplus C$. Let $\pi : A \longrightarrow C$ be the canonical projection. If $x, y \in G$ satisfy $Hx = Hy$, then there is some $h \in H$ such that $y = hx$. If $a \in A$, then we express $a = b + c$, where $b \in B$ and $c \in C$. Then we have

$$a(h^{-1}\pi h) = (ah^{-1})\pi h = (bh^{-1} + ch^{-1})\pi h = (ch^{-1})h = c = a\pi$$

and, hence,

$$\begin{aligned}
a(y^{-1}\pi y) &= a((x^{-1}h^{-1})\pi hx) = ((ax^{-1})(h^{-1}\pi h))x \\
&= ((ax^{-1})\pi)x = a(x^{-1}\pi x).
\end{aligned}$$

Given $f = Hg \in F = G/H$, we define an R-endomorphism $\vartheta(f) : A \longrightarrow A$ by $a\vartheta(f) = a(g^{-1}\pi g)$, if $a \in A$. Put now $\tau = \sum_{f \in F} \vartheta(f)$. If $x \in G$, then $Hx = t \in F$

and, if $a \in A$, we obtain

$$(a\tau)x = (\sum_{f \in F} a\vartheta(f))x = (\sum_{Hg \in F} a(g^{-1}\pi g)x = \sum_{Hg \in F} (ag^{-1}\pi g)x$$

$$= \sum_{Hg \in F} (ax)(x^{-1}g^{-1}\pi gx) = \sum_{f \in F} (ax)\vartheta(ft) = (ax)\tau,$$

because the mapping $f \mapsto ft, f \in F$ is a permutation of F. Therefore τ is an RG-endomorphism of A and so $A\tau$ is an RG-submodule of A. If $a \in A$, we have

$$na - a\tau = \sum_{f \in F} (a - a\vartheta(f)) = \sum_{Hg \in F} (a - a(g^{-1}\pi g))$$

$$= \sum_{Hg \in F} (ag^{-1} - a(g^{-1})\pi)g.$$

On the other hand, $ag^{-1} - a(g-1)\pi \in B$ and then $(ag^{-1} - a(g^{-1})\pi)g \in B$. Thus $na - a\tau \in B$.

Let $b \in B$. Then $bg^{-1} \in B$ and $(bg^{-1})\pi = 0$. Hence,

$$b\tau = \sum_{f \in F} b\vartheta(f) = \sum_{Hg \in F} ((b(g^{-1})\pi)g = 0.$$

If $a \in nA$, then $a = nu$, for some $u \in A$. Then $a - u\tau = nu - u\tau \in B$ so that $a \in B + A\tau$; that is, $nA \leq B + A\tau$.

If $b \in B \cap A\tau$, then $b = v\tau$, for some $v \in A$. Again, $nv - v\tau \in B$; that is, $nv \in B$. Therefore $0 = (nv)\tau = n(v\tau) = nb$. This means that $n(B \cap A\tau) = \langle 0 \rangle$. Hence, it suffices to define $E = A\tau$ to obtain the required result. □

Corollary 5.10. *Let G be a finite group and suppose that A is a $\mathbb{Z}G$-module whose additive group has finite special rank. If B is a $\mathbb{Z}G$-submodule admitting a $\mathbb{Z}$-complement, then there is a $\mathbb{Z}G$-submodule E such that $A/(B+E)$ and $B \cap E$ are finite.*

Proof. It follows immediately from Theorem 5.9 since a bounded abelian group of finite special rank is finite. □

The following simple consequence is due to D.I. Zaitsev [294] and only makes use of standard properties of divisible abelian groups.

Corollary 5.11. *Let G be a finite group and suppose that A is a $\mathbb{Z}G$-module, and B is a $\mathbb{Z}G$-submodule of A. If the underlying additive groups of A and B are divisible, then there exists a $\mathbb{Z}G$-submodule E such that $A = B + E$ and $n(B \cap E) = \langle 0 \rangle$, where $n = |G|$. In particular, if B is $\mathbb{Z}$-torsion-free, then $A = B \oplus E$.*

Corollary 5.12. *Let G be a finite group. Suppose that A is a $\mathbb{Z}G$-module and B is a $\mathbb{Z}G$-submodule of A. If the underlying additive group of A is finitely generated, then there exists a $\mathbb{Z}G$-submodule E such that $B \cap E$ and $A/(B \cap E)$ are finite.*

Proof. Since the factor-group A/B is finitely generated,

$$A/B = T/B \oplus C/B,$$

where T/B is finite and C/B is torsion-free. If $|T/B| = t$, then $t(A/B) = U/B$ is torsion-free. Moreover, the $\mathbb{Z}G$-submodule U has finite index. Since U/B is a free abelian group, $U = B \oplus E$ for some subgroup E. It suffices to apply Corollary 5.10 to the $\mathbb{Z}G$-submodule U. $\qquad\qquad\square$

Corollary 5.13. *Let G be a finite group. Suppose that A is a $\mathbb{Z}G$-module whose additive group is periodic such that $\Pi(A) \cap \Pi(G) = \emptyset$. If B is a $\mathbb{Z}G$-submodule of A admitting a $\mathbb{Z}$-complement, then there is a $\mathbb{Z}G$-submodule C such that $A = B \oplus C$.*

Corollary 5.14. *Let G be a finite group, F a field, A an FG-module, and B an FG-submodule of A. If char $F = 0$ or char$F \notin \Pi(G)$, then there is an FG-submodule C such that $A = B \oplus C$.*

Corollary 5.15. *Let A be an FG-module, where G is a finite group and F is a field. If char $F = 0$ or char $F \notin \Pi(G)$, then A is a semisimple FG-module.*

Given a group G and a ring R, the R-homomorphism $\omega : RG \longrightarrow R$ given by

$$\left(\sum_{g \in G} x_g g\right)\omega = \sum_{g \in G} x_g,$$

where all but finitely many x_g are zero (i. e. both sums are finite), is called *the unit augmentation of RG* or simply *the augmentation of RG*. We denote the kernel of ω by ωRG. It is a two-sided ideal called *the augmentation ideal of RG*. This ideal is generated by the elements $\{g - 1 \mid 1 \neq g \in G\}$.

Corollary 5.16. *Let A be an FG-module, where G is a finite group and F is a field. If char $F = 0$ or char $F \notin \Pi(G)$, then $A = C_A(G) \oplus A(\omega FG)$.*

Proof. Indeed, by Corollary 5.14, there exists an FG-submodule E such that

$$A = C_A(G) \oplus E.$$

If $a \in A$, then $a = c + u$, where $c \in C_A(G)$ and $u \in E$. Then

$$a(g - 1) = c(g - 1) + u(g - 1) = u(g - 1) \in E$$

for each element $g \in G$; that is, $A(\omega FG) \leq E$. By Corollary 5.14 again, we have the decomposition

$$E = A(\omega FG) \oplus V,$$

where V is an FG-submodule. If $v \in V$, then $v(g - 1) \in A(\omega FG) \cap V = \langle 0 \rangle$. It follows that $E = A(\omega FG)$. $\qquad\qquad\square$

Lemma 5.17. *Let G be a finite group and suppose that A is a $\mathbb{Z}G$-module of composition length 2. If $\Pi(A) \cap \Pi(G) = \emptyset$, then $A = C_A(G) \oplus A(\omega \mathbb{Z}G)$.*

Proof. Let B be a simple $\mathbb{Z}G$-submodule of A such that A/B is also simple. If $B = B(\omega\mathbb{Z}G)$ and $A/B = (A/B)(\omega\mathbb{Z}G)$, then $A = A(\omega\mathbb{Z}G)$. Therefore we can assume that either $A/B \neq (A/B)(\omega\mathbb{Z}G)$ or $B \neq B(\omega\mathbb{Z}G)$. We consider these cases separately.

Suppose first that $B \neq B(\omega\mathbb{Z}G)$. Since B simple, $B(\omega\mathbb{Z}G) = \langle 0 \rangle$; that is, $B \leq C_A(G)$. If $B \neq C_A(G)$, then $A = C_A(G)$ and all is done. Hence, we may assume that $B = C_A(G)$. It is not hard to see that B and A/B have to be finite elementary abelian groups. If $\Pi(B) \cap \Pi(A/B) = \emptyset$, then all is done. Suppose that the additive group of A is a p-group, where p is a prime. If $pA = \langle 0 \rangle$, we may think of A as an $\mathbb{F}_p G$-module and apply Corollary 5.16. Suppose that $pA \neq \langle 0 \rangle$. In this case $B = \Omega_1(A)$, the lower layer of A. The mapping $\rho : a \longrightarrow pa, a \in A$ is a $\mathbb{Z}G$-endomorphism, so that Im ρ and Ker ρ are $\mathbb{Z}G$-submodules of A such that Im $\rho \cong_{\mathbb{Z}G} A/\operatorname{Ker} \rho$. It follows that Im $\rho = B$ and Ker $\rho = B$; that is, $B \cong_{\mathbb{Z}G} A/B$. In particular, $(A/B)(\omega\mathbb{Z}G) = A/B$. Let $a \in A$ and $g \in G$. Then $ag = a + b$ for some $b \in B$. We have $ag^m = a + mb$ for every $m \in \mathbb{N}$. In particular, $ag^p = a$. Since $(|g|, p) = 1$, $g \in C_A(G)$. This contradiction shows that $pA = \langle 0 \rangle$.

Consider now the case in which $B = B(\omega\mathbb{Z}G)$ and $A(\omega\mathbb{Z}G) = B$. Considering again the mapping $\rho : a \mapsto pa, a \in A$, we obtain another time that $pA = \langle 0 \rangle$. In this case, it suffices to apply Corollary 5.16. $\square$

This implies immediately the following consequence.

Corollary 5.18. *Let G be a finite group and suppose that A is a $\mathbb{Z}G$-module of finite composition length. If $\Pi(A) \cap \Pi(G) = \emptyset$, then $A = C_A(G) \oplus A(\omega\mathbb{Z}G)$.*

Proposition 5.19. *Let G be a finite group and suppose that A is a $\mathbb{Z}G$-module whose additive group is periodic. If $\Pi(A) \cap \Pi(G) = \emptyset$, then $A = C_A(G) \oplus A(\omega\mathbb{Z}G)$.*

Proof. Let $\mathcal{L}$ be a local system of finite subgroups of A and let

$$\mathcal{Q} = \{LG \mid L \in \mathcal{L}\}.$$

Since G is finite, every member $B \in \mathcal{Q}$ is likewise finite. By Corollary 5.18, $B = C_B(G) \oplus B(\omega\mathbb{Z}G)$. If $C \in \mathcal{Q}$ satisfies $B \leq C$, then $C_B(G) \leq C_C(G)$ and $B(\omega\mathbb{Z}G) \leq C(\omega\mathbb{Z}G)$. These inclusions imply

$$C_A(G) = \bigcup_{B \in \mathcal{Q}} C_B(G) \text{ and } A(\omega\mathbb{Z}G) = \bigcup_{B \in \mathcal{Q}} B(\omega\mathbb{Z}G).$$

It follows that $A = C_A(G) \oplus A(\omega\mathbb{Z}G)$. $\square$

Chapter 6

Modules over Dedekind domains

Many chapters of this book are dedicated to modules over a group ring of the form DG, where D is a Dedekind domain. Thus we shall need some results about Dedekind domains and modules over Dedekind domains. Certain complications arise here — there is no single book that contains all the necessary material suitable for our specific purposes. Many of these results can be found only in journal articles, where they have been obtained for modules over wider classes of rings; but most of the time we only need particular cases of those general results. Therefore, it would be useful to present a modicum of such important information in this chapter. In subsequent chapters we will provide results as needed along with the appropriate references, and supply proofs for results that are not stated in the needed form (for example, the primary decompositions of periodic modules over Dedekind domains).

There are many books and papers dedicated to Dedekind domains and modules over Dedekind domains. Coming up with an exhaustive list would be very difficult. Below we mention some of them, which include proofs of the most frequently used and effective classic results: N. Bourbaki [20], Z.I. Borevich and A.R. Shafarevich [21], S.U. Chase [32], C. Curtis and I. Reiner [47, 48], I. Fleichsner [70], L. Fuchs and S.B. Lee [80], L. Fuchs and L. Salce [81], R. Gilmer [84], I. Kaplansky [121, 122, 123, 124, 125], G. Karpilovsky [131], M.D. Larsen and P.J. McCarthy [180], E. Matlis [183, 184, 185, 186, 187, 188], W. Narkiewicz [201], D.G. Northcott [206], D.S. Passman [219], O.F.G. Schilling [254], R. Sharp [263], and D.W. Sharpe and P. Vamos [265].

The idea of the Dedekind domain naturally arose within the process of the development of Algebraic Number Theory. In his famous Supplement XI to Dirichlet's book, R. Dedekind [51] proved that every ideal of the ring of integers of a field of algebraic numbers can be decomposed into prime factors and this factorization is unique. He created the fractional ideal and also proved that non-zero fractional ideals formed an abelian group. There are many definitions of Dedekind domains. We will consider some of them.

If R is a commutative ring, we recall that *the prime spectrum* $\mathrm{Spec}(R)$ *of R* (or simply *the spectrum of R*) is the set of all prime ideals of R.

Let R be an integral domain, and let K be the field of fractions of R. A non-zero R-submodule A of K is called *a fractional ideal of R* if there is some $a \in R$ such that $aA \leq R$. This means that all elements of A can be expressed with fractions having a as denominator, that is, all of them have a *common denominator*. Note that the fractional ideals of R contained in R itself are ordinary ideals of R and conversely. If A and B are fractional ideals, then the set of all finite sums $a_1 b_1 + \cdots + a_n b_n, a_i \in A, b_j \in B$ forms a fractional ideal, known as *the product AB of A and B*. For, if $aA \leq R$ and $bB \leq R$, then $abAB \leq R$. Obviously $A(BC) = (AB)C$, for fractional ideals A, B, C of R. This means that the set $\mathcal{FI}(R)$ of all fractional ideals of R is a commutative semigroup, which admits R itself as its identity element.

If A is a fractional ideal of R, then

$$A' = \{x \in K \mid xA \leq R\}$$

is a fractional ideal of R and $AA' \leq R$. We say that A is *invertible*, if $AA' = R$. In this case, A' is the unique fractional ideal of R satisfying $AA' = R$ and we write A^{-1} instead of A'. An integral domain D is called *a Dedekind domain*, if every fractional ideal of D is invertible, which is equivalent to the fact that the semigroup $\mathcal{FI}(D)$ is a group. Every principal ideal domain is a Dedekind domain and, in particular, the ring $\mathbb{Z}$ of all integers is a Dedekind domain.

Theorem 6.1. *Let D be a Dedekind domain. Then:*

(1) *D is noetherian.*

(2) *Every non-zero prime ideal of D is maximal.*

Let R be an integral domain and let L be a field including R, so that L includes the field of fractions K of R. We recall that an element $c \in L$ is said to be *integral over R* (or shortly *R-integral*) if there are $a_0, a_1, \cdots, a_{n_1} \in R$ such that

$$a_0 + a_1 c + \cdots + a_{n-1} c^{n-1} + c^n = 0.$$

In other words, c is a root of a monic polynomial with coefficients belong to R. It is well known that the set C of all elements of L which are integral over R is a subring. Let L be a field and R be a subring of L such that $1 \in R$. The subring of L consisting of all elements which are integrals over R is called *the integral closure of R in L*. The integral domain R is said to be *integrally closed* if R is its integral closure in the field of fractions.

The following result proves the full characterization of a Dedekind domain by means of these concepts and the conditions of Theorem 6.1.

Theorem 6.2. *Let D be an integral domain. Then D is a Dedekind domain if and only if the following conditions are satisfied.*

(1) *D is noetherian.*

(2) *Every non-zero prime ideal of D is maximal.*

(3) *D is integrally closed.*

As we can see every principal ideal domain is a Dedekind domain. Among principal ideal domains the following are the most important for applications in Group Theory: the ring $\mathbb{Z}$ of all integers, the ring $F\langle x \rangle$ of an infinite cyclic group $\langle x \rangle$ over a (finite) field F; the ring $\mathbb{Z}_{p^\infty}$ of integer p-adics; the ring $F[[x]]$ of power series over a (finite) field F. Observe also, that if R is an integral noetherian domain whose non-zero prime ideals are maximal, F is the field of fractions of R, and K is a finite field extension of F, then the integral closure of R in K is a Dedekind domain (see, for example, [20, Chapter 7.2, Corollary 2 of Proposition 5].

Beyond all doubt, the next results describe the most important features of Dedekind domains.

Theorem 6.3. *Every proper non-zero ideal of a Dedekind domain can be represented uniquely as a product of non-zero prime ideals (up to ordering of these)*

Corollary 6.4. *Let D be a Dedekind domain. Then the group $FI(D)$ is a free abelian group generated by the non-zero prime ideals of D.*

Given two ideals A and B of a Dedekind domain D, we say that A *is divisible by* B (or that B *divides* A), if there exists an ideal C of D such that $A = BC$.

Corollary 6.5. *Let D be a Dedekind domain. Then every non-zero ideal can be divisible only by finitely many distinct ideals.*

It follows that every fractional ideal A of a Dedekind domain D can be uniquely written in the form

$$A = P_1^{a(P_1)} \cdots P_r^{a(P_r)},$$

where $P_1, \ldots, P_r \in \mathrm{Spec}(D)$ and $a(P_1), \ldots, a(P_r) \in \mathbb{Z}$. Sometimes, it is more convenient to write this in the general form

$$A = \Pi_{P \in \mathrm{Spec}(D) \backslash \langle 0 \rangle} P^{a(P)},$$

where $a(P) \in \mathbb{Z}$ and $a(P) = 0$ for all but finitely many values of $a(P)$. Note that A is an ordinary ideal of D if and only if $a(P) \geq 0$ for all P.

Let A and B be (ordinary) ideals of the Dedekind domain D. By definition *the greatest common divisor of A and B* is the (unique) ideal $\mathrm{GCD}(A, B)$ of D that satisfies the following:

(i) $\mathrm{GCD}(A, B)$ divides A and B.

(ii) If C is an ideal of D and C divides A and B, then C divides $\mathrm{GCD}(A, B)$.

In fact, if

$$\langle 0 \rangle \neq A = \Pi_{P \in \mathrm{Spec}(D) \setminus \langle 0 \rangle} P^{a(P)}$$

and

$$\langle 0 \rangle \neq B = \Pi_{P \in \mathrm{Spec}(D) \setminus \langle 0 \rangle} P^{b(P)},$$

then A divides B if and only if $a(P) \leq b(P)$, for every $P \in \mathrm{Spec}(D) \setminus \langle 0 \rangle$. It follows that

$$\mathrm{GCD}(A, B) = \Pi_{P \in \mathrm{Spec}(D) \setminus \langle 0 \rangle} P^{c(P)},$$

where $c(P) = \max\{a(P), b(P)\}$.

The least common multiple of A and B is the unique ideal $\mathrm{LCM}(A, B)$ that satisfies the following properties:

(i) A and B divide $\mathrm{LCM}(A, B)$.

(ii) If C is an ideal of D and A and B divide C, then $\mathrm{LCM}(A, B)$ divides C.

It is easy to see that

$$\mathrm{LCM}(A, B) = \Pi_{P \in \mathrm{Spec}(D) \setminus \langle 0 \rangle} P^{d(P)},$$

where $d(P) = \max\{a(P), b(P)\}$.

Let R be a commutative ring. We say that the proper ideals A and B of R are *relatively prime* if $A + B = R$. Now we are in a position to obtain the other accustomed properties of a Dedekind domain.

Proposition 6.6. *Let D be a Dedekind domain. Then the following assertions hold.*

(1) *If A and B are fractional ideals, then B divides A if and only if $A \leq B$.*

(2) *If A is a fractional ideal, then there exists a principal fractional ideal aR such that $(aR)A^{-1} \leq R$.*

(3) *If A and B are ideals of R, then $\mathrm{GCD}(A, B) = A + B$ and $\mathrm{LCM}(A, B) = A \cap B$.*

(4) *If A and B are ideals of R, then $AB = (A + B)(A \cap B)$.*

(5) *If A and B are relatively prime ideals of R, then $AB = A \cap B$.*

Lemma 6.7. *Let R be a commutative ring and let A and B be ideals of R. If A and B are relatively prime, then so are A^k and B^m, for every $k, m \in \mathbb{N}$*

Lemma 6.8. *Let R be a commutative ring. The ideals $A_1, \ldots, A_r$ of R are pairwise relatively prime if and only if for every $1 \leq i \leq r$ we have that*

$$A_i + A_1 \cdots A_{i-1} A_{i+1} \cdots A_r = R.$$

Corollary 6.9. *Let D be a Dedekind domain. If the ideals $A_1, \ldots, A_r$ of D are pairwise relatively prime, then*

$$A_1 \cdots A_r = A_1 \cap \cdots \cap A_r.$$

Theorem 6.10 (Chinese Remainder Theorem). *Let D be a Dedekind domain. If $A_1, \ldots, A_k$ are pairwise, relatively prime, non-zero ideals of D and $a_1, \ldots, a_k \in D$, then there exists some $a \in D$ such that $a_i + A_i = a + A_i$, for each $1 \leq i \leq k$. Moreover, if b satisfies $b + A_i = a_i + A_i$, for each $1 \leq i \leq k$, then*

$$a + A_1 \cdots A_k = b + A_1 \cdots A_k.$$

Corollary 6.11. *Let D be a Dedekind domain. If $A_1, \ldots, A_k$ are pairwise, relatively prime, non-zero ideals of D, then*

$$D/(A_1 \cdots A_k) \cong D/A_1 \times \cdots \times D/A_k.$$

Corollary 6.12. *Let D be a Dedekind domain. If I is a non-zero ideal of D and $I = P_1^{m_1} \cdots P_k^{m_k}$, where $P_1, \ldots, P_k$ are different prime ideals of D, then*

$$D/I \cong D/P_1^{m_1} \times \cdots \times D/P_k^{m_k}.$$

As we have already mentioned, every principal ideal domain is a Dedekind domain. The following results show that Dedekind domains are very close to principal ideal domains.

Proposition 6.13. *Let D be a Dedekind domain and let Q be a non-zero prime ideal of D. Suppose that B is an ideal such that $B \geq Q^m$, for some $m \in \mathbb{N}$. Then there is some $k \leq m$ such that $B = Q^k$. In particular, if $y \in Q \setminus Q^2$, then $Q^t = y^t D + Q^n$, for every $n \in \mathbb{N}$ and each $t \leq n$.*

Proof. By Theorem 6.3,

$$B = \Pi_{P \in (Spec(D)) \setminus \langle 0 \rangle} P^{a(P)},$$

and then there exists some $k \in \mathbb{N}$ such that $B = Q^k A$, where A is an ideal of D. In particular, $B \leq Q^k$ by Proposition 6.6. If $k \geq m$, then $B \leq Q^k = Q^{k-m} Q^m \leq Q^m$; that is, $B = Q^m$ and we are done. Therefore $k \leq m$. Suppose that $A \neq D$ so that A and Q are relatively prime by Lemma 6.7 and Lemma 6.8. We have $BA^{-1} \leq Q^k \leq D$, whence $A^{-1} \leq B^{-1}$ and then $B \leq A$. It follows that $Q^m \leq A$. By Lemma 6.7, $A + Q^m = D$, which implies that $A = D$, a contradiction. Hence $A = D$, and then $B = Q^k$.

Since $y \in Q$, $yD \leq Q$. This and $Q^n \leq Q$ for each $n \in \mathbb{N}$ give that $yD + Q^n \leq Q$. By the result shown in the above paragraph, $yD + Q^n = Q^s$, for some $s \leq n$. We claim that $s = 1$. For, if $s > 1$, then

$$yD + Q^n \leq Q^s = Q^2 Q^{s-2} \leq Q^2$$

and so $yD \leq Q^2$, which contradicts the choice of y. Therefore $s = 1$, and so $yD + Q^n = Q$. Pick $x_1, x_2 \in Q$. Then $x_j = yz_j + u_j$, where $z_j \in D$ and $u_j \in Q^n$ ($j = 1, 2$). Thus

$$x_1 x_2 = (yz_1 + u_1)(yz_2 + u_2)$$
$$= y^2 z_1 z_2 + y z_1 u_2 + y z_2 u_1 + u_1 u_2 \in y^2 D + Q^n.$$

It follows that $Q^2 \leq y^2 D + Q^n$, and hence $Q^2 = y^2 D + Q^n$. By induction, $Q^t = y^t D + Q^n$, for every $t \leq n$. $\square$

Corollary 6.14. *Let A be a non-zero ideal of a Dedekind ring D. Then every ideal of D/A is principal.*

Proof. Let B be an ideal of D such that $B \geq A$. By Theorem 6.3,

$$A = P_1^{k_1} \cdots P_t^{k_t},$$

where $P_1, \ldots, P_t \in \mathrm{Spec}(D)$. By Corollary 6.12,

$$D/A \cong D/P_1^{k_1} \times \cdots \times D/P_t^{k_t}.$$

By Proposition 6.13, there exist some $a_1, \ldots, a_t \in D$ such that

$$B + P_j^{k_j} = a_j D + P_j^{k_j},$$

for every $1 \leq j \leq t$. Using Theorem 6.10, we find an element a such that

$$a + P_j^{k_j} = a_j + P_j^{k_j},$$

for every $1 \leq j \leq t$. Now it is not hard to see that $B = aD + A$ and then B/A is principal. $\square$

Corollary 6.15. *Let A be a non-zero ideal of a Dedekind domain D. There exist some elements a_1 and a_2 such that $A = a_1 D + a_2 D$.*

Proof. Let $0 \neq a_1 \in A$ and put $B = a_1 D$. By Corollary 6.14,

$$A/B = (a_2 + B)(D/B) = (a_2 D + B)/B,$$

for some $a_2 \in A$. Then $A = a_2 D + a_1 D$, as required. $\square$

Corollary 6.16. *Let A and B be non-zero ideals of a Dedekind domain D. Then there exists a non-zero ideal C such that $B + C = D$ and AC is principal.*

Proof. Since D contains no zero-divisors, $AB \neq \langle 0 \rangle$. By Corollary 6.14, $A/AB = (aD + AB)/AB$, for some $a \in A$. Since $aD \leq A$, by Proposition 6.6, $aD = AC$, for some ideal $C \neq \langle 0 \rangle$. Thus $AC + AB = A$ and, multiplying by A^{-1}, we obtain that $C + B = D$. $\square$

Corollary 6.17. *Let A and B be non-zero ideals of a Dedekind domain D. Then the D-modules D/A and B/AB are isomorphic. In particular, if $k \in N$, then D/A and A^k/A^{k+1} are isomorphic D-modules.*

Proof. By Corollary 6.16, there exists a non-zero ideal C such that $C + A = D$ and $BC = aD$, for some $a \in D$. Define the mapping $\phi : D \longrightarrow B/AB$ by $x\phi = ax + AB, x \in D$. We have

$$(x + y)\phi = a(x + y) + AB = ax + ay + AB$$
$$= (ax + AB) + (ay + AB) = x\phi + y\phi$$

and

$$(dx)\phi = a(dx) + AB = (ax)d + AB$$
$$= (ax + AB)d = (x\phi)d \text{ (for each } d \in D),$$

and so ϕ is a D-homomorphism. Furthermore,

$$aD + AB = CB + AB = (C + A)B = DB = B.$$

Thus Im $\phi = B/AB$. If $x \in$ Ker ϕ, then $ax \in AB$ and so $axC \leq ABC = aA$. It follows that $xC \leq A$. Since $A + C = D$, there are some elements $a_1 \in A$ and $c \in C$ such that $1 = a_1 + c$. Thus $x = x \cdot 1 = xa_1 + xc \in A$. Conversely, if $x \in A$, then $ax \in AB$ and so $x\phi = ax + AB = AB$. Therefore Ker $\phi = A$. $\square$

Let R be a ring and let A be an R-module. We define

$$t_R(A) = \{a \in A \mid \text{Ann}_R(a) \neq \langle 0 \rangle\}.$$

This subset has a nice structure provided R has no zero-divisors, as the following result shows.

Lemma 6.18. *Let R be an integral domain. If A is an R-module, then $t_R(A)$ is an R-submodule of A.*

Proof. Let $a_1, a_2 \in t_R(A)$. Put $I_1 = \text{Ann}_R(a_1)$ and $I_2 = \text{Ann}_R(a_2)$. Then $I_1 \cap I_2 \leq \text{Ann}_R(a_1 - a_2)$. Since R contains no zero-divisors, $I_1 I_2 \neq \langle 0 \rangle$. Since $I_1 I_2 \leq I_1 \cap I_2$, $\text{Ann}_R(a_1 - a_2) \neq \langle 0 \rangle$.

Given $a \in t_R(A)$ and $0 \neq x \in R$, if $u \in \text{Ann}_R(a)$, then

$$(ax)u = a(xu) = a(ux) = (au)x = 0x = 0.$$

It follows that $\text{Ann}_R(a) \leq \text{Ann}_R(ax)$, that is $ax \in t_R(A)$. $\square$

Let R be an integral domain. Then the submodule $t_R(A)$ of A is called *the R-periodic part of A. A is called *R-periodic* if $A = t_R(A)$ and *R-torsion-free* if $t_R(A) = \langle 0 \rangle$. Note that, for A arbitrary, then $t_R(A)$ is R-periodic and $A/t_R(A)$ is R-torsion-free.

The following concept is dual to the concept of R-annihilators of subsets of A. If $L \subseteq R$ is a subset of R, then *the A-annihilator of L* is

$$\text{Ann}_A(L) = \{a \in A \mid aL = \langle 0 \rangle\}.$$

If R is commutative, then it is customary that $\text{Ann}_A(L)$ is an R-submodule of A.

Lemma 6.19. *Let A be an R-module, where R is a commutative ring and let $\mathcal{W}$ be the set of all ideals H of R such that $\operatorname{Ann}_A(H) \neq \langle 0 \rangle$. If $\mathcal{W}$ has a maximal element $P \neq R$, then P is a prime ideal of R.*

Proof. All we have to show is that P is a prime ideal. Let $B = \operatorname{Ann}_A(P)$. Since R is a commutative ring, B is an R-submodule. Let x, y be elements of R such that $xy \in P$ and let $0 \neq b \in B$. Then $b(xy) = 0$. If $bx = 0$, then $xR \in \mathcal{W}$. In this case, $xR + P \leq \operatorname{Ann}_R(b)$ and, in particular, $\operatorname{Ann}_A(xR + P) \neq \langle 0 \rangle$; that is, $xR + P \in \mathcal{W}$. By the choice of P, it follows that $xR + P = P$; that is, $x \in P$. If $bx \neq 0$, then $\operatorname{Ann}_A(yR) \neq \langle 0 \rangle$. Repeating the previous arguments, we obtain $y \in P$. This means that P is a prime ideal. $\qquad\square$

Corollary 6.20. *Let R be a noetherian commutative ring and suppose that A is an R-module such that $t_R(A) \neq \langle 0 \rangle$. Then R has a non-zero prime ideal P such that $\operatorname{Ann}_A(P) \neq \langle \emptyset \rangle$.*

Let A be an R-module, where R is a commutative ring. We define *the R-assasinator of A* as the set $\operatorname{Ass}_R(A)$ consisting of all prime ideals P of R such that $\operatorname{Ann}_A(P) \neq \langle 0 \rangle$.

Now we finish this chapter showing the existence of the primary decomposition of certain modules over a Dedekind domain. Again, inspired by the ordinary layer of a p-group, we consider the following construction. Let I be an ideal of a commutative ring R. If $k \in \mathbb{N}$, we define

$$\Omega_{I,k}(A) = \{a \in A \mid aI^k = \langle 0 \rangle\}.$$

It is easy to see that $\Omega_{I,k}(A)$ is an R-submodule and

$$\Omega_{I,1}(A) \leq \Omega_{I,2}(A) \leq \cdots \leq \Omega_{I,k}(A) \leq \cdots.$$

The R-submodule $A_I = \bigcup \Omega_{I,k}(A)$ is said to be *the I-component of A*. If $A = A_I$, then A is said to be *an I-module*. If $I = P$ is a prime ideal, a P-module is generically called *a primary module*.

Lemma 6.21. *Let A be a D-module, where D is a Dedekind domain. If U and V are relatively prime ideals of D, then $A_U \cap A_V = \langle 0 \rangle$.*

Proof. Pick $u \in U$ and $v \in V$ such that $1 = u + v$. Suppose that $A_U \cap A_V \neq \langle 0 \rangle$. Then $\Omega_{U,1}(A) \cap \Omega_{V,1}(A) \neq \langle 0 \rangle$. If

$$0 \neq a \in \Omega_{U,1}(A) \cap \Omega_{V,1}(A),$$

then $a = a \cdot 1 = a(u + v) = au + av = 0$. This contradiction shows that $A_U \cap A_V = \langle 0 \rangle$. $\qquad\square$

Corollary 6.22. *Let A be a D-module, where D is a Dedekind domain. Suppose that U and V are two relatively prime ideals of D such that $A = A_U \oplus A_V$. If B is a submodule of A, then*

$$(A/B)_U = (A_U + B)/B, \quad (A/B)_V = (A_V + B)/B,$$

and $A/B = (A/B)_U \oplus (A/B)_V$.

Proof. Since $A = A_U \oplus A_V$,

$$A/B = (A_U + B)/B + (A_V + B)/B.$$

It easy to see that

$$(A_U + B)/B \leq (A/B)_U \text{ and } (A_V + B)/B \leq (A/B)_V.$$

By Lemma 6.21,

$$(A_U + B)/B \cap (A_V + B)/B = \langle 0 \rangle .$$

It follows that

$$A/B = (A_U + B)/B \oplus (A_V + B)/B$$

and then

$$(A/B)_U = (A_U + B)/B \text{ and } (A/B)_V = (A_V + B)/B. \qquad \square$$

Corollary 6.23. *Let D be a Dedekind domain. Suppose we are given a family $\mathcal{U}$ of proper, non-zero, pairwise, relatively prime, ideals of D. If A is a D-module, then*

$$\sum_{U \in \mathcal{U}} A_U = \bigoplus_{U \in \mathcal{U}} A_U.$$

Proof. It suffices to show the result for every finite subset $\mathcal{V} \subseteq \mathcal{U}$. Suppose that $|\mathcal{V}| = k$ and proceed by induction on k. If $k = 2$, then the result follows from Lemma 6.21. Let $k > 2$. If $V \in \mathcal{V}$, by induction

$$\sum_{U \in \mathcal{V} \setminus \{V\}} A_U = \bigoplus_{U \in \mathcal{V} \setminus \{V\}} A_U.$$

Suppose that

$$A_V \cap (\bigoplus_{U \in \mathcal{V} \setminus \{V\}} A_U) \neq \langle 0 \rangle .$$

Thus, for every $U \in \mathcal{V}$, there are some elements $a_U \in A_U$ such that

$$a_V = \sum_{U \in \mathcal{V} \setminus \{V\}} a_U$$

and $a_V \neq 0$. Then there exists some $V \neq W \in \mathcal{V}$ such that $a_W \neq 0$. Pick $r_W, r_V \in \mathbb{N}$ such that $a_V V^{r_V} = \langle 0 \rangle$ and $a_W U^{r_W} = \langle 0 \rangle$. By Lemma 6.7, $W^{r_W} + V^{r_V} = D$, and so there are $x_W \in W^{r_W}$ and $x_V \in V^{r_V}$ such that $1 = x_W + x_V$. Then

$$a_V = a_V \cdot 1 = a_V (x_W + x_V) = a_V x_W + a_V x_V$$

$$= a_V x_W = \sum_{U \in \mathcal{V} \setminus \{V\}} a_U x_W.$$

Since A_U is a D-submodule of A, $a_U x_W \in A_U$. But $a_W x_W = 0$, so that

$$\sum_{U \in V \setminus \{V\}} a_U x_W$$

contains in fact $k - 2 < k$ summands. By induction, $a_V = 0$. $\qquad \square$

Corollary 6.24. *Let D be a Dedekind domain. Suppose we are given a family $\mathcal{U}$ of proper, non-zero, pairwise, relatively prime, ideals of D. Let A be a D-module such that*

$$A = \bigoplus_{U \in \mathcal{U}} A_U.$$

If B is a D-submodule of A, then

$$A/B = \bigoplus_{U \in \mathcal{U}} (A_U + B)/B.$$

If A is a D-module over a Dedekind domain, then the set $\mathrm{Ass}_D(A) \neq \emptyset$ by Corollary 6.20, because D is noetherian by Theorem 6.1. Also, if A is D-periodic, then $\mathrm{Ass}_D(A)$ is a family of proper, non-zero, pairwise, relatively prime, ideals of D. Therefore, applying the above results, we obtain *the primary decomposition of A*.

Corollary 6.25. *Let D be a Dedekind domain. If A is a D-periodic D-module, then*

$$A = \bigoplus_{P \in \mathrm{Ass}_D(A)} A_P.$$

If B is a D-submodule of A, then $(A/B)_P = (A_P + B)/B$ and

$$A/B = \bigoplus_{P \in \mathrm{Ass}_D(A)} (A_P + B)/B.$$

Proof. Indeed, if $a \in A$, then $L = \mathrm{Ann}_D(a) \neq \emptyset$. Hence $aD \cong D/L$ and it suffices to apply Corollary 6.12 to obtain $a = \sum_{P \in \pi} A_P$, where $\pi = \mathrm{Ass}_D(A)$. $\qquad \square$

Chapter 7

The Kovacs–Newman theorem

A major part of this book is dedicated to artinian modules over group rings. Artinian modules are one of the oldest subjects of investigation in Algebra. They are named after E. Artin, who initiated the study of rings and modules with the minimal condition. The minimal condition (as well as the maximal condition) has shown to be very useful and truly effective in many branches of Algebra, particularly in Group Theory. A natural way of connecting groups and modules can be described as follows. Let G be a group, and suppose that G has an abelian normal subgroup A. If $H = G/A$, then H acts on A in the following way: if $h = gA \in H$ and $a \in A$, then we define $a \cdot h = a^g$. Since A is abelian, this definition only depends on h. If $n \in \mathbb{Z}$, we put

$$a^{nh} = (a^n)^h = (a^h)^n.$$

Every element $x \in \mathbb{Z}H$ can be expressed as a (finite) sum as

$$x = \sum_{h \in H} n_h h,$$

where $n_h \in \mathbb{Z}$ for every $h \in H$ and $n_h = 0$ for all but finitely many $h \in H$. Then, the law

$$a \cdot x = \prod_{h \in H} a^{n_h h}$$

(which will be briefly written by juxtaposition) transforms A into a $\mathbb{Z}H$-module. Since additive notations are widely employed in Module Theory, from now on we will use the additive notations and write ax instead a^x when we think of A as a $\mathbb{Z}H$-module and keep the multiplicative notations when we think of A as a normal subgroup of G.

Very often, the subgroup A satisfies some finiteness condition, which turns out to be linked with the minimal condition for G-invariant subgroups. In other words, the $\mathbb{Z}G$-module A is artinian.

In the theory of infinite groups, artinian modules first appeared in the study of metabelian groups with the minimal condition on normal subgroups (Min-n). The first interesting result in this direction can be found in the paper [33], in which V.S. Charin constructed an example of a metabelian group satisfying Min-n but not Min. Subsequent studies showed that this example enjoyed very typical properties. Almost simultaneously P. Hall began his study of soluble groups with the maximal condition on normal subgroups (the condition Max-n). Unfortunately, after this first step, V.S. Charin stopped his study of metabelian groups with Min-n, a strange event especially considering that a very intensive research of groups with finiteness conditions had been developed in the former USSR by that time. V.S. Charin was a student of S.N. Chernikov, to whom we owe one of the more important accomplishments in this area, namely, the investigation of groups with Min. Since then, in his next research, V.S. Charin [35] explored the question: *for what classes of groups does* Min-n *imply* Min*?*; but he did not come back to groups with Min-n anymore. The description of metabelian groups satisfying Min-n was obtained much later by B. Hartley and D. McDougall [110]. More precisely, B. Hartley and D. McDougall described artinian modules over Chernikov groups. That paper has stimulated investigations of artinian modules over group rings as well as many other research projects in this area. At this point, it is worth mentioning a fundamental distinction between the above topic and its *dual* theory of noetherian modules over group rings. P. Hall [95] showed that the group ring of a polycyclic-by-finite group over a noetherian ring is noetherian itself. This result is fundamental in the theory of noetherian modules: it led to the developing of the rich theory of group rings, which brought up many significant achievements in the study of noetherian modules over such rings. Among many other publications we can mention, for example, C.J.B. Brookes [26, 27, 28], C.J.B. Brookes and J.R. J. Groves [29], K.A. Brown [30, 31], J. Cornick and P.H. Kropholler [46], S. Donkin [57, 58], D.R. Farkas [68], K.W. Gruenberg [94], P. Hall [95, 96, 98], A.V. Jategaonkar [118, 120], P.H. Kropholler, P.A. Linnel and J.A. Moody [139], I. Musson [195, 196, 197], D.S. Passman [216, 217, 218], J.E. Roseblade [248, 249, 250, 251, 252], R.L. Snider [268].

Apart from this, in the case of artinian modules, the properties of a group ring are not very useful. A group ring of a non-polycyclic group with finiteness conditions (even one that is very close to being polycyclic) defies deep description, and there are not too many papers devoted to it (see S.D. Berman [10], S.D. Berman and A.G. Podgorny [11], S.D. Berman and N.I. Vishnyakova [12, 13], C.J.B. Brookes [25], B. Hartley [100], H. Heineken and I.J. Mohamed [112], D. Segal [256], A.V. Tushev [276, 277]). Therefore the study of artinian modules requires different approaches. We collect here some of the most important tools, which will be considered in detail throughout the book. One of these tools is the study of the semisimplicity conditions of artinian modules, whose main source is the celebrated theorem of Maschke. One can find many important results in this direction in the following papers: D.R. Farkas and R.L. Snider [69], B. Hartley [101, 102, 103, 107],

B. Hartley and D. McDougal [110], L.G. Kovacs and M.F. Newman [137], D.I. Zaitsev [294, 301].

Another direction, specifically available for artinian modules over group rings, is based on the Fitting lemma and detects direct decompositions of modules that are determined by a group action. This runs parallel to the study of direct decompositions of groups relative to formations (see R. Baer [8]), which allows us to unify the points of view on many results. In this direction, we can mention the following papers: Z.Y. Duan [59, 60, 61, 62, 63], Z.Y. Duan and M.J. Tomkinson [64], B. Hartley and M.J. Tomkinson [111], L.A. Kurdachenko, J. Otal and I.Ya. Subbotin [158], L.A. Kurdachenko, B.V. Petrenko and I.Ya. Subbotin [159, 160], D.I. Zaitsev [301, 302, 304, 307, 309, 310], D.I. Zaitsev and V.A. Maznichenko [313].

Finally, the description of artinian modules over group rings, perhaps, is the first problem in this list. Since the situation here is far from being satisfactory, it is better to reformulate this problem as follows: *for what groups G is the description of artinian modules over the group ring $\mathbb{Z}G$ possible?* As we have already noted, artinian modules over Chernikov groups have been described by B. Hartley and D. McDougall [110]. On the other hand, S.A. Kruglyak [140] pointed out that the description of finite modules (and hence artinian modules) over a free abelian group of 0-rank at most 2 is *wild* from the representation theory point of view. The following analogy could be useful at this point. The description of torsion-free abelian groups of finite 0-rank at most 2 is also *wild* even though the $\mathbb{Z}$-injective envelope of these groups is a direct sum of finitely many copies of the full rational group $(\mathbb{Q}, +)$. This leads to the question on the description of the injective envelope of artinian modules over group rings, a partial case of which is the description of simple modules over group rings. These topics have been reflected in many papers. We would like to mention some of them: D.R. Farkas and R.L. Snider [69], B. Hartley [101, 102, 103], B. Hartley and D. McDougall [110], A.V. Jategaonkar [119], L.A. Kurdachenko [148, 150], L.A. Kurdachenko and J. Otal [155], L.A. Kurdachenko, J. Otal and I.Ya. Subbotin [157], L.A. Kurdachenko and I.Ya. Subbotin [170], I. Musson [194, 195, 196, 197, 198, 199, 200], R.L. Snider [267, 269, 270], B.A.F. Wehrfritz [281, 282, 283].

We begin with the problem of existence of complements for submodules of modules over group rings. Let A be a module over a ring R. A submodule B of A is said *to have a complement* in A (or B *is a complemented submodule*) if there is an R-submodule C such that $A = B \oplus C$. In the theory of modules over a group ring RG, the following question arises very often. Let B be an RG-submodule of A such that $A = B \oplus C$, for some R-submodule C. *Under which conditions does B have an RG-complement?* It is well known that such an RG-complement does not always exist. In some cases, only a supplement exists: an RG-submodule E such that both $A/(B+E)$ and $B \cap E$ are *good* in some sense, they are both finite. One of the most celebrated positive results for modules over finite groups is the theorem of Maschke. In Chapter 5, we considered general forms of this result. Hence, the question on the existence of analogous theorems in some classes of infinite groups appears to be very plausible. Now we consider other useful results, which imply, in particular, semisimplicity criteria. The next one is the Kovacs–Newman theorem [137]. We formulate this result in a slightly generalized form, namely, we extend the scalar ring on $\mathbb{Z}$ to an arbitrary Dedekind domain. For this reason, some results on modules over Dedekind domains will be needed.

Let R be a ring and let A be an R-module. If $0 \neq x \in R$, we say that A *is x-divisible* if given an element $a \in A$ there exists some $c \in A$ such that $cx = a$, that is, $A = Ax$. A module A is said to be R-*divisible* if A is x-divisible for every $0 \neq x \in R$. The concept of a divisible module is tightly connected with the concept of the injective module, which is dual to the concept of the projective module. A module E over a ring R is called *injective* if for every R-monomorphism $\beta : A \longrightarrow B$ and every R-homomorphism $\alpha : A \longrightarrow E$, there exists an R-homomorphism $\gamma : B \longrightarrow E$ such that $\beta\gamma = \alpha$. Now we formulate some useful properties of injective and divisible modules. By the same reasons mentioned in Chapter 6, and for the reader's convenience, we will supply some of them with suitable proofs. The proof of all omitted results can be directly found in the books cited in Chapter 6.

Theorem 7.1 (Baer's criterion). *Let R be a ring and let E be an R-module. Then E is an injective module if and only if for every right ideal J of R and for every R-homomorphism $\sigma : J \longrightarrow E$ there exists an R-homomorphism $\sigma^* : R \longrightarrow E$ such that $x\sigma^* = x\sigma$ for each $x \in J$.*

Proposition 7.2. *Let R be a ring.*

(1) *Let $\{E_\lambda \mid \lambda \in \Lambda\}$ be a family of R-modules. Then $\prod_{\lambda \in \Lambda} E_\lambda$ is injective if and only if E_λ is injective for every $\lambda \in \Lambda$.*

(2) *Let $\{E_\lambda \mid \lambda \in \Lambda\}$ be a family of R-modules. If $\bigoplus_{\lambda \in \Lambda} E_\lambda$ is injective, then E_λ is injective for every $\lambda \in \Lambda$.*

(3) *Let $\{E_\lambda \mid \lambda \in \Lambda\}$ be a family of R-modules such that set Λ is finite. If E_λ is injective for each $\lambda \in \Lambda$, then $\bigoplus_{\lambda \in \Lambda} E_\lambda$ is injective.*

(4) *An R-module A is injective if and only if A has an R-complement in every R-module $E \geq A$, that is there exists an R-module L such that $E = A \oplus L$.*

The next theorem elucidates the significance of injective modules.

Theorem 7.3. *Let R be a ring. Then every R-module can be embedded in some injective R-module.*

Let A be a module over a ring R. Then an R-module E is said to be *an injective envelope* of A if E satisfies the following conditions:

(IE1) $A \leq E$;

(IE2) E is injective; and

(IE3) if B is an injective R-module such that $A \leq B \leq E$, then $B = E$.

An R-module C is called *an essential extension of an R-submodule A if* $S \cap A \neq \langle 0 \rangle$ *for every non-zero R-submodule S of C.* An essential extension C of a module A is called *maximal*, if there is no essential extension E of A and monomorphism $\phi : C \longrightarrow E$ satisfying $a\phi = a$ for every $a \in A$ and $C\phi \neq E$.

Theorem 7.4. *Let A be a module over a ring R. Then there exists an R-module E satisfying the following equivalent conditions:*

(1) *E is injective and E is an essential extension of A.*

(2) *E is a maximal essential extension of A.*

(3) *E is a minimal injective extension of A.*

Moreover, if U and V are minimal injective extensions of A, then there exists an isomorphism $\vartheta : U \longrightarrow V$.

In other words, for every R-module A there exists an injective envelope and this injective envelope is unique (up to isomorphism).

The R-injective envelope of A is denoted by $\mathrm{IE}_R(A)$.

Proposition 7.5. *Let R be a ring and suppose that $y \in R$ is not a right zero-divisor. If E is an injective R-module, then E is y-divisible.*

Proof. Given $0 \neq e \in E$, we define a mapping $\phi : Ry \longrightarrow E$ by $(xy)\phi = ex$, $x \in R$. By the choice of y, ϕ is a well-defined R-homomorphism and $\mathrm{Ker}\,\phi \neq \langle 0 \rangle$. Therefore there exists an R-homomorphism $\vartheta : R \longrightarrow E$ such that $(xy)\vartheta = (xy)\phi = ex$. We have that

$$e = e\phi = y\vartheta = (1 \cdot y)\vartheta = (1\vartheta)y$$

so that $e \in Ey$. Therefore $E = Ey$, as required. $\square$

We also will consider other properties of divisible modules.

Lemma 7.6. *Let R be an integral domain.*

(1) *If E is an R-divisible R-module and C is an R-submodule of E, then E/C is R-divisible too.*

(2) *If E_λ is an R-divisible R-module for every $\lambda \in \Lambda$, then $\bigoplus_{\lambda \in \Lambda} E_\lambda$ and $\prod_{\lambda \in \Lambda} E_\lambda$ are also R-divisible.*

Proof. (1) Let $e \in E$ and $x \in R$. There is some $a \in E$, such that $ax = e$. Then $e + C = ax + C = (a + C)x$, that is, $(E/C)x = E/C$.

(2) Suppose that $(e_\lambda)_{\lambda \in \Lambda} \in \prod_{\lambda \in \Lambda} E_\lambda$, where $0 \neq x \in R$. For every $\lambda \in \Lambda$, there is some $a_\lambda \in E_\lambda$, such that $a_\lambda x = e_\lambda$. Therefore $((a_\lambda)_{\lambda \in \Lambda})x = (e_\lambda)_{\lambda \in \Lambda}$ that is,

$$\left(\prod_{\lambda \in \Lambda} E_\lambda \right) x = \prod_{\lambda \in \Lambda} E_\lambda.$$

Suppose now that $S =\mathrm{Supp}((e_\lambda)_{\lambda\in\Lambda})$ is finite. If $\lambda \in \Lambda$, we define

$$b_\lambda = \begin{cases} a_\lambda, & \text{if } \lambda \in S, \\ 0, & \text{if } \lambda \notin S. \end{cases}$$

Then $\mathrm{Supp}((b_{\lambda\in\Lambda})_{\lambda\in\Lambda}) = S$. In particular, $(b_\lambda)_{\lambda\in\Lambda} \in \bigoplus_{\lambda\in\Lambda} E_\lambda$. By construction, we have that $((b_\lambda)_{\lambda\in\Lambda})x = (e_\lambda)_{\lambda\in\Lambda}$, and hence $(\bigoplus_{\lambda\in\Lambda} E_\lambda)x = \bigoplus_{\lambda\in\Lambda} E_\lambda$. □

Proposition 7.7. *Let R be an integral domain and let E be an injective R-module. Then E is an R-divisible module.*

Proof. Let $0 \neq x \in R$ and $e \in E$. We define the mapping $\phi : Rx \longrightarrow E$ by $(ux)\phi = eu$, $u \in R$. Clearly, ϕ is an R-homomorphism. Since E is R-injective, by Theorem 7.1, there is an R-homomorphism $\phi^* : R \longrightarrow E$ extending ϕ. We have

$$e = x\phi = x\phi^* = (1 \cdot x)\phi^* = (1\phi^*)x.$$

Thus $Ex = E$, i. e., E is R-divisible. □

We now consider an abelian group as a Z-module. An abelian group A is said to be *divisible*, if A is a $\mathbb{Z}$-divisible module; that is, $A = kA$ for each $0 \neq k \in \mathbb{Z}$. Proposition 7.7 has the following converse.

Proposition 7.8. *Let R be an integral domain and let E be an R-module. If E is R-divisible and R-torsion-free, then E is an injective module.*

Proof. Let L be an ideal of R and let $\phi : L \longrightarrow E$ be an R-homomorphism. We assume that $L \neq \langle 0 \rangle$, and then we pick an element $0 \neq x \in L$. Since E is R-divisible, there is an element $e \in E$ such that $ex = x\phi$. If $y \in L$, then

$$(y\phi)x = (yx)\phi = (xy)\phi = (x\phi)y = (ex)y$$
$$= e(xy) = e(yx) = (ey)x.$$

It follows that $(y\phi - ey)x = 0$, which means that $y\phi = ey$ since E is R-torsion-free. Define now the mapping $\gamma : R \longrightarrow E$ by $ug = eu$ for every $u \in R$. Then $y\gamma = ey = y\phi$, for each $y \in L$, so γ is an extension of ϕ. By Theorem 7.1, E is injective. □

We need to have a careful look at the case in which R is a Dedekind domain. This case is extremely important for our purposes. In this case, as we will show, the concepts of injectivity and divisibility are coincide.

Theorem 7.9. *Let D be a Dedekind domain and let E be a D-module. If E is divisible, then E is injective.*

Proof. Let L be an ideal of D. Again we assume that $L \neq \langle 0 \rangle$. Let $\phi : L \longrightarrow E$ be a D-homomorphism. Since D is a Dedekind domain, there exists L^{-1}, and so there are some $x_1, \ldots, x_k \in L$ and $y_1, \ldots, y_k \in L^{-1}$ such that

$$1 = x_1 y_1 + \cdots + x_k y_k.$$

There is no loss if we assume that $x_1, \ldots, x_k \neq 0$. Since E is D-divisible, there are some $e_1, \ldots, e_k \in E$ such that $x_j \phi = e_j x_j$ for every $1 \leq j \leq k$. Pick $x \in L$. Since $x y_j \in LL^{-1} = D$,

$$x\phi = (x \cdot 1)\phi = (xx_1 y_1 + \cdots + xx_k y_k)\phi = (x_1\phi)xy_1 + \cdots + (x_k\phi)xy_k$$
$$= (e_1 x_1)xy_1 + \cdots + (e_k x_k)xy_k = (e_1 x_1 y_1 + \cdots + e_k x_k y_k)x.$$

Since
$$x_1 y_1 + \cdots + x_k y_k \in LL^{-1} = D,$$
$$e = e_1 x_1 y_1 + \cdots + e_k x_k y_k \in E.$$

Define the mapping $\phi^* : D \longrightarrow E$ by $y\phi^* = ey$ whenever $y \in D$. It is easy to see that ϕ^* is a D-homomorphism and $x\phi^* = ex = x\phi$, and so ϕ^* is an extension of ϕ. Then it suffices to apply Theorem 7.1. □

Corollary 7.10. *Let R be a principal ideal domain and let E be an R-module. Then E is injective if and only if E is R-divisible.*

Corollary 7.11 (R. Baer). *An abelian group A is divisible if and only if A is an injective $\mathbb{Z}$-module.*

Let D be a Dedekind domain and let A be a D-module. We say that a D-divisible module E is *a minimal divisible module for A* if E contains A and every proper D-submodule B of E containing A is not D-divisible.

Corollary 7.12. *Let D be a Dedekind domain.*

(1) *Each D-module A can be embedded in a divisible D-module.*

(2) *Let A be a D-module and let E be a divisible D-module such that $A \leq E$. Then E includes a minimal divisible submodule for A.*

(3) *Let A be a D-module. Then there exists a minimal divisible module for A, and two minimal divisible modules for A are isomorphic.*

A minimal divisible module for A (unique up to an isomorphism) will be called *the divisible envelope of A*.

Lemma 7.13. *Let D be a Dedekind domain and let A be a D-module. Suppose that $\{E_\lambda\}_{\lambda \in \Lambda}$ is a family of D-divisible submodules of A. Then $E = \sum_{\lambda \in \Lambda} E_\lambda$ is also D-divisible.*

Proof. Indeed, a sum of submodules is a homomorphic image of the direct sum of them, and then it suffices to apply Lemma 7.6. □

Corollary 7.14. *Let D be a Dedekind domain and let A be a D-module. If A has D-divisible submodules, then A has a unique largest D-divisible submodule, say div (A).*

Proof. In fact, div (A) is the sum of all D-divisible submodules of A. □

If A has no D-divisible submodules, then we put $\mathrm{div}(A) = \langle 0 \rangle$. The submodule $\mathrm{div}\,(A)$ defined above is called *the D-divisible part of A.*

Lemma 7.15. *Let R be a ring and let A and B be R-modules.*

(1) *If there are R-homomorphisms $\alpha : A \longrightarrow B$ and $\beta : B \longrightarrow A$ such that $\alpha\beta = \varepsilon_A$, then α is a monomorphism, β is an epimorphism, and $B = \mathrm{Im}\,\alpha \oplus \mathrm{Ker}\,\beta$. In particular, A is isomorphic to a direct summand of B.*

(2) *If $B = A_1 \oplus A_2$ and $A_1 \cong A$, then there exists an R-monomorphism $\alpha : A \longrightarrow B$ and an R-epimorphism $\beta : B \longrightarrow A$ such that $\alpha\beta = \varepsilon_A$.*

Proof. (1) If $b = a\alpha \in \mathrm{Im}\,\alpha \cap \mathrm{Ker}\,\beta$, then

$$a = a\varepsilon_A = a\alpha\beta = b\beta = 0$$

and so $b = 0$. Hence $\mathrm{Im}\,\alpha \cap \mathrm{Ker}\,\beta = \{0\}$. On the other hand, if $b \in B$, then $b - b\beta\alpha \in \mathrm{Ker}\,\beta$ since

$$(b - b\beta\alpha)\beta = b\beta - b\beta\varepsilon_A = 0.$$

Since $(b\beta)\alpha \in \mathrm{Im}\,\alpha$, $b \in \mathrm{Im}\,\alpha + \mathrm{Ker}\,\beta$, that is

$$B = \mathrm{Im}\,\alpha + \mathrm{Ker}\,\beta$$

and hence

$$B = \mathrm{Im}\,\alpha \oplus \mathrm{Ker}\,\beta,$$

as required.

(2) This is obvious. □

Theorem 7.16. *Let D be a Dedekind domain. Suppose that A is a D-module including two submodules E and B, which satisfy the following conditions:*

(1) *E is D-divisible.*

(2) *$B \cap E = \langle 0 \rangle$.*

Then there exists a D-submodule C such that $B \leq C$ and $A = E \oplus C$.

Proof. Let $\psi_1 : E \longrightarrow E \oplus B$ and $\psi_2 : E \oplus B \longrightarrow A$ be the canonical embeddings. Then the mapping $\alpha : E \oplus B \longrightarrow E$ given by $(e + b)\alpha = e$, where $e \in E$ and $b \in B$, is a D-homomorphism. By Proposition 7.7, E is injective. Hence there is a D-homomorphism $\beta : A \longrightarrow E$ such that $\psi_2\beta = \alpha$. Then

$$\psi_1\psi_2\beta = \psi_1\alpha = \varepsilon_E.$$

By Lemma 7.15,

$$A = \mathrm{Im}\,\psi_1\psi_2 \oplus \mathrm{Ker}\,\beta.$$

Obviously, $\mathrm{Im}\,\psi_1\psi_2 = E$. If $C = \mathrm{Ker}\,\beta$, then $A = E \oplus C$. Note that, if $b \in B$, then

$$b\beta = (b\psi_2)\beta = b\alpha = 0$$

and so $B \leq \mathrm{Ker}\,\beta = C$. □

Corollary 7.17. *Let D be a Dedekind domain and let A be a D-module. Then $A = \operatorname{div}(A) \oplus B$, where B is a D-submodule including no D-divisible submodules.*

Let F be the field of fractions of the Dedekind domain D. If $a \in F$ and $0 \neq x \in D$, we have $a = (ax^{-1})x$. This means that the D-module F is divisible. Also, note that if A is a D-module, by Corollary 7.12, there exists a divisible module E such that $E \geq A$.

Lemma 7.18. *Let D be a Dedekind domain. Suppose that A is a D-module and put $T = t_D(A)$. If A is divisible, then T is likewise divisible.*

Proof. If $a \in T$, then $\operatorname{Ann}_D(a) = J \neq \langle 0 \rangle$. Pick $0 \neq x \in D$. Since A is divisible, there is some $b \in A$ such that $bx = a$. If $0 \neq y \in J$, since $xy \neq 0$ and $0 = ay = (bx)y = b(xy)$, we have $0 \neq xy \in \operatorname{Ann}_D(b)$; that is, $b \in T$. Thus T is divisible, as required. $\qquad \square$

Corollary 7.19. *Let D be a Dedekind domain. Suppose that A is a D-module and E is a minimal divisible module for A.*

(1) *If A is D-periodic, then E is likewise D-periodic.*

(2) *If A is D-torsion-free, then E is likewise D-torsion-free.*

Proof. Put $T = t_D(E)$. By Lemma 7.18, T is divisible. If A is D-periodic, then $A \leq T$ and the minimality of E assures that $E = T$. If A is D-torsion-free, then

$$A \cap T = \langle 0 \rangle .$$

By Theorem 7.16, there exists a D-submodule $E_1 \geq A$ such that

$$E = E_1 \bigoplus T.$$

Since $E_1 \cong E/T$, E_1 is divisible by Lemma 7.6. By the choice of E again, $E = E_1$. Thus the result is proved. $\qquad \square$

Lemma 7.20. *Let D be a Dedekind domain. Suppose that $C = cD$ is a cyclic D-module and that E is a minimal divisible module for C. If C is D-torsion-free, then $E \cong_D F$, where F is the field of fractions of D.*

Proof. Since C is D-torsion-free, the mapping $\phi : D \longrightarrow C$, defined by the rule $x\phi = cx, x \in D$, is a D-homomorphism. Corollary 7.19 implies that E is D-torsion-free. Let $c_1 \in C$, $0 \neq y \in D, e_1, e_2$ be elements of E such that $e_1 y = c_1 = e_2 y$. Then

$$0 = e_1 y - e_2 y = (e_1 - e_2)y$$

and, since E is D-torsion-free,

$$e_1 - e_2 = 0,$$

that is, $e_1 = e_2$. This assures that if $c_1 \in C$ and $0 \neq y \in D$, there is a unique $e \in E$ such that $ey = c_1$. If $c_1 = cx$ for some $x \in D$, then put $e = c\frac{x}{y}$.

Extend the mapping ϕ to a mapping $\phi_1 : F \longrightarrow E$ by the rule $\frac{x}{y}\phi_1 = c\frac{x}{y}$. Let $x_1 \in C$, $0 \neq y_1 \in D$ and $e_1 = c\frac{x_1}{y_1}$. Then

$$ey = cx, \quad e_1 y_1 = cx_1$$

and we obtain

$$eyy_1 = cxy_1, \quad e_1 y_1 y = cx_1 y$$

and

$$(e + e_1)yy_1 = cx_1 y + cxy_1 = c(x_1 y + xy_1).$$

It follows that

$$e + e_1 = c\frac{x_1 y + xy_1}{yy_1}.$$

Hence

$$(\frac{x}{y} + \frac{x_1}{y_1})\phi_1 = (\frac{x_1 y + xy_1}{yy_1})\phi_1 = c\frac{x_1 y + xy_1}{yy_1}$$

$$= e + e_1 = c\frac{x}{y} + c\frac{x_1}{y_1} = (\frac{x}{y})\phi_1 + (\frac{x_1}{y_1})\phi_1.$$

Let $u \in D$. Put $e_2 = c\frac{ux}{y}$. Then

$$e_2 y = cux = (cx)u = (ey)u = (eu)y.$$

It follows that $(e_2 - eu)y = 0$ and, then $e_2 = eu$. Thus

$$e_2 = (u\frac{x}{y})\phi_1 = eu = c\frac{x}{y}u = (\frac{x}{y})\phi_1 u.$$

This shows that ϕ_1 is a D-homomorphism. Let $\frac{x}{y} \in \text{Ker } \phi_1$, that is $c\frac{x}{y} = 0$. Then

$$cx = c\frac{x}{y}y = 0,$$

i.e., $x \in \text{Ann }_D(C) = \langle 0 \rangle$. This means that $\frac{x}{y} = 0$ and therefore $\text{Ker } \phi_1 = \langle 0 \rangle$. As $\text{Im } \phi_1$ is a divisible submodule of E and $C \leq \text{Im } \phi_1$, by the choice of E, we conclude that $E = \text{Im } \phi_1$. Hence $E \cong_D F$. $\qquad\square$

Let D be a Dedekind domain. If A is a simple D-module, then there exists a maximal ideal P of D such that $A \cong D/P$. By Corollary 6.17, D/P^k and P/P^{k+1} are isomorphic as D-modules for every $k \in \mathbb{N}$. In particular, for every $k \in \mathbb{N}$ the D-module D/P^k can be embedded in the D-module D/P^{k+1}. Therefore we may consider the injective limit of the family of D-modules $\{D/P^k \mid k \in \mathbb{N}\}$. We put

$$C_{P\infty} = \varinjlim \{D/P^k \mid k \in \mathbb{N}\}.$$

The D-module $C_{P\infty}$ is called a Prüfer P-module. By construction, $C_{P\infty}$ is a P-module such that $\Omega_{P,k}(C_{P\infty}) \cong_D D/P^k$ and

$$\Omega_{P,k+1}(C_{P\infty})/\Omega_{P,k}(C_{P\infty}) \cong (D/P^{k+1})/(P/P^{k+1}) \cong D/P$$

for every $k \in \mathbb{N}$. Hence, if C is a proper D-submodule of $C_{p\infty}$, then there exists some $k \in \mathbb{N}$ such that $C = \Omega_{P,k}(C_{P\infty})$. In fact, if $b \notin \Omega_{P,k}(C_{P\infty})$, then $C = bD$. We also note that the Prüfer P-module $C_{P\infty}$ is monolithic and its monolith is $\Omega_{P,1}(C_{P\infty})$.

Lemma 7.21. *The Prüfer P-module is divisible.*

Proof. This has been established in the book by L.A. Kurdachenko, J. Otal and I.Ya. Subbotin [157, Lemma 5.1]. □

Corollary 7.22. *Let D be a Dedekind domain. Suppose that A is a simple D-module and let E be a minimal divisible module for A. Then $E \cong_D C_{P\infty}$, where $P = \mathrm{Ann}_D(A)$.*

Proof. By Corollary 7.19, E is a D-periodic module. By Corollary 6.25, $E = \bigoplus_{Q \in \pi} E_Q$, where E_Q is the Q-component of E and $\pi = \mathrm{Ass}\ _D(A)$. In particular, $E = E_P \oplus C$, where C is the P'-component of E, that is $C = \bigoplus_{Q \neq P} E_Q$. Then Lemma 7.6 and the isomorphism $E_P \cong E/C$ at once give that E_P is divisible. By the choice of E, $E = E_P$ since $A \leq E_P$. Hence E is a P-module. By Proposition 6.13, if $y \in P \setminus P^2$, then $P = yD + P^2$. Let $0 \neq a_1 \in A$. Since E is divisible, there are elements $a_2, a_3, \ldots, a_n, \ldots \in A$ such that

$$a_{n+1}x = a_n,$$

for every $n \in \mathbb{N}$. Obviously, $\mathrm{Ann}\ _D(a_2) = \mathrm{Ann}\ _D(a_2 D) \neq P$ and so $\mathrm{Ann}\ _D(a_2 D) = P^{k_1}$, where $k_1 > 1$. Thus $a_2 D \cong D/P^{k_1}$ and, in particular, $A_2 = a_2 D$ contains copies of submodules isomorphic to

$$D/P \cong P^{k_1-1}/P^{k_1} \cong A, \ldots, D/P^{k_1-1} \cong P/P^{k_1}.$$

Put $A_3 = a_3 D$. As above, $\mathrm{Ann}\ _D(a_3) = \mathrm{Ann}\ _D(A_3) = P^{k_2}$ and $k_2 > k_1$. Proceeding in this way, if $\mathrm{Ann}\ _D(a_n) = P^{k_{n-1}}$, we have that $k_1 < k_2 < \cdots < k_n < \cdots$. Therefore $\bigcup_{n \in \mathbb{N}} A_n$ has a submodule B which is isomorphic to

$$\varinjlim \{D/P^k \mid k \in \mathbb{N}\}.$$

Hence $B = E$. □

Lemma 7.23. *Let R be a ring. Given a family of R-modules $\{A_\lambda \mid \lambda \in \Lambda\}$, suppose that, for every $\lambda \in \Lambda$, E_λ is an essential extension of A_λ. Then $\bigoplus_{\lambda \in \Lambda} E_\lambda$ is an essential extension of $\bigoplus_{\lambda \in \Lambda} A_\lambda$.*

Proof. Put

$$A = \bigoplus_{\lambda \in \Lambda} A_\lambda \text{ and } E = \bigoplus_{\lambda \in \Lambda} E_\lambda.$$

Given $0 \neq e \in E$, $e = \sum_{\lambda \in \Lambda} e_\lambda$, where $\varsigma = \mathrm{Supp}\ ((e_\lambda)_{\lambda \in \Lambda})$ is finite. Let $\lambda \in \varsigma$ such that $0 \neq e_\lambda \in E_\lambda$. Since E_λ is an essential extension of A_λ, there is some $x_1 \in R$ such that $0 \neq e_\lambda x_1 \in A_\lambda$. If $\mu \in \varsigma \setminus \{\lambda\}$ satisfies $e_\mu x_1 = 0$, then

$0 \neq ex_1 = e_\lambda x_1 \in A$, that is $eR \cap A \neq \langle 0 \rangle$. Therefore we may assume that there exists some $\mu \in \varsigma \setminus \{\lambda\}$ such that $e_\mu x_1 \neq 0$. Since E_μ is an essential extension of A_μ, there is some $x_2 \in R$ such that

$$0 \neq e_\mu x_1 x_2 \in A_\mu.$$

Note that $e_\lambda x_1 x_2 \in A_\lambda$. If $\nu \in \varsigma \setminus \{\lambda, \mu\}$ satisfies $e_\nu x_1 x_2 = 0$, then

$$0 \neq ex_1 x_2 = e_\lambda x_1 x_2 + e_\mu x_1 x_2 \in A$$

and so $eR \cap A \neq \langle 0 \rangle$. Proceeding in the same way, we obtain after finitely many steps, that $eR \cap A \neq \langle 0 \rangle$. It means that E is an essential extension of A, as required. $\square$

Corollary 7.24. *Let D be a Dedekind domain. If $\{A_\lambda \mid \lambda \in \Lambda\}$ is a family of D-modules and, for every $\lambda \in \Lambda$, E_λ is a D-divisible envelope of A_λ, then $\bigoplus_{\lambda \in \Lambda} E_\lambda$ is a D-divisible envelope of $\bigoplus_{\lambda \in \Lambda} A_\lambda$.*

Proof. By Lemma 7.6, $U = \bigoplus_{\lambda \in \Lambda} E_\lambda$ is divisible. By Corollary 7.12, U contains the divisible envelope E of $A = \bigoplus_{\lambda \in \Lambda} A_\lambda$. By Theorem 7.16, there exists a D-submodule B such that $U = E \oplus B$. In particular, $B \cap A = \langle 0 \rangle$ and then, by Lemma 7.23, U is an essential extension of A. This means that $B = \langle 0 \rangle$ and so $E = U$. $\square$

Let D be a Dedekind domain, and let A be a D-module. If P is a maximal ideal of D, let A_P be the P-component of A. Since $\text{Ann}_D(\Omega_{P,1}(A_P)) = P$, $\Omega_{P,1}(A_P)$ may be considered as a vector space over the field D/P. By definition, the corresponding dimension $r_P(A) = \dim_{D/P}(\Omega_{P,1}(A_P))$ is called *the P-rank of A*. Note that $r_P(A)$ is an invariant of the module A.

Let R be an integral domain, and let A be an R-module. If B is a maximal free subset of A, then $|B|$ is called *the torsion-free R-rank* of A or the 0-rank of the R-module A and it is denoted by $r_0(A)$. It is worth mentioning that this invariant for R-modules is analogous to the dimension of a vector space on a field.

Theorem 7.25. *Let D be a Dedekind domain and let F be the field of fractions of D. If E is a divisible D-module and $\pi = \text{Ass}_D(E)$, then $E = (\bigoplus_{P \in \pi} E_P) \oplus U$ and*

(1) *E_P is the P-component of E;*

(2) *$E_P = \bigoplus_{\lambda \in \Lambda(P)} C_\lambda$, where C_λ is the Prüfer P-module for every $\lambda \in \Lambda(P)$;*

(3) *$U = \bigoplus_{\delta \in \Delta} U_\delta$, where $U_\delta \cong_D F$, for every $\delta \in \Delta$.*

Moreover, $|\Lambda(P)| = r_p(E)$ and $|\Delta| = r_0(E)$.

Proof. Let $T = t_D(E)$. By Lemma 7.18, T is divisible, and, by Theorem 7.16, there exists a D-submodule U such that $E = T \oplus U$. Further, by Corollary 6.25, $T = \bigoplus_{P \in \pi} E_P$, where E_P is the P-component of E and $\pi = \text{Ass}_D(E)$. Since E_P is divisible by Lemma 7.6, there exists a set $\Lambda(P)$ such that $\Omega_{P,1}(E_P) = \bigoplus_{\lambda \in \Lambda(P)} M_\lambda$, where each M_λ is a simple D-module; that is, $M_\lambda \cong D/P$ for every $\lambda \in \Lambda(P)$. By Corollary 7.12, E_P contains a divisible envelope $C(P)$ of $\Omega_{P,1}(E_P)$ and, by Corollary 7.24,

$$C(P) = \bigoplus_{\lambda \in \Lambda(P)} C_\lambda,$$

where, for every $\lambda \in \Lambda(P)$, each C_λ is a divisible envelope of M_λ, and $C_\lambda = C_{P^\infty}$ (by Corollary 7.22). Moreover, by Theorem 7.16, there exists a submodule $A(P)$ such that

$$E_P = C(P) \oplus A(P).$$

Since $\Omega_{P,1}(E_P) \leq C(P)$ and E_P is an essential extension of $\Omega_{P,1}(E_P)$, $A(P) = \langle 0 \rangle$; that is, $E_P = C(P)$. By the definition of $\Lambda(P)$, $|\Lambda(P)| = r_P(A)$.

On the other hand, since $U \cong E/T$, $r_0(E) = r_0(U)$. Let $\mathcal{S}$ be a maximal D-free subset of U and define $V = \bigoplus_{a \in \mathcal{S}} aD$. If $u \in U$, then $uD \cap V \neq \langle 0 \rangle$, and so U is an essential extension of V. Since $r_0(U) = |\mathcal{S}|$, $r_0(E) = |\mathcal{S}|$. Since $C \cong E/T$, by Lemma 7.6, U is divisible. By Corollary 7.12, U contains a D-divisible envelope W of V, and, by Corollary 7.24, $W = \bigoplus_{a \in \mathcal{S}} V_a$, where each summand V_a is a D-divisible envelope of aD. By Lemma 7.20, $V_a \cong_D F$. Finally, by Theorem 7.16, there exists a submodule M such that $U = W \oplus M$ and, since $V \leq W$ and U is an essential extension of V, $M = \langle 0 \rangle$; that is, $U = W$. Thus the theorem is proved. $\qquad\square$

Corollary 7.26. *Let D be a Dedekind domain and let F be the field of fractions of D. Suppose that A is a D-module and let E be a divisible envelope of A.*

(1) *$E = \left(\bigoplus_{P \in \pi} E_P \right) \oplus U$, where if $\pi = \mathrm{Ass}_D(E)$, $E_P = \bigoplus_{\lambda \in \Lambda(P)} C_\lambda$, and C_λ is the Prüfer P-module for every $\lambda \in \Lambda(P)$.*

(2) *$U = \bigoplus_{\delta \in \Delta} U_\delta$, where $U_\delta \cong_D F$, for every $\delta \in \Delta$.*

(3) *$|\Lambda(P)| = r_P(A)$ and $|\Delta| = r_0(A)$.*

The above result finishes our preliminary work, and we are now ready to prove the Kovacs-Newman theorem. The next lemma is the basis for that proof. We remark that this lemma is originally shown for the ring $D = \mathbb{Z}$, although we can slightly modify its proof in such a way that it remains true for an arbitrary Dedekind domain.

Lemma 7.27 (L.G. Kovacs and M.F. Newman [137]). *Let D be a Dedekind domain, G a periodic central-by-finite group, A a DG-module. Suppose that A satisfies the following conditions:*

(i) *A is a monolithic DG-module with monolith M.*

(ii) *$\mathrm{Ann}_D(A) = P \neq \langle 0 \rangle$.*

(iii) *Either $\mathrm{char}(D/P) = 0$ or $\mathrm{char}(D/P) \notin \Pi(G)$.*

Then:

(1) $A = \bigoplus_{\lambda \in \Lambda} A_\lambda$, *where either* $A_\lambda \cong D/P^n$ *or* A_λ *is a Prüfer P-module for every* $\lambda \in \Lambda$. *In particular, A is D-periodic and* $\mathrm{Ass}_D(A) = P$.

(2) *If E is a proper DG-submodule of A, then there exists some* $k \in \mathbb{N}$ *such that* $E = \Omega_{P,k}(A)$.

Proof. (a) Put

$$C = \zeta(G), \ S = DC, \ L = \Omega_{P,1}(A) \text{ and } A_0 = t_D(A).$$

Since the P-component of A contains the monolith, by Corollary 6.25, A_0 is exactly the P-component of A. Given some $x \in \mathrm{Ann}_S(M)$, we choose a finite subgroup $C_x \leq C$ such that $x \in DC_x$. By Theorem 5.9, there exists a DC_x-submodule M_1 such that $L = M \oplus M_1$. Then

$$Lx = Mx \oplus M_1 x \leq M_1 x \leq M_1.$$

If $Lx \neq \langle 0 \rangle$, then $M \leq Lx \leq M_1$, a contradiction. Hence $Lx = \langle 0 \rangle$; that is, $\mathrm{Ann}_S(M) = \mathrm{Ann}_S(L)$.

(b) Put $T = \mathrm{Ann}_S(M)$ so that T is an ideal of S. Since $MP = \langle 0 \rangle$, $SP \leq T$. By Theorem 5.5,

$$M = M_0 \oplus M_0 g_1 \oplus \cdots \oplus M_0 g_m,$$

where M_0 is a simple S-module and $g_1, \ldots, g_m \in G$. It is rather easy to prove that $T = \mathrm{Ann}_S(M_0)$. Since M_0 is a simple S-module, there exists some $a \in A$ such that $M_0 = aS$ and $\mathrm{Ann}_S(a) = T$ so that $M_0 \cong S/T$. Moreover, T is a maximal ideal of S, i.e. the factor-ring S/T is a field. If $b \in L$ and $bS \neq \langle 0 \rangle$, then $T \leq \mathrm{Ann}_S(b)$ since $\mathrm{Ann}_S(L) = \mathrm{Ann}_S(M) = T$. Since T is a maximal ideal, $T = \mathrm{Ann}_S(b)$. This means that bS is a simple S-module.

(c) Let B be a maximal S-submodule of L under $B \cap M = \langle 0 \rangle$. If $b \in L \setminus B$, then B does not include bS. Since bS is a simple S-submodule, $bS \cap B = \langle 0 \rangle$. By the choice of B, $(bS + B) \cap M \neq \langle 0 \rangle$, and so there are some $x \in S$ and $b_1 \in B$ such that $0 \neq b_2 = b_1 + bx \in M$. Then $bx = b_2 - b_1 \in M + B$. If $bx = 0$, then $b_1 = b_2$. Since $M \cap B = \langle 0 \rangle$, $b_1 = b_2 = 0$, a contradiction. Hence $bS \cap (M + B) \neq \langle 0 \rangle$. It follows that $bS \leq B + M$, since bS is simple. In other words, $L = M \oplus B$.

Let F be a finite subgroup of G such that $G = CF$. By Theorem 5.9, there exists a DG-submodule B_1 such that $L = M \oplus B_1$. Since M is the DG-monolith of A, $L = M$.

(d) Pick $y \in P \setminus P^2$ and $a \in M$. Suppose that there is some $n \in \mathbb{N}$ such that $xy^n = a$ has solutions but $xy^{n+1} = a$ has not. If $0 \neq b \in M$, then there is some $u \in DG$ such that $bu = a$. If there is some $b_1 \in A$ such that $b_1 y^{n+1} = b$, then

$$(b_1 u)y^{n+1} = (b_1 y^{n+1})u = bu = a,$$

a contradiction. By Proposition 6.13, $P = Dy + P^{n+1}$, and therefore either $A_0 = \Omega_{P,n}(A)$ or A_0 is y-divisible. In the latter, A_0 is D-divisible, cf. L.A. Kurdachenko, J. Otal and I.Ya. Subbotin [157, Proposition 7.2].

For every $j \in \mathbb{N}$, the mapping $\phi_j : a \mapsto ay^j$, $a \in \Omega_{P,j+1}(A)$, is a *DG*-endomorphism. Hence Im ϕ_j and Ker ϕ_j are *DG*-submodules; moreover Im $\phi_j \leq \Omega_{P,1}(A) = M$. If $A_0 \neq \Omega_{P,j+1}(A)$, then Im $\phi_j \neq \langle 0 \rangle$, and so Im $\phi_j = M$. In this case, Ker $\phi_j = \Omega_{P,j}(A)$, and so $\Omega_{P,j+1}(A)/\Omega_{P,j}(A) \cong M$. In particular, this factor is a simple *DG*-module. Therefore, if E is a *DG*-submodule of A_0, then either $E = \Omega_{P,j}(A)$ for some $j \in \mathbb{N}$ or $E = A_0$.

(e) If Ann $_D(A_0) = P^t$, then

$$A = \bigoplus_{\lambda \in \Lambda} A_\lambda,$$

where $A_\lambda \cong D/P^t$, for every $\lambda \in \Lambda$ (see D. W. Sharpe and P. Vamos [265, Theorem 6.14]). If A is *D*-divisible, then, by Theorem 7.16, $A = \bigoplus_{\delta \in \Delta} B_\delta$, where B_δ is a Prüfer P-module for every $\delta \in \Delta$. In the first case, there exists some *D*-submodule U such that $A = A_0 \oplus U$ (see I. Kaplansky [123]). In the second one, by Theorem 7.25, there exists a *D*-submodule U_0 such that $A = A_0 \oplus U_0$. Therefore, in any case, A_0 is a direct summand of A; that is, there exists a *D*-submodule W such that $A = A_0 \oplus W$.

(f) Suppose that Ann $_D(A_0) = P^t$. If $A \neq A_0$, then the complement W of A_0 in A is a non-zero torsion-free *D*-submodule. Moreover,

$$AP^t = A_0 P^t \oplus WP^t = WP^t \leq W,$$

and, in particular, AP^t is a non-zero torsion-free *DG*-submodule. However, $M \leq AP^t$, a contradiction that shows that $A = A_0$.

(g) Now suppose that A_0 is *D*-divisible. Assume also that M is the *S*-monolith of A. If $0 \neq a \in A$, then aS is a non-zero *S*-submodule, and so $M \leq aS$. Thus, there is some $x \in S$ such that $0 \neq ax \in M$. Pick a finite subgroup C_x such that $x \in DC_x$. Since A_0 is *D*-complemented in A, by Theorem 5.9, there is a DC_x-submodule V such that

$$k_x A \leq A_0 + V \text{ and } k_x(A_0 \cap V) = \langle 0 \rangle,$$

where $k_x = |C_x|$.

Suppose that char $D = p > 0$ so that char$(D/P) = p$. It follows that $(k_x, p) = 1$ and, then $k_x A = A$ and $A_0 \cap V = \langle 0 \rangle$. In other words, $A = A_0 \oplus V$.

Suppose now that char $D = 0$ but char$(D/P) = p > 0$. In this case, the additive group of A_0 is a *p*-group. Since $(k_x, p) = 1$, $A_0 \cap V = \langle 0 \rangle$; that is, $A_0 + V = A_0 \oplus V$. The mapping $\psi : a \mapsto k_x a$, $a \in A$, is a DC_x-endomorphism of A so that Ker ψ and Im ψ are DC_x-submodules. Since the $\mathbb{Z}$-periodic part of A is A_0,

$$\text{Ker } \psi = \langle 0 \rangle, \ Im \psi = k_x A \leq A_0 \oplus V,$$

$$\text{and}$$

$$A_0 = k_x A_0 \leq \text{Im } \psi.$$

Therefore

$$\operatorname{Im} \psi = A_0 \oplus (V \cap \operatorname{Im} \psi),$$

and we notice that $V \cap \operatorname{Im} \psi$ is a DC_x-submodule. Since $A \cong \operatorname{Im} \psi$, we obtain $A = A_0 \oplus V_1$, where $A_0 = A_0 \psi^{-1}$ and $V_1 = (V \cap \operatorname{Im} \psi)\psi^{-1}$.

Finally, suppose that $\operatorname{char}(D/P) = 0$. Consider $\Omega_{P,1}(A)$ as a vector space over the field D/P. Since $\operatorname{char}(D/P) = 0$, $\Omega_{P,1}(A)$ is $\mathbb{Z}$-divisible. By the isomorphism

$$\Omega_{P,1}(A) \cong \Omega_{P,j+1}(A)/\Omega_{P,j}(A)$$

for every $j \in \mathbb{N}$, A_0 is likewise $\mathbb{Z}$-divisible. Therefore $A_0 = k_x A_0$. Since the factor-module A/A_0 is D-torsion-free and $\operatorname{char} D = 0$, A/A_0 is $\mathbb{Z}$-torsion-free. Hence A is $\mathbb{Z}$-torsion-free. It follows that the mapping

$$\psi : a \longrightarrow k_x a, \ a \in A$$

is then a DC_x-monomorphism. Moreover, $A_0 \cap V = \langle 0 \rangle$. We find

$$A \cong \operatorname{Im} \psi = A_0 \oplus (V \cap \operatorname{Im} \psi)$$

again, which proves that A_0 is also DC_x-complemented in A.

Therefore, in any case, $A = A_0 \oplus V_2$, for some DC_x-submodule V_2. We claim that $A_0 x \neq \langle 0 \rangle$. Otherwise $ax \in Ax = A_0 x \oplus V_2 x = V_2 x \leq V_2$. Since $ax \in M$ and $M \leq A_0$, this leads to a contradiction, which shows our claim.

Put $B = \operatorname{Ann}_A(x)$. Since $x \in DC$, B is a DG-submodule of A. Since $A_0 x \neq \langle 0 \rangle$, $B \cap A_0$ is a proper DG-submodule of A_0. We have already proved that $B \cap A_0 = \Omega_{P,j}(A)$, for some $j \in \mathbb{N}$. It follows that $\operatorname{Ann}_D(B \cap A_0) = \langle 0 \rangle$, and then $B = (B \cap A_0) \oplus B_1$, where B_1 is a D-torsion-free D-submodule provided $B_1 \neq \langle 0 \rangle$ (see I. Kaplansky [123]). Then $BP^j = B_1 P^j \leq B_1$ and, in particular, BP^j is D-torsion-free; this conditions assures that $BP^j = \langle 0 \rangle$. For, otherwise $M \leq BP^j$ and so $M \leq B_1$, a contradiction. Then $B \leq A_0$. Since $x \in S$, $A_0 x$ is a DG-submodule and, since $A_0 x \neq \langle 0 \rangle$, $M \leq A_0 x$. Then $ax \in A_0 x$; that is, $ax = bx$ for some $b \in A_0$. Thus $(a-b)x = 0$, and so $a - b \in B \leq A_0$. It follows that $a \in A_0$. Therefore $A = A_0$ also holds in this case.

(h) Now suppose that A_0 is D-divisible but M is not S-monolith. By Theorem 5.5, there are a simple S-module M_0 and elements $g_1, \dots, g_m \in G$ such that

$$M = M_0 \oplus M_0 g_1 \oplus \cdots \oplus M_0 g_m.$$

Define

$$M_1 = M_0 g_1, \dots, M_m = M_0 g_m.$$

For every $0 \leq j \leq m$, choose R_j to be a maximal S-submodule of A under $\oplus_{s \neq j} M_s \leq R_j$ and $R_j \cap M_j = \langle 0 \rangle$. Then every A/R_j is an S-monolithic module with S-monolith $(M_j + R_j)/R_j$. Since $\operatorname{Ann}_D(M_j) = \operatorname{Ann}_D(M_0) = P$, by applying the result shown in the last paragraphs, we deduce that A/R_j is D-periodic and Ass $_D(A/R_j) = \{P\}$. Put

$$X = R_0 \cap R_1 \cap \cdots \cap R_m.$$

If $X \neq \langle 0 \rangle$, then $X \cap M = \langle 0 \rangle$ and, as we showed above in (c), $M = \Omega_{P,1}(A)$. In particular, X is D-torsion-free. By Remak's theorem

$$A/X \hookrightarrow A/R_0 \oplus A/R_1 \oplus \cdots \oplus A/R_m,$$

and so A/X is D-periodic. It follows that A/Xg_j is likewise D-periodic, for every $0 \leq j \leq m$. If

$$X_1 = X \cap Xg_1 \cap \cdots \cap Xg_m,$$

then, by Remak's theorem, again,

$$A/X_1 \hookrightarrow A/X \oplus A/Xg_1 \oplus \cdots \oplus A/Xg_m,$$

which shows that A/X_1 is likewise D-periodic. Since X_1 is a DG-submodule such that $X_1 \leq X$, X_1 is also D-torsion-free. Consequently, $X_1 = \langle 0 \rangle$ and hence A is D-periodic in this case.

Thus, the proof is now completed. □

Theorem 7.28 (L.G. Kovacs and M.F. Newman [137])**.** *Let D be a Dedekind domain, G a periodic central-by-finite group, A a DG-module. Suppose that B is a DG-submodule of A satisfying the following conditions:*

(1) *B is DG-monolithic with monolith M.*

(2) *$\mathrm{Ann}_D(M) = P \neq \langle 0 \rangle$.*

(3) *Either $\mathrm{char}(D/P) = 0$ or $\mathrm{char}(D/P) \notin \Pi(G)$.*

If B is D-complemented in A, then B is DG-complemented in A.

Proof. By hypothesis, there exists a D-submodule C of A such that $A = B \oplus C$. Let E be a maximal DG-submodule of A under $E \cap M = \langle 0 \rangle$. If there exists some $j \in \mathbb{N}$ such that $BP^j = \langle 0 \rangle$, then $AP^j = CP^j$. In particular, $AP^j \cap B = \langle 0 \rangle$. Clearly, we may choose E such that $AP^j \leq E$. By Lemma 7.27, A/E is D-periodic, and Ass $_D(A/E) = \{P\}$. Since B also satisfies the conditions of Lemma 7.27, there exists a decomposition

$$B = \bigoplus_{\lambda \in \Lambda} B_\lambda,$$

where either $B_\lambda \cong D/P^t$ for every $\lambda \in \Lambda$, or B_λ is a Prüfer P-module for every $\lambda \in \Lambda$. In the first case, by the choice of E, we have $(A/E)P^j = \langle 0 \rangle$. By Lemma 7.27, there exists a decomposition

$$A/E = \bigoplus_{\delta \in \Delta} C_\delta,$$

where $C_\delta \cong D/P^t$ for every $\delta \in \Delta$. Since

$$\Omega_{P,1}(A/E) = (M+E)/E = \Omega_{P,1}((B+E)/E),$$

$A/E = (B+E)/E$, and so $A = B \oplus E$.

Suppose now that B is D-divisible. By Lemma 7.27, A/E is likewise D-divisible. Since

$$\Omega_{P,1}(A/E) = \Omega_{P,1}((B+E)/E),$$

as above,

$$A/E = (B+E)/E \text{ and } A = B \oplus E,$$

as required   □

Corollary 7.29. *Let D be a Dedekind domain, G a periodic central-by-finite group, A a DG-module. Suppose that B is a DG-submodule of A satisfying the following conditions:*

(1) $B = B_1 \oplus \cdots \oplus B_n$, *where every B_j is DG-monolithic with monolith M_j.*

(2) *For every $1 \leq j \leq n$, $\mathrm{Ann}_D(M_j) = P_j \neq \langle 0 \rangle$.*

(3) *Either $\mathrm{char}(D/P_j) = 0$ or $\mathrm{char}(D/P_j) \notin \Pi(G)$, for every $1 \leq j \leq n$.*

If B has a D-complement, then B has a DG-complement.

Corollary 7.30. *Let F be a field, G a periodic central-by-finite group, A an artinian FG-module. If $\mathrm{char} F = 0$ or $\mathrm{char} F \notin \Pi(G)$, then A is a semisimple FG-module.*

Chapter 8

Hartley's classes of modules

This chapter deals with some specific criteria of semisimplicity that were obtained by B. Hartley in [101, 103]. In these papers, Hartley considered a specific class of modules that is wider than artinian modules over Chernikov groups. We will consider here some results of [101] only.

Let us begin with some technical lemmas.

Lemma 8.1 ([101]). *Suppose that $r, m, n, k \in N$ satisfy $m^2 \mid r^n - 1$. Then*

$$1 + r^n + \cdots + r^{n(k-1)} \equiv k(\mathrm{mod}\ m^2).$$

Proof. We have that

$$1 + r^n + \cdots + r^{n(k-1)} - k = (r^{n1} - 1) + \cdots + (r^{n(k-1)}) - 1$$

and, for each $1 \le j \le k - 1$,

$$r^{nj} - 1 = (r^n - 1)(r^{n(j-1)} + \cdots + 1), \ j \ge 1$$

which is divisible by m^2. □

Let G be a group, H a subgroup of G, R a ring and A an RH-module. We define

$$A^G = A \otimes_{RH} RG.$$

We consider now the effect of inducing simple modules from a subgroup of an abelian group to the whole group.

Lemma 8.2 ([101]). *Let G be a periodic abelian group and suppose that K is a subgroup of G such that G/K is a Chernikov locally cyclic group. Let F be a finite field such that $\mathrm{char}\ F \notin \Pi(G)$. Then there exists a subgroup $H \ge K$ satisfying the following properties:*

(1) H/G_0 is finite.

(2) *If A is a simple FH-module such that $C_H(A) = K$, then A^G is a simple FG-module.*

Proof. It suffices to consider the case $K = \langle 1 \rangle$. Having dealt with this case, we may be assured that there is a finite subgroup $H/K \leq G/K$ such that if A is a simple $F(H/K)$-module and $C_{H/K}(A) = \langle 1 \rangle$, then $A^{G/K}$ is a simple $F(G/K)$-module. In this case, let V be a simple FH–module such that $C_H(V) = K$. We may naturally view V as a faithful simple $F(H/K)$-module, and, then the induced $F(G/K)$-module $V^{G/K}$ remains simple. We may think of this module as an FG-module so that, as consequence of the definitions, it is isomorphic to V^G. Thus V^G is simple.

Therefore, we suppose that $K = \langle 1 \rangle$. Then G is a Chernikov locally cyclic group, and we may assume that G is infinite. Suppose $\Pi(G) = \{p_1, \ldots, p_k\}$. Then

$$G = G_1 \times \cdots \times G_k,$$

where G_j is a cyclic p_j-subgroup or a Prüfer p_j-subgroup, $1 \leq j \leq k$. Let H be any finite subgroup containing all elements of G of order dividing $p_1^2 \cdots p_k^2$. The group G has an ascending chain

$$H = H_0 \leq H_1 \leq \cdots \leq H_j \leq \cdots$$

of finite cyclic subgroups such that

$$G = \bigcup_{j \in \mathbb{N}} H_j,$$

and all the indexes $|H_{j+1} : H_j| = q_j$ are prime. Since $q_j \in \{p_1, \ldots, p_k\}$, $q_j \in \Pi(H)$, and hence q_j^2 divides $|H_j|$. Therefore H_j has elements of order q_j^2, and then

$$(q_j)^2 \text{ divides } |H_0| \text{ for every } j \in \mathbb{N}. \tag{8.1}$$

Now we proceed to choose H in a particular way. Let

$$L = \mathrm{Dr}_{p \in \Pi(G)} \, \Omega_{p,2}(G).$$

Suppose that $|F| = r = p^t$ and let n be the smallest natural number such that $|L|$ divides $r^n - 1$; the existence of n is guaranteed since $(p, |L|) = 1$. Let H be the unique largest subgroup of G of order dividing $r^n - 1$. Then $L \leq H$, and so n is the smallest natural number such that $|H|$ divides $r^n - 1$. Put $q_0 = n$. We show by induction on $j \in \mathbb{N}$ that if $n_j = q_0 q_1 \cdots q_j$, then H_j is the largest subgroup of G of order dividing $r^{n_j} - 1$ and n_j is the smallest natural number j such that $|H_j|$ divides $r^j - 1$.

Suppose that we have already shown this for some $j \geq 0$. Let X be any finite subgroup including H such that $|X|$ divides $r^{n_{j+1}} - 1$. Then

$$r^{n_{j+1}} - 1 = r^{n_j q_{j+1}} - 1 = (r^{n_j} - 1)(r^{n_j(q_{j+1}-1)} + \cdots + 1),$$

and so the index $|X : H_j|$ divides

$$r^{n_j(q_{j+1}-1)} + \cdots + 1.$$

Let s be a prime divisor of $|X : H_j|$. By [1], s^2 divides $|H_0|$, and so s^2 divides $r^{n_j} - 1$. By Lemma 8.1,

$$r^{n_j(q_{j+1}-1)} + \cdots + 1 \equiv q_{j+1} (\mathrm{mod}\ s^2). \tag{8.2}$$

Since s divides the left-hand side of (8.2), we find that $s = q_{j+1}$. It also follows, that $(q_{j+1})^2$ does not divide that left-hand side of (8.2), and hence either $|X : H_j| = 1$ or $|X : H_j| = q_{j+1}$. Since G has at most only one subgroup of any given finite order, it follows that $X \leq H_{j+1}$. By Lemma 8.1, (8.1) assures that (8.2) holds with $s = q_{j+1}$. Therefore $|H_{j+1}|$ divides $r^{n_{j+1}} - 1$, and so H_{j+1} itself is the largest subgroup of G with this property. Furthermore, if $|H_{j+1}|$ divides $r^d - 1$, then, by induction, d has the form $d = n_j u$, for some $u \in \mathbb{N}$. Then q_{j+1} divides

$$r^{n_j(u-1)} + \cdots + 1 \equiv u (\mathrm{mod}\ q_{j+1}^2),$$

and hence q_{j+1} divides u. This completes the inductive proof.

Let now F_j be the finite field of order r^{n_j}. Since $|H_j|$ divides $(r^{n_j} - 1)$, there is a monomorphism $\phi : H_j \longrightarrow U(F_j)$. The additive group of F_j is made into an FH_j-module by the action $ah = a(h\phi)$, where $a \in F_j$ and $h \in H_j$. Moreover, $C_{H_j}(F_j) = \langle 1 \rangle$. Since $|H_j| \mid r^d - 1$ if and only if $n_j \mid d$, it follows that $H_j \phi$ generates the field F_j; and so we obtain in this way a simple FH_j-module. Since every simple module with trivial centralizer arises in this way, we have just seen

$$\dim_F A = n_j, \text{ if } A \text{ is any simple } FH_j\text{-module}$$
$$\text{such that } A \text{ has an identity centralizer in } H_j. \tag{8.3}$$

If A is such a module, then the induced module $A^{H_{j+1}}$ is an FH_{j+1}-module of dimension $n_j q_{j+1} = n_{j+1}$ over F. Since the FH_j-module $A^{H_{j+1}}$ is a direct sum of copies of A, every element of prime order acts fixed-point freely on $A^{H_{j+1}}$. Consequently, every simple submodule of $A^{H_{j+1}}$ is faithful for H_{j+1}; that is, its corresponding centralizer is trivial. By (8.3), $A^{H_{j+1}}$ is a simple FH_{j+1}-module. Iterating the above argument, we find that A^{H_d} is simple for every $d > j$. Since A^G is the union of the FH_d-modules A^{H_d} for every $d > j$, A^G is simple. It suffices to take $j = 0$. $\qquad \square$

Further we will use some ring-theoretical results.

Lemma 8.3. *Let L be a nil-ideal of a ring R, and let $\psi : R \longrightarrow R/L$ the canonical epimorphism. If a is an element of R such that $a\vartheta$ is an idempotent, then there is some $b \in R$ such that $aba = e$ is an idempotent and $a\psi = e\psi$.*

Proof. We have

$$0 = a\psi - (a\psi)^2 = a\psi - (a^2)\psi = (a - a^2)\psi,$$

and so $a - a^2 \in L$. Then there exists some $n \in \mathbb{N}$ such that $(a-a^2)^n = 0$. Obviously, we may assume that $n \geq 2$. Moreover, $(1 - a)^n = 1 - ac$, for some $c \in R$ such that $ac = ca$. In particular,

$$1 - (ac)\psi = (1 - ac)\psi = ((1 - a)^n)\psi = (1 - a\psi)^n = 1 - a\psi,$$

and then $(ac)\psi = a\psi$. Thus

$$0 = (a - a^2)^n = a^n(1 - a)^n = a^n(1 - ac);$$

that is, $a^n = a^n ac$. It follows that $a^n = a^n(ac)^j$ for every $j \geq 1$. Making $j = n$, we obtain $a^n = a^n(ac)^n = a^{2n}c^n$, since $ac = ca$. Put $e = (ac)^n$. Then

$$e^2 = ((ac)^n)^2 = (ac)^{2n} = a^{2n}c^n c^n = a^n c^n = (ac)^n = e$$

and so e is an idempotent. Further,

$$e\psi = ((ac)^n)\psi = ((ac)\psi)^n = (a\psi)^n = a\psi.$$

Since $n \geq 2$, $e = a(a^{n-2}c^n)a$, as required. $\square$

Let R be a ring, e_1, e_2 idempotents of R. We say that e_1, e_2 are *orthogonal* if $e_1 e_2 = e_2 e_1 = 0$. Let e be a non-zero idempotent of the ring R. We say that e is a *primitive idempotent* if e cannot be expressed in a sum of two non-zero orthogonal idempotents.

If R is a ring, then $J(R)$ denotes the *Jacobson radical* of R.

Lemma 8.4. *Let R be an artinian ring.*

(1) *If e is a non-zero idempotent of R, then the R-module eR is directly indecomposable if and only if e a primitive idempotent.*

(2) *If e is a primitive idempotent, then eR includes an unique maximal R-submodule $eJ(R)$; in particular, $eR/eJ(R)$ is a simple R-module.*

Proof. (1) If $e = e_1 + e_2$ and e_1, e_2 are non-zero orthogonal idempotents, then $eR = e_1 R + e_2 R$. Let $a \in e_1 R \cap e_2 R$. Then there are $x_1, x_2 \in R$ such that $a = e_1 x_1 = e_2 x_2$. Thus $e_1 x_1 = e_1 e_1 x_1 = e_1 e_2 x_1 = 0$, and so $eR = e_1 R \oplus e_2 R$. Conversely, suppose that R_1, R_2 are R-submodules of R such that $eR = R_1 \oplus R_2$. Since R_1, R_2 are R-submodules, $e_1 e_2, e_2 e_1 \in R_1 \cap R_2 = \langle 0 \rangle$. Clearly, e_1, e_2 are idempotents, so that e_1, e_2 are orthogonal idempotents.

(2) Consider $R/J(R)$ as a right R-module, and let $\mathfrak{M}$ be the set of maximal right ideals of R. We claim that $\mathfrak{M}$ is finite. Otherwise, $\mathfrak{M}$ has a countably infinite subset $\{M_n \mid n \in \mathbb{N}\}$ such that

$$M_1 \cap \cdots \cap M_n \neq M_1 \cap \cdots \cap M_n \cap M_{n+1},$$

for every $n \in \mathbb{N}$. Hence we may construct an infinite strictly descending chain of right ideals

$$M_1 > M_1 \cap M_2 > \cdots > M_1 \cap \cdots \cap M_n > \cdots$$

that contradicts the fact that R is artinian. Thus the set $\mathfrak{M}$ is finite, and it follows that

$$R/J(R) \hookrightarrow \prod_{M \in \mathfrak{M}} R/M.$$

Since each R/M is a simple R-module, $R/J(R)$ is a semisimple R-module of finite composition length; Corollary 4.3 also yields that

$$(eR + J(R))/J(R) \cong_R (eR)/(eR \cap J(R))$$

is a semisimple R-module. In particular, $eR \cap J(R) \geq J(eR)$. The mapping $\phi : x \mapsto ex, x \in R$ is an R-endomorphism; therefore $eR/eJ(R)$ is likewise a semisimple R-module, so that $eJ(R) \geq J(eR)$. On the other hand, $J(R)\phi \leq J(R\phi)$, and thus $eJ(R) = J(eR)$. Since $J(R)$ is a two-sided ideal, $eJ(R) \leq J(R)$. Hence $eJ(R) = J(R) \cap eR$. Since eR contains an idempotent e, $eR \neq eJ(R)$. Let B be an R-submodule of eR such that $eJ(R)$ does not include B. Then $J(R)$ does not include B, and $(B + J(R))/J(R)$ is a non-zero R-submodule of the semisimple module $R/J(R)$. By Corollary 4.3, $(B + J(R))/J(R)$ is a direct summand of $R/J(R)$. Hence $(B + J(R))/J(R)$ contains a non-zero idempotent, say $f + J(R)$. Since R is artinian, $J(R)$ is a nilpotent ideal (see D. S. Passman [219, Lemma 5.11, Lemma 5.1]). By Lemma 8.3, there exists an idempotent $0 \neq g \in B$ such that $f+J(R) = g+J(R)$. In other words, we may assume that f is an idempotent. Since $f + B \subseteq eR$, there exists some $y \in R$ such that $f = ey$. Then $ef = e^2 y = ey = f$. Put $z = fe \in B$. Since $ez = efe = fe = fee = ze$, $z^2 = f(ef)e = f(fe) = fe = z$. It follows that

$$e = z + (e - z), \qquad \text{and}$$
$$z(e - z) = ze - z^2 = (fe)e - z^2 = fe - z = z - z = 0;$$
$$(e - z)z = ez - z^2 = ze - z = z - z = 0;$$
$$(e - z)^2 = (e^2 - ze) - (ez - z^2) = e^2 - fee = e - fe = e - z.$$

Thus $e = z + (e - z)$ is a sum of two orthogonal idempotents. Moreover, since

$$0 \neq f = f^2 = (ef)(ef) = e(fe)f = e(zf),$$

$z \neq 0$. Since e is a primitive idempotent, $e - z = 0$; that is, $e = z$. But in this case $e \in B$, thus $B = eR$. It follows that $eJ(R)$ contains every proper submodule of eR. This means that $eR/eJ(R)$ is a simple R-module, as required. $\qquad \square$

Lemma 8.5 ([101]). *Let G be an abelian group and suppose that G has a finite subgroup K such that G/K is a Chernikov locally cyclic group. Let F be a finite field such that* char $F \notin \Pi(G)$. *Then there exists a finite subgroup $H \geq K$ satisfying the following properties:*

(1) *If e is a primitive idempotent in $B = FH$ such that $C_H(eB) = K$, then e is a primitive idempotent in $R = FG$, and eR is a minimal ideal of R such that $C_G(eR) = K$.*

(2) *If A is a simple R-module such that $C_G(A) = K$, then $A \cong_R eR$, for some primitive idempotent $e \in B$ such that $C_H(eB) = K$.*

Proof. We choose a subgroup H satisfying the properties (1) and (2) of Lemma 8.2. Thus H is finite because K is finite.

(1) Since char $F \notin \Pi(H)$, B is a semisimple FH-module by Corollary 5.15. In particular $J(FH) = \langle 0 \rangle$. By Lemma 8.4, eB is a simple FH-submodule. Now it is immediate from the definition that eR is isomorphic as right R-module to a module obtained by including the FH-module eB up to G. Since $C_H(eB) = K$, by the choice of H and Lemma 8.2, eR is a simple R-module; and so is a minimal ideal of R. By Lemma 8.4, e is a primitive idempotent of FG. Since $eR \cong (eB)^G$, it is straightforward to obtain that $C_G(eR) = K$.

(2) Since A is a simple R-module, there exists some maximal ideal M of R such that $A \cong_R R/M$. By Corollary 5.15, the FH-module A is semisimple. Let C be a simple FH-submodule of A. By Lemma 5.4, there is a subset S of G such that

$$A = \bigoplus_{x \in S} Cx.$$

Since G is abelian, $Cx \cong_{FH} C$. It follows that A and every FH-submodule of A is a direct sum of copies of C, and hence $C_H(C) = K$. Since B is finite dimensional and semisimple, by Corollary 4.3,

$$B = (B \cap M) \oplus D_1 \oplus \cdots \oplus D_k,$$

where $D_1, \ldots, D_k$ are simple FH-submodules of B. Given $1 \leq j \leq k$, by Lemma 8.4, there exists a primitive idempotent e_j such that $D_j = e_j B$. Since $e_j B$ are pairwise non-isomorphic as B-modules, it follows that $k = 1$. Now $B/(B \cap M)$ is naturally isomorphic to an FH-submodule of the FH-module A. Hence, if $e = e_1$, then $eB \cong C$ and $C_H(eB) = K$. By (1), eR is a minimal ideal of R. Since M does not include eR, $R = M \oplus eR$, and hence $R/M \cong_R eR$. □

Lemma 8.6 ([101]). *Let G be a Chernikov abelian group having a locally cyclic subgroup of finite index. Suppose that F is a finite field such that char $F \notin \Pi(G)$ and let A be a simple FG-module such that $C_G(A)$ is finite. Then there is an uniquely determined primitive idempotent $e \in FG$ such that $A \cong_{FG} e(FG)$. This correspondence determines a bijection between the isomorphism classes of the simple FG-modules A with $C_G(A)$ finite and the primitive idempotents in FG.*

Proof. The factor-group $G/C_G(A)$ is locally cyclic (see L.A. Kurdachenko, J. Otal and I.Ya. Subbotin [157, Theorem 2.3]). Thus, the existence of e follows from Lemma 8.5. Since G is abelian, eFG and $(1 - e)FG$ are mutual annihilators in

FG, whence, if e_1 is an idempotent in FG. Therefore, if $e_1 \in FG$ is an idempotent such that $e_1 FG \cong_{FG} eFG$, we obtain

$$(1 - e)FG = (1 - e_1)FG \text{ and } eFG = e_1 FG.$$

Since both e and e_1 are the identity of the subring eFG, we have $e = e_1$.

Finally, if e is any primitive idempotent of FG, then $e \in FH$, for some finite subgroup H of G and since $eFG \cong (eFH)^G$, we have thar $C_G(eFG) \leq H$. □

Lemma 8.7 ([101]). *Let G be a periodic abelian group such that $\Pi(G)$ is finite, and suppose that K is a finite subgroup of G. If F is a finite field, then the number of isomorphism types of simple FG-modules A such that $C_G(A) = K$ is finite.*

Proof. We clearly may assume that $K = \langle 1 \rangle$. Then G is a locally cyclic p'-group, where $p =$char F (see L.A. Kurdachenko, J. Otal and I.Ya. Subbotin [157, Theorem 2.3]). Since $\Pi(G)$ is finite, G is Chernikov. By Lemma 8.5, $A \cong eFG$, where e is a primitive idempotent in FH and H is a finite subgroup satisfying the conditions of Lemma 8.2. Since H is finite, the set of such primitive idempotents is finite. This shows the result. □

Let R be a ring, G a group, A an RG-module. The set $\mathcal{S}$ of RG-submodules (called *the terms of* $\mathcal{S}$) of A is called *a series of* A if it satisfies the following conditions:

(i) $\mathcal{S}$ is linearly ordered by inclusion.

(ii) If $0 \neq a \in A$, then there are terms of $\mathcal{S}$ that do not contain a, and the union of all such terms is a term $V(a)$ of $\mathcal{S}$.

(iii) If $0 \neq a \in A$, then there are terms of $\mathcal{S}$ that contain a, and the intersection of all such terms is a term $\Lambda(a)$ of $\mathcal{S}$.

(iv) Each term of $\mathcal{S}$ has the form $V(a)$ or $\Lambda(a)$, for some $0 \neq a \in A$.

(v) $\langle 0 \rangle, A \in \mathcal{S}$.

Thus a belongs to $\Lambda \setminus V(a)$. The corresponding factor-module $\Lambda(a)/V(a)$ is called *a factor of* $\mathcal{S}$. If $\mathcal{S}$ and $\mathcal{S}_1$ are series of A, $\mathcal{S}_1$ is said to be *a refinement of* $\mathcal{S}$ if every term of $\mathcal{S}$ is a term of $\mathcal{S}_1$. A series that has no refinement other than itself is called *a composition series* (more precisely, *an RG-composition series*). Clearly, in such series every factor is a simple RG-module. As in groups, every series of submodules can be refined to a composition series.

Lemma 8.8 ([101]). *Let G be a periodic abelian group, and let F be a field with char $F = p$, where, if $p > 0$, then $p \notin \Pi(G)$. Suppose that A is an FG-module and put $L = \mathrm{Ann}_{FG}(A)$. If A has a composition series whose factors fall into finitely many isomorphism types, then FG/L is a direct sum of finitely many fields, and A is a semisimple FG-module.*

Proof. Let $\mathcal{S}$ be a composition series of submodules of A. Suppose that each factor of $\mathcal{S}$ is isomorphic to one of the finitely many pairwise non-isomorphic FG-modules $X_1, \ldots, X_n$. For each $1 \leq j \leq n$, we put $L_j = \mathrm{Ann}_{FG}(X_j)$ and

$$U = L_1 \cap \cdots \cap L_n.$$

Let

$$\phi : FG/U \longrightarrow (FG/L_1) \oplus \cdots \oplus (FG/L_n) = B$$

be the homomorphism given by $(x + L)\phi = (x + L_1, \ldots, x + L_n)$, $x \in FG$.

Clearly, ϕ is both a ring homomorphism and an FG-homomorphism. Since the FG-modules $A/L_1, \ldots, A/L_n$ are simple and pairwise non-isomorphic, every FG-submodule of B is a direct sum of some of them and, since the image of FG/U in B projects onto each summand FG/L_j, it must be the whole of B. Hence $FG/U \cong B$ as rings. Since every X_j is a simple FG-module, FG/L_j is a field and so FG/U is a direct sum of finitely many fields.

Obviously, $U \geq L$, and we claim that equality holds. This will show that FG/L is a direct sum of finitely many fields. Hence, it is a semisimple artinian ring. Therefore A, which is naturally an FG/L-module, is semisimple.

Let $u \in U$. Then there is a finite subgroup H of G such that $u \in FH$. By Corollary 5.15, the FH-module A is semisimple. Let C be a simple FH-submodule of A, and choose $0 \neq c \in C$ arbitrary. Then $c \in \Lambda(c) \setminus V(c)$, where $\Lambda(c), V(c) \in \mathcal{S}$. Since C is a simple FH-submodule, $\Lambda(c) \geq C$ and $V(c) \cap C = \langle 0 \rangle$, so that C is isomorphic to an FH-submodule of $\Lambda(c)/V(c)$. Since $u \in U \cap FH$, it follows that $Cu = \langle 0 \rangle$. Since A is a direct sum of copies of the FH-submodule C, it follows that $Au = \langle 0 \rangle$. Therefore $u \in L$, and $U = L$ as claimed. $\square$

Lemma 8.9 ([101]). *Let G be a Chernikov abelian group and suppose we have an infinite set $\{K_\lambda \mid \lambda \in \Lambda\}$ of distinct subgroups. Then there exists a subgroup $B \leq G$ such that*

(1) *B is contained in only finitely many subgroups K_λ; and*

(2) *$B = \bigcup_{n \in \mathbb{N}} B_n$, where*

$$B_1 \leq B_2 \leq \cdots \leq B_n \leq \cdots$$

and each B_n is contained in infinitely many subgroups K_λ.

Proof. We proceed by induction on the sum of the ranks of the Sylow p-subgroups of G. Since G contains finitely many elements of prime order, it follows that some of such elements lie in infinitely many subgroups K_λ. Then there exist a non-identity subgroup X_1 and an infinite subset $\Lambda_1 \subseteq \Lambda$ such that $X_1 \leq K_\lambda$, for every $\lambda \in \Lambda_1$. This means that infinitely many subgroups K_λ/X_1 are distinct. Applying the same argument, there exist a non-identity subgroup X_2/X_1 and an infinite subset $\Lambda_2 \subseteq \Lambda_1$ such that $X_2/X_1 \leq K_\lambda/X_1$, for every $\lambda \in \Lambda_2$. Iterating, we obtain a tower of subgroups of G

$$\langle 1 \rangle = X_0 < X_1 < \cdots < X_n < \cdots,$$

each of which is contained in infinitely many of the subgroups K_λ. Put $X = \bigcup_{n \in \mathbb{N}} X_n$. If X lies in finitely many subgroups K_λ, then it suffices to define $B = X$ and $B_n = X_n$, for every $n \in \mathbb{N}$. Otherwise, the result follows by applying induction to G/X. □

Let G be a locally finite group. Suppose that F is a field of characteristic $p \geq 0$, and let A be an FG-module. Then A is said to be *an $\mathcal{M}_c$-module over FG* if for each p'-subgroup H of G the set of centralizers in A of subgroups of H satisfies the minimal condition (B. Hartley [101]) where, as usual, by $0'$ we mean the set of all primes. It is straightforward that A is an $\mathcal{M}_c$-module over FG if and only if each p'-subgroup H of G has a finite subgroup K such that $C_A(H) = C_A(K)$. Note that every artinian FG-module is trivially an $\mathcal{M}_c$-module.

Theorem 8.10 ([101]). *Let G be a Chernikov group. Suppose that F is a finite field, and let A be an $\mathcal{M}_c$-module over FG. If $\operatorname{char} F \notin \Pi(G)$, then A is semisimple FG-module.*

Proof. First we suppose that G is abelian. Let $\mathcal{S}$ be any composition series of A. If $0 \neq a \in A$, then we put $K_a = C_G(\Lambda(a)/V(a))$. Suppose, if possible, that infinitely many distinct subgroups of G lie in $\mathcal{K} = \{K_a \mid 0 \neq a \in A\}$. By Lemma 8.9, there exists an ascending chain of subgroups

$$B_1 \leq B_2 \leq \cdots \leq B_n \leq \cdots$$

such that if $B = \bigcup_{j \in \mathbb{N}} B_j$, then B is contained in finitely many subgroups K_a whereas each B_j is contained in infinitely many of them. Since A is an $\mathcal{M}_c$-module, there is a finite subgroup E of B such that $C_A(B) = C_A(E)$. Then there exists some $j \in \mathbb{N}$ such that $E \leq B_j$, and so there is some $a \in A$ such that $E \leq K_a$, but K_a does not include B. If C and D are FG-submodules of A such that $C \geq D$ and we put $C_1/D = C_{C/D}(E)$, by Corollary 5.15, there exists some FE-submodule C_2 such that $C_1 = C_2 \oplus D$. By the choice of C_1, $C_2 \leq C_A(E)$, and therefore $C_{C/D}(E) = (C_C(E) + D)/D$. Since $E \leq K_a$, the above argument gives that

$$\Lambda(a)/V(a) = C_{\Lambda(a)/V(a)}(E) = (C_{\Lambda(a)}(E) + V(a))/V(a)$$
$$= (C_{\Lambda(a)}(B) + V(a))/V(a) \leq C_{\Lambda(a)/V(a)}(B).$$

Hence, $B \leq K_a$, a contradiction. Consequently, only finitely many subgroups of G occur among $\mathcal{K}_a$. This and Lemma 8.7 give that the composition factors $\Lambda(a)/V(a)$ fall into finitely many isomorphism types as FG-modules. By Lemma 8.8, A is a semisimple FG-module.

In the general case, let G_0 be a normal abelian subgroup of G having finite index. Then A is an $\mathcal{M}_c$-module over FG_0, and so A is a semisimple FG_0-module. Let C be a non-zero FG-submodule of A. Since C is an FG_0-submodule, by Corollary 4.3, there exists an FG_0-submodule D such that $A = C \oplus D$. By Theorem 5.9, there exists an FG-submodule U such that $A = C \oplus U$. Applying Corollary 4.3 again, we obtain that A is a semisimple FG-module, as required. □

Lemma 8.11 (A.D. Gardiner, B. Hartley and M.J. Tomkinson [82, Lemma 3.2]).
Let G be a locally finite group, and suppose that F is a field with char $F = p > 0$.
If A is a simple FG-module such that $C_G(A) = \langle 1 \rangle$, then $O_p(G) = \langle 1 \rangle$.

Proof. Assume the contrary, and choose $1 \neq x \in O_p(G)$. Put $X = \langle x \rangle$, and let a
be an element of A such that $ax \neq a$. Define $A_0 = \langle a \rangle^X$. Since X is finite, A_0 is a
finite subgroup of the additive group of A. If $0 \neq u \in A_0$, then $uFG = A$. Hence,
there exists a finite subgroup $G_u \leq G$ such that $u(FG_u) \geq A_0$. Put

$$\langle H = x, G_u \mid 0 \neq u \in A_0 \rangle.$$

Then H is finite and $C = A_0 FH$ is finite dimensional over F. Since $u(FG_u) \geq A_0$,
every non-zero element of A_0 generates C as FH-submodule. Thus if we choose an
FH-submodule C_1 of C maximal under $C_1 \cap A_0 = \langle 0 \rangle$, then C_1 is a maximal FH-
submodule of C and C/C_1 is a simple FH-module. Now $x \in O_p(G) \cap H \leq O_p(H)$.
Since C/C_1 is a simple FH-module, cf. K. Doerk and T.O. Hawkes [56, Theorem
10.5], $O_p(H) \leq C_H(C/C_1)$. However $A_0 \cong (A_0 + C_1)/C_1$ as X-groups, and so x
acts as the identity on A_0. This contradiction proves the result. $\square$

  If A is a module over a ring R, then the intersection of all maximal R-
submodules of A is called *the Frattini submodule of A* and denoted by $Fratt_{FG}(A)$.

Theorem 8.12 ([101]). *Let G be a Chernikov group. Suppose that F is a finite field,
and let A be an $\mathcal{M}_c$-module over FG. Then A is a semisimple FG-module if and
only if $Fratt_{FG}(A) = \langle 0 \rangle$.*

Proof. If A is a semisimple FG-module, obviously $Fratt_{FG}(A) = \langle 0 \rangle$. Conversely,
suppose that $Fratt_{FG}(A) = \langle 0 \rangle$. We clearly may assume that $C_G(A) = \langle 1 \rangle$. If char
$F = p$ and $L = \omega F O_p(G)$ is the augmentation ideal of the group-ring $F O_p(G)$,
by Lemma 8.11, $AL \leq M$, for every maximal FG-submodule M. Hence, $AL \leq
Fratt_{FG}(A) = \langle 0 \rangle$. It follows that $O_p(G) = \langle 1 \rangle$.
  Let H be a normal abelian subgroup of G of finite index in G. Since $O_p(H) =
\langle 1 \rangle$, $p \notin \Pi(H)$. Clearly, A is an $\mathcal{M}_c$-module over FH, and so, by Theorem 8.10,
A is a semisimple FH-module. Let B be a simple FH-submodule of A. Choose T
to be a transversal to H in G, and put

$$C = \sum_{x \in T} Bx.$$

By Lemma 5.4, Bx is a simple FH-module for every $x \in T$. Since T is finite,
by Lemma 1.1 and Lemma 1.2, C satisfies Max-FH and Min-FH. But C is an
FG-submodule, so it actually satisfies Max-FG and Min-FG. If M is a maximal
FG-submodule of A, then either $M \geq C$, or $A = M + C$. In the second case, we
note that

$$C/(C \cap M) \cong_{FG} A/M$$

is a simple FG-module, and so $C \cap M$ is a maximal FG-submodule of C. It follows that $Fratt_{FG}(C) = \langle 0 \rangle$. Since C satisfies Min-FG, there are finitely many maximal submodules $M_1, \ldots, M_n$ of C such that

$$M_1 \cap \cdots \cap M_n = \langle 0 \rangle .$$

By Remak's theorem,
$$C \hookrightarrow C/M_1 \oplus \cdots \oplus C/M_n.$$

By Corollary 4.4, C is a semisimple FG-submodule. It follows that $C \leq \mathrm{Soc}_{FG}(A)$. From the choice of C we obtain that $A = \mathrm{Soc}_{FG}(A)$, that is A is a semisimple FG-module. $\qquad\square$

It is worth mentioning that Theorem 8.10 and Theorem 8.12 can fail if $\Pi(G)$ is infinite. Specific examples of this were constructed by B. Hartley in [101]. For the reader's convenience, we quote them here but we have omitted their proofs.

Theorem 8.13 ([101]). *Let p be a prime. Then there exists a group G satisfying the following conditions:*

(1) *$G = Dr_{q \in \pi} \langle g_q \rangle$, where $|g_q| = q$ and π is an infinite set of primes.*

(2) *there are $2^{\aleph_0}$ pairwise non-isomorphic simple FG-modules which are faithful for G, where $F = \mathbb{F}_p$ is the finite field of p elements.*

(3) *the ring FG contains no primitive idempotents and no minimal ideals.*

(4) *There is an $\mathcal{M}_c$-module A over FG such that $Fratt_{FG}(C) = \langle 0 \rangle$ but A is not semisimple.*

B. Hartley [103] continued the study of $\mathcal{M}_c$-modules over group-rings. In this case, the underlying group has a periodic abelian subgroup of finite index. We also quote the main result of the above paper, omitting the proof as well. We will need the following definition.

Suppose that G is a locally finite group, F is a field and A is an FG-module. Then A is called *a strong locally semisimple module* if G has a finite subgroup K such that A is a semisimple FH-module whenever H is a finite subgroup such that $H \geq K$ (B. Hartley [103]).

Theorem 8.14. *Let G be a periodic abelian-by-finite group. Suppose that F is a field, and let A be an $\mathcal{M}_c$-module over FG. Then the following three conditions are equivalent:*

(1) *$Fratt_{FG}(A) = \langle 0 \rangle$.*

(2) *A is a strong locally semisimple module.*

(3) *every composition factor of A has an FG-complement.*

Chapter 9

The injectivity of some simple modules

In this chapter we consider a more general topic than the semisimplicity of modules over group rings. In a semisimple module all submodules are complemented (Corollary 4.3), and so, in particular, every simple submodule has a complement. On the other hand, if a simple module A has a complement in every module E including A, then A is injective (see Proposition 7.2). The commutative rings R such that every simple R-module is injective have been already described. By a result due to I. Kaplansky, such rings are necessarily von Neumann regular rings. We are not going to study the non-commutative case in depth here. We will focus on modules over group rings and our objectives in this chapter are some important results that were obtained by D.R. Farkas and R.L. Snider [69] and B. Hartley [107]. If A is a simple RG-module where R is a commutative ring, then $L = \mathrm{Ann}_R(A)$ is a maximal ideal of R, and we may think of A as an FG-module where $F = R/L$ is a field. Therefore we will focus on modules over a group ring FG, where F is a field.

The first problem that arises here is the characterization of the injectivity of the trivial FG-module. A field F can be thought as a right FG-module under the trivial action. Specifically, if $a \in F$ and $x \in FG$, then $ax = a\omega(x)$, where ω denotes the augmentation map. If H is a subgroup of the group G, then, by considering maps from ideals of FH to F, we can easily show the following result.

Lemma 9.1. *Let F be a field, and let G be a group. If H is a subgroup of G and F is FG-injective, then F is likewise FH-injective.*

Lemma 9.2 (D.R. Farkas and R.L. Snider [69]). *Let G be a finite group. Then a field F is FG-injective if and only if $\mathrm{char}\, F \notin \Pi(G)$.*

Proof. If char $F \notin \Pi(G)$, by Corollary 5.15, FG is semisimple; then it is clear that F is FG-injective.

Suppose that char $F = p > 0$, and F is FG-injective. Suppose that the result is false, and let x be an element of G of order p. By Lemma 9.1, F is $F\langle x \rangle$-injective. Let L be the augmentation ideal of $F\langle x \rangle$. We recall that L admits

$\{x - 1, x^2 - 1, \ldots, x^{p-1} - 1\}$ as F-basis. Define the mapping $\phi : L \longrightarrow F$ by

$$(\alpha_1(x^1 - 1) + \cdots + \alpha_{p-1}(x^{p-1} - 1))\phi = \alpha_1 1 + \cdots + \alpha_{p-1}(p - 1),$$

and extend ϕ to FG. Then there exists some $c \in F$ such that $1 = (x - 1)\phi = c(x - 1) = 0$, a contradiction. $\qquad\square$

Theorem 9.3 (G.O. Michler and O.E. Villamayor [193], O.E. Villamayor [278]). *Let G be a group. Then a field F is FG-injective if and only if G is locally finite and char $F \notin \Pi(G)$.*

Proof. Suppose that F is FG-injective but G is not locally finite. Then there exist $g_1, \ldots, g_n \in G$ such that $G_1 = \langle g_1, \ldots, g_n \rangle$ is infinite. Put

$$R = \underbrace{FG \times \cdots \times FG}_{n},$$

and define a mapping $\phi : FG \longrightarrow R$ by

$$x\phi = ((g_1 - 1)x, \ldots, (g_n - 1)x), \quad x \in FG.$$

Clearly, ϕ is an FG-homomorphism. If $z \in \mathrm{Ker}\ \phi$, then

$$(g_j - 1)z = 0 \text{ or } g_j z = z \text{ for every } j, 1 \le j \le n.$$

Therefore,

$$g_1 z = \cdots = g_n z = z,$$

and it follows that $gz = z$ for every $g \in G_1$. Thus the coefficients of z are constant on the left cosets of G_1. Since Supp z is finite and G_1 infinite, we have $z = 0$. Hence $\mathrm{Ker}\ \phi = \langle 0 \rangle$, and ϕ is a monomorphism. Let $\omega : FG \longrightarrow F$ be the augmentation map of FG. Since F is FG-injective, there exists a mapping $\psi : R \longrightarrow F$ such that $\omega = \phi\psi$. If $1 \le j \le n$, let

$$\nu_j = (0, \ldots, 0, 1, 0, \ldots, 0) \in R$$

be the n-tuple with all zeros unless a unique 1 is in the place j, and define $\lambda_j = \nu_j \psi$. Then

$$1 = 1\omega = 1(\phi\psi) = (1\phi)\psi = (g_1 - 1, \ldots, g_n - 1)\psi$$
$$= ((g_1 - 1)\nu_1 + \cdots + (g_n - 1)\nu_n)\psi = (g_1 - 1)\lambda_1 + \cdots + (g_n - 1)\lambda_n = 0,$$

a contradiction. This contradiction shows that G is locally finite. The other condition follows from Lemma 9.1 and Lemma 9.2.

Conversely, suppose that G is a locally finite group and char $F \notin \Pi(G)$. Let L be a non-zero right ideal of FG, and let $\phi : L \longrightarrow F$ a non-zero mapping. Pick $u \in L$ such that $u\phi \ne 0$. Since G is locally finite, $H = \langle \mathrm{Supp}\ u \rangle$ is finite and $L \cap FH \ne \langle 0 \rangle$. By Lemma 9.2, the restriction of ϕ to $L \cap FH$ can be extended to FH; that is, there exists some $\lambda \in F$ such that $a\phi = \lambda a = \lambda(aw)$ for every $a \in L \cap FH$. Pick $b \in L$. There exists a finite subgroup S such that Supp $b \subseteq S$ and $H \leq S$. As above, there exists some $\mu \in F$ such that $b\phi = \mu(bw)$ and $u\phi = \mu(uw)$. Then $(\lambda - \mu)(uw) = 0$. But $(uw) \neq 0$ since $u\phi \neq 0$. Therefore $\lambda = \mu$. Thus $b\phi = \lambda b$ and ϕ can be extended. By the criterion of Baer (Theorem 7.1), F is FG-injective, as required. $\qquad \square$

In a natural way, Theorem 9.3 leads us to the following class of algebras. Let A be an algebra over a field F. Then A is said to be *a locally Wedderburn algebra* if every finite set of elements in A generates a semisimple subalgebra having finite dimension over F.

Now we study the conditions of injectivity of simple FG-modules. Let R be a ring and A be a simple R-module, and put $E = \mathrm{End}_R(A)$. If $\phi \in \mathrm{End}_R(A)$, then Ker ϕ and Im ϕ are R-submodules of A. Therefore, if $\phi \neq \langle 0 \rangle$, then Im $\phi = A$ and Ker $\phi = \langle 0 \rangle$. Hence, every non-zero endomorphism of a simple R-module A is an automorphism. In other words, E is a division ring. Thus we can naturally assume A to be in a vector space over E. On the other hand, let G be a group and let B be an RG-module. If $x \in RG$, then the mapping $\tau_x : b \longrightarrow bx$, $b \in B$ is an R-endomorphism, and the mapping $\Psi : x \longrightarrow \tau_x$, $x \in RG$ is a ring homomorphism of RG in the ring $\mathrm{End}_R(B)$ of all R-endomorphisms of B.

Lemma 9.4 (D.R. Farkas and R.L. Snider [69]). *Let W be a locally Wedderburn algebra over a field F. Suppose that A is a simple W-module and put $E = \mathrm{End}_W(A)$. If $\dim_E(A)$ is finite, then A is injective.*

Proof. Suppose that the result is false. By the criterion of Baer (Theorem 7.1), there exist a right ideal L of W and a W-homomorphism $\phi : L \longrightarrow A$, which cannot be lifted to W. Given a subalgebra V of W, we put

$$D(V) = \{a \in A \mid ax = x\phi \text{ for every } x \in L \cap V\}.$$

Let $\mathfrak{M}$ be the family of all finite dimensional semisimple subalgebras of W. Our assumption yields

$$\bigcap_{V \in \mathfrak{M}} D(V) = \emptyset.$$

Since all modules over such an algebra are injective, $D(V) \neq \emptyset$, for each $V \in \mathfrak{M}$. We pick $U \in \mathfrak{M}$ such that the dimension d of $L/(L \cap U)$ is minimal. Since $\bigcup_{V \in \mathfrak{M}} D(V) = \emptyset$, there exists a subalgebra $V_1 \in \mathfrak{M}$ such that $D(V_1)$ does not include $D(U)$. If $V_2 \in \mathfrak{M}$ contains a basis for U and V_1, then $U \leq V_2$ implies

$$\emptyset \neq D(V_2) \subseteq D(U) \text{ and } D(V_2) \neq D(U).$$

Thus, if $\mu \in D(V_2)$, then

$$D(V_2) = \mu + \mathrm{Ann}_A(L \cap V_2) \subseteq \mu + \mathrm{Ann}_A(L \cap U)$$

and

$$\mu + \mathrm{Ann}_A(L \cap V_2) \neq \mu + \mathrm{Ann}_A(L \cap U) = D(U),$$

contradicting the minimality of d. $\square$

The next result is due to B. Hartley and leads us in a natural way to the class of abelian-by-finite groups.

Lemma 9.5 (B. Hartley [107]). *Let A be a simple FG-module, where G is an abelian-by-finite group and F is a field. If $E = \mathrm{End}_{FG}(A)$, then $\dim_E(A)$ is finite.*

Proof. Let $K = (FG)\Psi$ be the image of FG in $\mathrm{End}_F(A)$. Consider first the case in which G is abelian. Then τ_x is an FG-endomorphism for every $x \in FG$. Given $\sigma \in E$, we choose a fixed non-zero element a of A. Since A is a simple FG-module, there exists some $z \in FG$ such that $a\sigma = az$. If b is an arbitrary element of A, then there exists some $x \in FG$ such that $b = ax$. Therefore

$$b\sigma = (ax)\sigma = (a\sigma)x = (az)x = a(zx) = a(xz) = (ax)z = bz,$$

and it follows that $\sigma = z\Psi$ so that $E = K$. In particular, K is a field.

In the general case, let H be an abelian normal subgroup of G of finite index, and put $R = (FH)\Psi$. By Theorem 5.5, there are finitely many simple FH-submodules $B_1, \ldots, B_n$ such that

$$A = B_1 \oplus \cdots \oplus B_n.$$

For each $1 \leq j \leq n$, we put $L_j = \mathrm{Ann}_R(B_j)$. By the result shown in the above paragraph, each R/L_j is a field and clearly

$$L_1 \cap \cdots \cap L_n = \mathrm{Ann}_R(A) = \langle 0 \rangle.$$

Therefore R is a direct sum of finitely many fields; so, in particular, it is an artinian ring. Let S be a transversal to H in G. Then $K = \sum_{x \in S} xR$ and so, in particular, K satisfies the minimal condition for R-submodules and *a fortiori* for right ideals. Since K is a simple ring with the minimal condition for right ideals, the classical Wedderburn–Artin structure theorem (see, for example, [66, Theorem 7.9]) shows that there are a positive integer r and a division ring D such that $K \cong M_r(D)$, the ring of $r \times r$-matrices over D. Moreover, A is isomorphic to the natural $M_r(D)$-module of r-row vectors with entries in D. Of course, this module has dimension r over its endomorphism ring D, which proves the general case. $\square$

Let F be a field of characteristic $p \geq 0$. Following B. Hartley [107], we say that a locally finite p'-group G is *restricted* if every simple FG-module has finite dimension over its endomorphism ring (as usual, $0'$ denotes the set of all primes). Our next goal is to prove here an important result due to B. Hartley, in which it is shown that a restricted group is abelian-by-finite. To deal with this, some technical results are needed.

Lemma 9.6 ([69]). *Let S be a ring, and let R be a subring of S. Suppose that A is a simple R-module, B is a simple S-module, and $\phi : A \longrightarrow B$ is an R-monomorphism. Put $E = \mathrm{End}_R(A)$ and $D = \mathrm{End}_S(B)$. If X is a subset of A linearly independent over E, then $X\phi$ is linearly independent over D.*

Proof. Suppose that
$$(a_1\phi)\delta_1 + \cdots + (a_n\phi)\delta_n = 0,$$
where $a_1, \ldots, a_n$ are different elements of X and $\delta_1, \ldots, \delta_n \in D$. Applying the Jacobson-Chevalley Density Theorem (see, for example, C. Faith [67, Theorem 19.22]), for every $1 \leq j \leq n$, there exists some $u_j \in R$ such that $a_j u_j = a_j$, whereas $a_k u_j = 0$, provided $k \neq j$. Then
$$0 = ((a_1\phi)\delta_1) + \cdots + (a_n\phi)\delta_n)u_k$$
$$= a_1 u_k \phi \delta_1 + \cdots + a_n u_k \phi \delta_n = a_k \phi \delta_k.$$

Since $a_k\phi \neq 0$ and D is a division ring, $\delta_k = 0$, establishing the result. □

Lemma 9.7 ([107]). *Let F be a field with char $F = p$, and suppose that G is a restricted locally finite p'-group. If V is a subgroup of G and U is a normal subgroup of V, then V/U is likewise restricted.*

Proof. Let A be a simple $F(V/U)$-module. We can think of A as a simple FV-module by allowing U to act trivially; this does not affect the endomorphism dimension. Then $A \cong FV/M$, for some maximal right ideal M of FV. Since $M(FG)$ is a proper right ideal of FG, there is a maximal right ideal L of FG including M, and necessarily $L \cap FV = M$. Thus we have an FV-monomorphism of $A \cong FV/M$ into the simple FG-module FG/L. By assumption, the latter has finite endomorphism dimension. Lemma 9.6 implies that A also has finite endomorphism dimension. □

Now we consider some needed facts about linear groups over division rings.

Lemma 9.8 ([107]). *Let F be a field, and suppose that D is a division algebra over F and A is a vector space of finite dimension over D. Let G be a locally finite p'-subgroup of $GL(D, A)$. Then:*

(1) *For each prime q, the Sylow q-subgroups of G are Chernikov.*

(2) *if for every subgroup H of finite index the DH-module A is indecomposable, then every finite abelian normal subgroup of G is cyclic.*

Proof. We think of F as naturally embedded in the ring of D-endomorphisms of A. Let R be the subring of $\mathrm{End}_D(A)$ generated by F and G. Thus, R is a homomorphic image of FG.

(1) Let Q be a finite elementary abelian q-subgroup of G, and let K be the image of FQ in R. Then K is a finite dimensional F-algebra. We may write
$$1 = e_1 + \cdots + e_t,$$

where $e_1, \ldots, e_t$ are orthogonal primitive idempotents in K. Then

$$A = Ae_1 \oplus \cdots \oplus Ae_t,$$

and we note that each summand is a non-zero D-subspace. Therefore $t \leq n = \dim_D(A)$. Furthermore,

$$K = Ke_1 \oplus \cdots \oplus Ke_t,$$

and the summands are fields. We note that the projection of Q into each summand has order 1 or q, and $|Q|$ divides q^t, $q^t | q^n$. In particular, every elementary abelian q-subgroup of G is finite. By a result due to S. N. Chernikov [43, Theorem 4.1] every Sylow q-subgroup of G is Chernikov.

(2) Suppose the contrary, and let U be a finite noncyclic abelian normal subgroup of G. Denote by L the subring of $\operatorname{End}_D(A)$ generated by F and U. Then L is not a field, and so there exist non-zero orthogonal idempotents $e_1, e_2 \in L$ such that $1 = e_1 + e_2$. It follows that

$$A = Ae_1 \oplus Ae_2$$

and the latter two summands are non-zero D-subspaces. Let $C = C_G(U)$. Then G/C is finite, and, since C commutes with e_1 and e_2, Ae_1 and Ae_2 are C-invariant. Thus A is a direct sum of two non-zero C-invariant subspaces. This contradiction proves this result.   □

Before dealing with the proof of the basis theorem, we require another result about modules for normal subgroups. Let H be a normal subgroup of a group G and let U be an FH-module. If $g \in G$, by definition, *the conjugate FH-module U^g* has U as underlying space with the H-action

$$(u, x) \to u(gxg^{-1}), \text{ where } u \in U \text{ and } x \in H.$$

The stabilizer of U (or *the inertia group of U*) is the set

$$S = \{g \in G \mid U^g \cong_{FH} U\}.$$

Clearly, S is a subgroup of G and there exists a bijection between the isomorphism classes of conjugates of U and the set of the right cosets Sg, $g \in G$. Indeed these facts are all well known (see B. Huppert [115]).

Lemma 9.9 ([107]). *Let F be a field, and suppose that H is a normal subgroup of a restricted group G. If U is a simple FH-module, then the stabilizer S of U in G has finite index in G.*

Proof. If $|G : S|$ is infinite, then U has infinitely many pairwise non-isomorphic conjugates under G. As in Lemma 9.7, U can be embedded in a simple FG-module A. If $g \in G$, then $U^g \cong_{FH} Ug$; so the FH-module A has infinitely many homogeneous components. Each one of these is invariant under the ring $E = \operatorname{End}_{FG}(A)$. Thus, $\dim_E(A)$ is infinite, and G cannot be restricted.   □

The next proposition is the key step of the proof of the basis result. We show it independently because the result is interesting in its own way.

Proposition 9.10 ([107]). *Let F be a field with char$F = p$. Suppose that G is a locally finite p'-group. If G is not abelian-by-finite, then there exists a simple FG-module A that satisfies one of the following equivalent conditions:*

(1) *$G/C_G(A)$ is not abelian-by-finite.*

(2) *$(FG)\Psi$ satisfies no non-trivial polynomial identity.*

Proof. We first show that (1) and (2) are equivalent. Indeed, if $Q = G/C_G(A)$ is abelian-by-finite, then (see D.S. Passman [216, Theorem 5.1]) the algebra FQ satisfies the standard identity; hence, so does the image of FG in $\text{End}_F(A)$, because this is a homomorphic image of FQ. Conversely, suppose that $R = (FG)\Psi$ satisfies a non-trivial polynomial identity. Since R is a primitive ring, by a result due to Kaplansky (see D. S. Passman [216, Theorem 6.4]), R can be embedded in a ring of matrices over a field, which is a field extension of F. By B.A.F. Wehrfritz [280, Corollary 9.4], $G/C_G(A)$ is abelian-by-finite.

Thus, it suffices to show (1). First of all, we note that G has a countable subgroup H that is not abelian-by-finite (D.S. Passman [216, Lemma 13.2]). Applying the argument of Lemma 9.7, every simple FH-module B can be embedded in a simple FG-module A. Moreover, if $H/C_H(B)$ is not abelian-by-finite, neither is $G/C_G(A)$. Therefore, replacing G by H, we may assume that G is countable. Then G has an ascending chain of finite subgroups

$$H_1 \leq H_2 \leq \cdots \leq H_n \leq \cdots$$

such that

$$G = \bigcup_{n \in \mathbb{N}} H_n.$$

We consider all the sequences

$$[\mathcal{S}] \quad S_1 \longrightarrow S_2 \longrightarrow \cdots \longrightarrow S_n \longrightarrow \cdots ,$$

in which each S_n is a minimal two-sided ideal of FH_n, and the mapping into S_n is the monomorphism induced by the natural projection of FH_n onto S_n resulting from writing FH_n as a direct sum of minimal ideals. Let r_n be the smallest natural number m such that S_n satisfies the standard identity s_m. Clearly

$$r_1 \leq r_2 \leq \cdots \leq r_n \leq \cdots ,$$

and the argument falls into two cases.

Case 1. *There exists a sequence $[\mathcal{S}]$ for which $r_n \longrightarrow \infty$.* We note that $[\mathcal{S}]$ induces

$$[\mathcal{S}_1] \quad A_1 \longrightarrow A_2 \longrightarrow \cdots \longrightarrow A_n \longrightarrow \cdots ,$$

in which every A_n is a minimal right ideal of S_n (and so of FH_n), and the mapping into A_{n+1} is an FH_n-monomorphism. Having obtained A_n, we map it into S_{n+1} through the sequence $[\mathcal{S}]$, and then project onto a suitable minimal right ideal A_{n+1} of S_{n+1}. The injective limit of the sequence $[\mathcal{S}_1]$ is a simple FG-module A, and the image $(FG)\Psi$ of FG in $\mathrm{End}_F(A)$ cannot satisfy a standard identity, since it includes isomorphic copies of the S_n for each $n \in \mathbb{N}$. If $Q = G/C_G(A)$ is abelian-by-finite, then FQ satisfies an standard identity; hence, so does $(FG)\Psi$, a contradiction. This finishes **Case 1**.

Suppose now that for each $j \geq 1$ and for a minimal (two-sided) ideal U_j of FH_j, there exists some $k \geq j$ and a minimal ideal U_k of FH_k such that there is a monomorphism $U_j \longrightarrow U_k$ and U_k fails to satisfy some standard identity satisfied by U_j. Then we may clearly construct a sequence $[\mathcal{S}]$ to which **Case 1** applies. In the contrary case, by omitting some of the H_n and re-indexing the remainders, we may suppose that H_1 and S_1 satisfy the following:

Case 2. *For every sequence $[\mathcal{S}]$ beginning with S_1, we have $r_n = r$ for each $n \in \mathbb{N}$.*

As indicated above, any sequence $[\mathcal{S}]$ induces a sequence $[\mathcal{S}_1]$. If $n \in \mathbb{N}$, we put $E_n = \mathrm{End}_{FH_n}(A_n)$. Clearly, $E_n = \mathrm{End}_{S_n}(A_n)$. Let d_n be the dimension of A_n over E_n. By Lemma 9.6, the image of an E_n-basis of A_n in A_{n+1} is E_{n+1}-independent, and so

$$d_1 \leq d_2 \leq \cdots \leq d_n \leq \cdots .$$

Applying the Jacobson–Chevalley Density Theorem (see, for example, C. Faith [67, Theorem 19.22]), we get that S_n is isomorphic to a ring of $d_n \times d_n$ matrices over E_n, which includes a ring of $d_n \times d_n$ matrices over F. Since each S_n satisfies the standard identity s_r, the Amitsur–Levitzki theorem (D.S. Passman [216, Lemma 4.2]) shows that d_n are bounded by a number depending only on r. By considering a sequence $[\mathcal{S}]$ in which the least such bound is attained by omitting the first few terms and re-indexing the remainders, we may suppose

$$d_1 = \cdots = d_n = \cdots = d,$$

for every sequence $[\mathcal{S}]$ starting with S_1.

Consider now a sequence $[\mathcal{S}]$ and its induced sequence $[\mathcal{S}_1]$. Let e be a central idempotent of FH_1 generating S_1, that is, the identity of the ring S_1. If $n \geq 1$, we know that the image in A_n of an E_1-basis of A_1 is an E_n-basis of A_n. Therefore the projection of e in S_n fixes by right multiplication an E_n-basis of A_n and it has to be the identity of S_n. Write

$$FH_n = T_1 \oplus \cdots \oplus T_s,$$

where $T_1, \ldots, T_s$ are minimal ideals and the ordering is such that

$$e = a_1 + \cdots + a_t,$$

where $t \leq s$ and $0 \neq a_j \in T_j$ for every $1 \leq j \leq t$. Then each T_j can be taken as the nth-term S_n of a sequence $[\mathcal{S}]$. Therefore a_j is the identity of each such T_j, and each such T_j satisfies s_r. Hence, e is central in FH_n and eFH_n satisfies s_r. It readily follows that e is a central idempotent of FG and so eFG satisfies s_r.

From D.S. Passman [215], $|G : FC(G)|$ is finite and $[FC(G), FC(G)]$ is finite. Since G is a p'-group, FG is semisimple, and then for every $1 \neq g \in [FC(G), FC(G)]$ there exists a simple FG-module A_g such that $g \notin C_G(A_g)$. Let Y be a finite subset such that $[FC(G), FC(G)] = \langle Y \rangle$. If every group $G/C_G(A_g)$ has an abelian subgroup $V_g/C_G(A_g)$ of finite index, then putting

$$W = FC(G) \cap \left(\bigcap_{g \in Y} V_g \right),$$

we obtain that

$$[W, W] \leq [FC(G), FC(G)] \bigcap \left(\bigcap_{g \in Y} C_G(A_g) \right) = \langle 1 \rangle .$$

Since W has finite index in G, we get a contradiction to the assumption that G cannot be abelian-by-finite. Thus, one of the groups $G/C_G(A_g)$ cannot be abelian-by-finite, and all is done. □

Now we consider some particular cases of the mentioned theorem, which are needed for its proof. The next result was originally proved by R. L. Snider in [267], but we prefer to follow the proof given by B. Hartley in [107].

Proposition 9.11. *Let F be a field with char $F = p$, and suppose that G is a restricted p'-group. If G is locally nilpotent, then G is abelian-by-finite.*

Proof. Suppose the contrary. By Lemma 9.7 and Proposition 9.10, we may assume that there exists a simple FG-module A such that $C_G(A) = \langle 1 \rangle$. Put $E =$ End $_{FG}(A)$. Since G is restricted, $\dim_E(A)$ is finite. We have

$$G = \text{Dr } _{q \in \Pi(G)} G_q,$$

where G_q is the Sylow q-subgroup of G. By Lemma 9.8, each G_q is Cernikov; then, in particular, it is abelian-by-finite. Therefore, there is an infinite subset $\pi \subseteq \Pi(G)$ such that G_q is not abelian for every $q \in \pi$. Since each one of these G_q has a finite non-abelian subgroup, by Lemma 9.7 we may assume that

$$G = \text{Dr } _{q \in \Pi(G)} G_q,$$

where $\Pi(G)$ is infinite and each Sylow q-subgroup G_q is finite non-abelian. We may even assume that $2 \notin \Pi(G)$ and every proper subgroup and factor group of G_q is abelian. If $q \in \Pi(G)$, it follows that G_q is nilpotent of class 2, $Z_q = \zeta(G_q)$ is cyclic, and $|[G_q, Gq]| = q$. Moreover, G_q has a non-cyclic maximal abelian normal subgroup U_q (see D. Gorenstein [91, Theorem 5.4.10]). Let V_q be a subgroup of U_q, maximal under $V_q \cap Z_q = \langle 1 \rangle$. Then U_q/V_q is cyclic, $V_q \neq \langle 1 \rangle$. Moreover, if

$$U = \text{Dr } _{q \in \Pi(G)} U_q \text{ and } V = \text{Dr } _{q \in \Pi(G)} V_q,$$

then U/V is locally cyclic. This means that there exists a simple FU-module B such that $V = C_U(B)$ (see L.A. Kurdachenko, J. Otal and I.Ya. Subbotin [157, Chapter 2]). Let $H = N_G(V)$ so that

$$H = \mathrm{Dr}\ _{q \in \Pi(G)} H \cap G_q.$$

Clearly, H contains the stabilizer S of B in G. Since $V_q \cap Z_q = \langle 1 \rangle$, V_q is not normal in G_q; hence, $G_q \neq H \cap G_q$ for all $q \in \Pi(G)$. Therefore, $|G : H|$ is infinite, and $|G : S|$ is infinite. By Lemma 9.9, G cannot be restricted, a contradiction. $\square$

Proposition 9.12 ([107]). *Let F be a field with char $F = p$. Suppose that G is a periodic restricted p'-group. If G is metabelian, then G is abelian-by-finite.*

Proof. Suppose the contrary. By Lemma 9.7 and Proposition 9.10, we assume that there exists a simple FG-module A such that $C_G(A) = \langle 1 \rangle$. Put $E =\mathrm{End}\ _{FG}(A)$. Since G is restricted, $\dim_E(A) = n$ is finite and then G can be embedded into the linear group $GL_n(E)$. We may suppose that n is minimal subject to the existence of a counterexample, which can be embedded in this way. If G has a subgroup L of finite index such that $A = A_1 \oplus A_2$, where A_1 and A_2 are non-zero L-invariant submodules, then by Lemma 9.7, $L/C_L(A_1)$ and $L/C_L(A_2)$ are restricted so that, by the minimality of n, each $L/C_L(A_j)$ is abelian-by-finite, $1 \leq j \leq 2$. Therefore, L is abelian-by-finite, and so is G; a contradiction that shows that this case cannot occur.

By Lemma 9.8, the Sylow q-subgroups of G are Chernikov for every prime q. It also follows that every finite abelian normal subgroup of G is cyclic. Let H be the smallest normal subgroup of G such that G/H is locally nilpotent. Since G is metabelian, H is certainly abelian. The Sylow q-subgroup H_q of H is Cernikov, and $\Omega_m(H_q)$ is finite, for every $m \in \mathbb{N}$. Since the latter is normal in G, it has to be cyclic. Therefore H_q, and hence H is locally cyclic. Then there exists a monomorphism $\chi : H \longrightarrow U(K)$, where K is the algebraic closure of F. Put $M = F[H\chi]$. Since every element of $H\chi$ is algebraic over F, M is a subfield of K. We may extend χ by linearity over F to the whole group ring FH. If $a \in M$ and $v \in FH$, the action $(a, v) \to a(v\chi)$ makes M into an FH-module, which clearly is simple (see, for example, L.A. Kurdachenko, J. Otal and I.Ya. Subbotin [157, Chapter 2]). Let S be the stabilizer of the FH-module M in G. By Lemma 9.9, $|G : S|$ is finite. We are going to show how the FH-module structure of M can be extended to an FS-module structure in such a way that the resulting module has infinite endomorphism dimension, then obtaining a contradiction. A crucial consideration for this is the fact that G splits over H (B. Hartley [104]); that is, there exists a subgroup B such that $G = HB$ and $H \cap B = \langle 1 \rangle$.

If $g \in S$, the conjugate module gMg^{-1} is FH-isomorphic to M. Thus, there exists an F-linear isomorphism $\psi : M \longrightarrow M$ such that

$$(av)\psi = a\psi(g^{-1}vg),$$

for every $a \in M$ and $v \in FH$. Thus,

$$(av\chi)\psi = a\psi(g^{-1}vg)\chi,$$

if $a \in M$ and $v \in FH$. Since the multiplication by any non-zero element of M is an FH-automorphism of M, we may assume that $1\psi = 1$. Putting $a = 1$, we obtain that

$$v\chi\psi = (g^{-1}vg)\chi,$$

for every $v \in FH$; hence we have that

$$(av\chi)\psi = a\psi v\chi\psi,$$

for every $a \in M$ and $v \in FH$. Since χ is surjective, it follows that ψ is an element of the Galois group Γ of M over F. Further, ψ is uniquely determined (in terms of g) by

$$(av\chi)\psi = a\psi(g^{-1}vg)\chi \text{ and } 1\psi = 1,$$

via $v\chi\psi = (g^{-1}vg)\chi$. We define $\psi = g\phi$ so that ϕ is a mapping from S into Γ. Since $v\chi\psi = (g^{-1}vg)\chi$, we readily obtain that ϕ is a homomorphism.

Let Σ be the translation subgroup of the group of all F-linear automorphisms of M. Thus Σ consists of all maps $y^* : a \mapsto ay$, where $a \in M$ and $0 \neq y \in M$. If $\gamma \in \Gamma$, then $\gamma^{-1}y^*\gamma = (y\gamma)^*$, and so Γ normalizes Σ. Since G splits over H, S likewise splits over H. Actually, if B is a complement of H in G (as indicated above), $B \cap S$ is a complement of H in S. Then the mapping $\theta : hb \mapsto (h\chi)^* b\phi$, where $h \in H$ and $b \in B \cap S$, is a well-defined map of S into the group $\Sigma\Gamma$ of semilinear transformations of M. Let $h' \in H$ and $b' \in V$. Then

$$hb = h'b' = h(bh'b^{-1})(bb').$$

Thus

$$(hbh'b')\theta = (h\chi)^*((bh'b^{-1})\chi)^*(b\phi)(b'\phi)$$
$$= (h\chi)^*(b\phi)((bh'b^{-1})\chi)b\phi)^*(b'\phi) = (h\chi)^*(b\phi)(h'\chi)^*(b'\phi).$$

Matching $v = bh'b^{-1}$ and $\psi = b\phi$ in $v\chi\psi = (g^{-1}vg)\chi$, we obtain

$$(h\chi)^*(b\phi)(h'\chi)^*(b'\phi) = ((hb)\theta)((h'b')\theta).$$

Consequently, θ is in fact a homomorphism; by means of it, we can view M as an FS-module. The action of FH remains unchanged, and so M is certainly simple. Let ν be any FS-endomorphism of M. If $w \in FH$, then

$$w\chi\nu = (1w\chi)v = 1vw\chi.$$

Since χ is surjective, ν is induced by the multiplication by the element $y = 1\nu$ of M. If $b \in B \cap S$ and $a \in M$, then

$$(ab\phi)y = (ab\phi)\nu = (a\nu)(b\phi) = (ay)(b\phi).$$

Putting $a = 1$, we find that y is fixed by the element $b\phi \in \Gamma$. Since this holds for every $b \in B \cap S$, we find that ν is induced by the multiplication by an element of the fixed field P of $(B \cap S)\phi$. It is easy to see that such multiplication induces an FS-endomorphism of M, and so End $_{FS}(M)$ can be identified with P. Thus, it suffices to show that M has infinite dimension over P. This claim will follow if we prove that $(B \cap S)\phi$ is infinite. To see this, pick $b \in B \cap S$. Then $b \in \mathrm{Ker}\ \phi$ if and only if $(w\chi)(b\phi) = w\chi$, for all $w \in FH$. Matching $\psi = b\phi$ in $w\chi\psi = (g^{-1}wg)\chi$, we see that the above assertion is equivalent to $(b^{-1}wb)\chi = w\chi$, for all $w \in FH$, or, since χ is F-linear, to the same condition with $w \in H$. Thus, $b \in \mathrm{Ker}\ \phi$ if and only if $[w, b]\chi = 1$, for all $w \in H$. Since χ is a monomorphism of H, we conclude that $\mathrm{Ker}\,(\phi\,|_{B \cap C}) = C_{B \cap C}(H) = C$. Since H is abelian and C is locally nilpotent, $HC = H \times C$ is likewise locally nilpotent. By Lemma 9.7, HC is restricted, and by Proposition 9.11, HC is abelian-by-finite. If $|(B \cap S) : C|$ is finite, then $|S : HC|$ is finite. Since $|G : S|$ is finite, $|G : HC|$ is finite. This implies that G is abelian-by-finite, which was assumed not to be so. Hence, C has infinite index in $B \cap S$; so that $(B \cap S)\phi$ is infinite, and we are done. $\qquad\square$

The next result was proved by R.L. Snider in [267], but we follow the proof given by B. Hartley in [107].

Proposition 9.13. *Let F be a field with* char $F = p$. *Suppose that G is a restricted p'-group. If the conjugacy classes of G are finite (that is, G is an FC-group), then G is abelian-by-finite.*

Proof. Suppose the contrary. If $Z = \zeta(G)$, recall that G/Z is residually finite (see M.J. Tomkinson [272, Theorem 1.9]). If G/Z has an abelian subgroup U/Z of finite index, then U is nilpotent of class at most 2 and restricted. By Proposition 9.11, U is abelian-by-finite, and so is G. This contradiction shows that G/Z cannot be abelian-by-finite. This and Lemma 9.7 assure that we may assume that G is residually finite.

Let L_1 be an arbitrary finite normal subgroup of G. Since G is residually finite, there exists a normal subgroup H of G of finite index such that $H \cap L_1 = \langle 1 \rangle$. By our assumption, H cannot be abelian and then H has a finite non-abelian G-invariant subgroup L_2. Applying the above argument to $L_1 \times L_2$ and proceeding in this way, we find that G contains the direct product

$$L = L_1 \times L_2 \times \cdots \times L_n \times \cdots$$

of a countably infinite number of finite non-abelian subgroups L_n, $n \in \mathbb{N}$. By Lemma 9.7, we may assume that

$$G = L = \mathrm{Dr}_{n \in \mathbb{N}} L_n,$$

and, furthermore, that every proper subgroup of each L_n is abelian. Then L_n is metabelian (see D.J.S. Robinson [242, 9.1.9]), so that we have a metabelian counterexample G. It suffices to apply Proposition 9.12 to get a contradiction. $\qquad\square$

Lemma 9.14 ([107]). *Let F be a field with* char $F = p$. *If G is a restricted locally finite p'-group, and $FC(G)$ is finite, then G is finite.*

Proof. Put $Q = G/FC(G)$. Since $FC(G)$ is finite, we actually have $FC(Q) = \langle 1 \rangle$. We claim that Q is finite. Otherwise, it is easy to see that Q has a countably infinite subgroup H such that $FC(H) = \langle 1 \rangle$. By Lemma 9.7, H is restricted. On the other hand, by a result due to E. Formanek and R.L. Snider [71], the group ring FH is primitive. Applying the Jacobson–Chevalley Density Theorem (see C. Faith [67, Theorem 19.22]) to a faithful simple FH-module (which necessarily has finite dimension n over its endomorphism ring E), we find that $FH \cong M_n(E)$. In particular, FH is simple. Consideration of the augmentation ideal leads us to the contradiction $H = \langle 1 \rangle$. $\qquad\square$

We are now in a position to show one of the relevant results of this chapter.

Theorem 9.15 ([107]). *Let F be a field with* char $F = p$ *and suppose that G is a locally finite p'-group. Then every simple FG-module has finite endomorphism dimension if and only if G is abelian-by-finite.*

Proof. By Lemma 9.5, every periodic abelian-by-finite group is restricted.

Conversely, assume that G is restricted but G is not abelian-by-finite. Let L be the locally nilpotent radical of G. By Lemma 9.7 and Proposition 9.11, L includes an abelian normal subgroup L_0 of finite index. Form the product R of the abelian normal subgroups of finite index of L. Since $|L : L_0|$ is finite, we have

$$R = L_0 L_1 \cdots L_n,$$

for some normal abelian-by-finite subgroups $L_0, L_1, \ldots, L_n$ having finite index in L. Clearly,

$$L_1 \cap \cdots \cap L_n \le \zeta(L),$$

so that $|L : \zeta(L)|$ is finite. Hence L has a characteristic abelian subgroup Z of finite index. By our assumption, the factor-group G/Z is infinite. By Lemma 9.7 and Lemma 9.14, $D/Z = FC(G/Z)$ is infinite, and, by Lemma 9.7 and Proposition 9.13, D/Z includes an abelian normal subgroup U/Z of finite index. Then U/Z is infinite, and U is metabelian. By Lemma 9.7 and Proposition 9.12, U includes an abelian normal subgroup V of finite index. Since L contains every subnormal abelian subgroup of G, $V \le L$. Therefore, $U/(U \cap L)$ is finite. Since L/Z is finite, so is $(U \cap L)/Z$. It follows that U/Z is finite. This contradiction establishes the result. $\qquad\square$

Now we go back to the study of the injectivity of simple modules. As we did above, we consider modules over locally Wedderburn algebras. Concretely, we consider a locally Wedderburn algebra W of countable dimension over a field F and a simple W-module A with endomorphism ring E. By Lemma 9.4, A is injective if $\dim_E(A)$ is finite; however, we need to study the case in which $\dim_E(A)$ is infinite. We suppose this is the case. Since $A \cong_W W/M$, where M is a maximal right ideal of W, $\dim_F(A)$ is countably infinite. Let $\{w_n \mid n \in \mathbb{N}\}$ and $\{a_n \mid n \in \mathbb{N}\}$ be the basis of W and A, respectively. We recall the construction introduced by B. Hartley in [107]. We begin by constructing recursively quintuples $(W_n, A_n, B_n, C_n, D_n)$ satisfying the following conditions:

(a) W_n is a finite dimensional semisimple subalgebra of W such that $W_{n-1} \leq W_n$ and $\{w_1, \ldots, w_n\} \subseteq W_n$.

(b) A_n is a finite dimensional W_n-submodule of A such that $A_{n-1} \leq A_n$, $D_{n-1} \leq A_n$ and $\{a_1, \ldots, a_n\} \subseteq A_n$.

(c) B_n and C_n are W_n-submodules of A_n such that B_n is simple and $A_n = B_n \oplus C_n$.

(d) D_n is a simple W_n-submodule of A such that the E-span $A_n E$ of A_n does not include D_n, and the homogeneous component $\overline{D_n}$ of D_n in the W_n-module A has infinite E-dimension.

(e) If $E_n =$ End $_{W_n}(B_n)$ and $\pi_n : A_n \longrightarrow B_n$ is the projection associated with the decomposition $A_n = B_n \oplus C_n$, then $B_{n-1}\pi_n \neq \langle 0 \rangle$, and $(A_{n-1}\pi_n)E_n$ does not include $D_{n-1}\pi_n$.

(f) There exists a W_{n-1}-monomorphism of D_{n-1} into D_n.

In (d), as in other places, the homogeneous component of D_n means the sum of all W_n-submodules of A isomorphic to D_n. Of course, this is an E-submodule.

We start the construction with $n = 1$. Let W_1 be any finite dimensional semisimple subalgebra of W containing w_1. Let A_1 be any finite dimensional W_1-submodule of A containing a_1. Let B_1 be any simple submodule of A_1. Since W_1 is finite dimensional and semisimple, there exists some W_1-submodule C_1 of A_1 such that $A_1 = B_1 \oplus C_1$. Note that $A_1 E$ is finite dimensional over E, and W_1-invariant. Let X_1 be the sum of $A_1 E$, and all those homogeneous components of the W_1-module A that are finite dimensional over E. Then $\dim_E(X_1)$ is finite, and there exists a simple W_1-submodule D_1 of A such that X_1 does not include D_1. Clearly, (d) holds for D_1, and we have just established (a)–(d) in the case $n = 1$. Now suppose that $n \geq 1$, and we have already constructed $(W_j, A_j, B_j, C_j, D_j)$ for every $j \leq n$ satisfying the properties (a)-(f). By (d), there exists some $d \in D_n$, which is E-independent of A_n. By the Jacobson–Chevalley Density Theorem (see C. Faith [67, Theorem 19.22]), there exists some $w \in W$ such that $A_n w = \langle 0 \rangle$ and $0 \neq dw \in B_n$. Pick a finite dimensional semisimple subalgebra W_{n+1} of W such that $W_n \leq W_{n+1}$, $w_{n+1} \in W_{n+1}$ and $w \in W_{n+1}$. Let A_{n+1} be a finite dimensional W_{n+1}-submodule of A such that $A_n \leq A_{n+1}$, $D_n \leq A_{n+1}$ and $a_{n+1} \in A_{n+1}$. Then (a) and (b) hold for $j = n+1$. We can clearly choose a direct decomposition $A_{n+1} = B_{n+1} \oplus C_{n+1}$ such that B_{n+1} is simple and the projection $\pi_{n+1} : A_{n+1} \longrightarrow B_{n+1}$ is non-zero. This gives (c). Since π_{n+1} is a monomorphism, and $w \in W_{n+1}$, we have $0 \neq (dw)\pi_{n+1} = d\pi_{n+1}w$. Thus w does not annihilate $D_n\pi_{n+1}$, whereas it clearly annihilates $A_n\pi_{n+1}$ because it annihilates A_n itself. It follows that $(A_n\pi_{n+1})E_{n+1}$ does not include $D_n\pi_{n+1}$, which proves (c).

It remains to construct D_{n+1}. As in the case $n = 1$, $\dim_E(A_{n+1}E)$ is finite, and $A_{n+1}E$ is W_{n+1}-invariant. Adding to it the sum of those homogeneous components of the W_{n+1}-module A that are finite dimensional over E, we obtain a finite dimensional W_{n+1}-invariant E-subspace X_{n+1} of A. Since, by (d), $\overline{D_n}$ has infinite E-dimension, X_{n+1} does not include $\overline{D_n}$. Therefore X_{n+1} does not include some simple W_n-submodule $S \cong D_n$. We have $A = X_{n+1} \oplus T$, where

$$T = \bigoplus_{\lambda \in \Lambda} T_\lambda,$$

and every T_λ is a simple W_{n+1}-submodule. It follows that there is some λ such that the projection of S into T_λ is non-zero. We put $D_{n+1} = T_\lambda$. Then (d) and (f) hold for D_{n+1}, and the construction is done.

It follows from (a) and (b) that

(g) $W = \bigcup_{n \in \mathbb{N}} W_n$ and $A = \bigcup_{n \in \mathbb{N}} A_n$.

It is worth mentioning that condition (d) is only required to assure that the induction may be carried out. In fact, we will not make use of it any longer.

Lemma 9.16 ([107]). *Suppose that B_{n+1} and D_{n+1} are not isomorphic, and they form the direct sum $Y_{n+1} = D_{n+1} \oplus A_{n+1}$. Then Y_{n+1} has a W_n-submodule $L_n \cong D_n$ such that $L_n \cap A_{n+1} = \langle 0 \rangle$, and no W_n-complement to A_n in $L_n + A_n$ lies in a W_{n+1}-complement to A_{n+1} in Y_{n+1}.*

Proof. It will be convenient to identify A_{n+1} with the external direct sum $B_{n+1} \oplus C_{n+1}$ (see (c)). In this case A_n is identified with the submodule

$$\{(0, a\pi_{n+1}, a\rho_{n+1}) \mid a \in A_n\} \leq D_{n+1} \oplus B_{n+1} \oplus C_{n+1}$$

where ρ_{n+1} is the projection of A_{n+1} onto C_{n+1} associated with the decomposition (c). First, we notice that each W_{n+1}-complement to A_{n+1} in Y_{n+1} is isomorphic to D_{n+1}, and, since B_{n+1} and D_{n+1} are not isomorphic, projects trivially into D_{n+1}. Next, we identify D_n in some way with a W_n-submodule of D_{n+1} (which is possible by (f)). Because of (e), there certainly exists a W_n-monomorphism ϕ of D_n into B_{n+1} such that $A_n \pi_{n+1}$ does not include $D_n \phi$. Let $L_n = \{(a, a\phi, 0) \mid a \in D_n\}$. Clearly L_n is a W_n-submodule of Y_{n+1} such that $L_n \cap A_{n+1} = \langle 0 \rangle$. Choose an element $a \in D_n$ such that $a\phi \notin A_n \pi_{n+1}$. Then any complement Z to A_n in $L_n + A_n$ contains an element of the form

$$(a, a\phi + a\pi_{n+1}, a\rho_{n+1}) \text{ with } a \in A_n.$$

By the choice of a, the second component of this element is non-zero. Therefore, the projection of Z into B_{n+1} is non-zero and cannot lie in a W_{n+1}-complement to A_{n+1} in Y_{n+1}. $\qquad\square$

Lemma 9.17 ([107]). *Suppose that $B_n \cong D_n$ for every $n \in N$ and form the direct sum $Y_{n+1} = B_{n+1} \oplus A_{n+1}$. Then Y_{n+1} includes a W_n-submodule $M_n \cong B_n$ such that $M_n \cap A_{n+1} = \langle 0 \rangle$, and no complement to A_n in $M_n + A_n$ lies in a complement to A_{n+1} in Y_{n+1}.*

Proof. In this case, we will identify Y_{n+1} with the external direct sum $B_{n+1} \oplus B_{n+1} \oplus C_{n+1}$, A_{n+1} with the set of elements with zero as the first coordinate, and A_n with $\{(0, a\pi_{n+1}, a\rho_{n+1}) \mid a \in A_n\}$. Since $B_n \cong D_n$ for every $n \in \mathbb{N}$, and we have the property (e), we obtain

(h) There is a W_n-monomorphism $\phi_{n+1} : B_n \longrightarrow B_{n+1}$ such that $(A_n\pi_{n+1})E_{n+1}$ does not include $B_n\phi_{n+1}$.

Let
$$M_n = \{(b\pi_{n+1}, b\phi_{n+1}, 0) \mid b \in B_n\}.$$

By (e), π_n maps B_n monomorphically; thus $M_n \cap A_{n+1} = \langle 0 \rangle$, and M_n is a W_n-submodule isomorphic to B_n. Now any complement to A_{n+1} in Y_{n+1} has the form

$$\{(b, b\psi, b\psi') \mid b \in B_{n+1}\},$$

where $\psi \in E_{n+1}$ and ψ' is a W_{n+1}-homomorphism of B_{n+1} in C_{n+1}). Choose $d \in B_n$ such that $d\phi_{n+1} \notin (A_n\pi_{n+1})E_{n+1}$. The existence of d is assured by (h). Then any complement to A_n in $M_n + A_n$ contains an element of the form

$$(d\pi_{n+1}, d\phi_{n+1} + a\pi_{n+1}, a\rho_{n+1})$$

with $a \in A_n$. If this element lies in a complement to A_{n+1} in Y_{n+1}, then we can identify it as an element of the form $(b, b\psi, b\psi')$, obtaining $d\pi_{n+1} = b$ and $d\phi_{n+1} + a\pi_{n+1} = b\psi$. Substituting for b in the second equation and rearranging, we get
$$d\phi_{n+1} = d\pi_{n+1}\psi - a\pi_{n+1} \in (A_n\pi_{n+1})E_{n+1}.$$

But this contradicts the choice of d. Therefore, no complement to A_n in $M_n + A_n$ lies in a complement to A_{n+1} in Y_{n+1}.   □

Now we are ready to establish the second main result of this chapter.

Theorem 9.18 ([107]). *Let W be a locally Wedderburn algebra of countable dimension over a field F and let A be a simple W-module with endomorphism ring E. Then exactly one of the following statements holds:*

(1) $\dim_E(A)$ *is finite, and A is injective; or*

(2) $\dim_E(A)$ *is infinite, and A can be embedded in an indecomposable W-module of composition length 2.*

Proof. Consider
$$\Xi = \{n \in \mathbb{N} \mid B_{n+1} \cong D_{n+1}\}.$$

We split the proof into two cases.

Case 1. *The set $\mathbb{N} \setminus \Xi$ is infinite.*
In this case, we form the direct sum $Y_n = D_n \oplus A_n$ as in Lemma 9.16. If B_{n+1} is not isomorphic to D_{n+1}, then we embed Y_n as a W_n-submodule of

Y_{n+1} embedding D_n as L_n and embedding A_n in A_{n+1} by the natural inclusion. Otherwise, using (f), we embed D_n in D_{n+1} in any way and again embed A_n in A_{n+1} by the natural inclusion. Taking the direct limit of the direct system so obtained and making suitable identifications, we obtain a W-module Y that is the union of the sequence

$$Y_1 \leq Y_2 \leq \cdots \leq Y_n \leq \cdots,$$

where each Y_n is a W_n-submodule and Y contains A in such a way that $A \cap Y_n = A_n$. Furthermore, $Y_n = D_n \oplus A_n$, and for infinitely many $n \in \mathbb{N}$, no complement to A_n in Y_n lies in a complement to A_{n+1} in Y_{n+1}. Now

$$Y/A = \bigcup_{n \in \mathbb{N}} (A + D_n)/A,$$

and

$$(A + D_n)/A \cong_{W_n} D_n/(A \cap D_n) \cong_{W_n} D_n.$$

Since the W_n-module D_n is simple, Y/A is a simple W-module. Thus Y has composition length 2. Now let T be any non-zero W-submodule of Y, and choose n such that $Y_n \cap T \neq \langle 0 \rangle$. If $A_n \cap T \neq \langle 0 \rangle$, then $T \geq A$. Otherwise, $Y_n \cap T$ is a complement to A_n in Y_n. But we can arrange that no complement to A_n in Y_n lies in a complement to A_{n+1} in Y_{n+1}. Therefore, the W_{n+1}-submodule of A_{n+1} generated by $Y_n \cap T$ meets A_{n+1} non-trivially. It follows that $A \cap T \neq \langle 0 \rangle$ and so $T \geq A$. Consequently, A is the unique simple submodule of Y, which is indecomposable.

If **Case 1** does not hold, then by suppressing finitely many terms in our sequence and re-indexing the rest, we reduce to

Case 2. $B_n \cong D_n$ *for all* $n \in \mathbb{N}$.

The proof of the theorem in **Case 2** now follows from Lemma 9.17, exactly as it did with **Case 1**. $\square$

The above result allows us to completely describe countable groups G, for which every simple FG-module is injective.

Theorem 9.19 ([107]). *Let F be a field of characteristic $p \geq 0$, and let G be a countable group. Then every simple FG-module is injective if and only if G is a periodic abelian-by-finite p'-group.*

Proof. Suppose that every simple FG-module is injective. If G is not a locally finite p'-group, then the FG-module F, which is clearly simple, is not injective by Theorem 9.3. Thus we must consider only the case, when G is a locally finite p'-group. Suppose that G is not abelian-by-finite. By Theorem 9.15, there exists a simple FG-module A that has infinite endomorphism dimension. Clearly the group algebra FG is a locally Wedderburn algebra of countable dimension over F. By Theorem 9.18, A can be embedded in an indecomposable FG-module of composition length 2. This shows that A cannot be injective, a contradiction.

Conversely, if G is a periodic abelian-by-finite group, and A is a simple FG-module, then by Lemma 9.5, A has finite endomorphism dimension. By Lemma 9.4, A is injective. $\square$

Theorem 9.19 was formerly obtained by D.R. Farkas and R.L. Snider [69] under the additional assumption that F contains all roots of identity if char $F = 0$.

Chapter 10

Direct decompositions in artinian modules

In the first chapter of this book we considered a celebrated result due to H. Fitting known as the Fitting lemma (Proposition 1.5). This result raises very interesting questions about artinian modules over group rings concerned with direct decompositions of modules that are determined by a group action. Since we may obtain very fruitful results studying this topic and make them available in the scope of this book, we develop them here. To deal with this, it is convenient to formulate the Fitting lemma in the following way.

Fitting lemma. *Let A be an RG-module of finite composition length, where R is a ring and G is a finite nilpotent group. Then $A = C \oplus E$, where the RG-chief factors U/V of C (respectively of E) satisfy $G = C_G(U/V)$ (respectively, $G \neq C_G(U/V)$).*

This fact raises related issues such as the discovery of complements to the upper RG-hypercenter (see the definition below) and the studying of certain extensions of modules that are very near to modules of finite composition length. These problems have found very important applications in the study both of groups and modules with finiteness conditions. Moreover, they are also related to the question of the existence of complements for some residuals in groups. We quote some papers dedicated to these problems: Z.Y. Duan [60, 61, 63], D.Y. Duan and M.J. Tomkinson [64], B. Hartley and M.J. Tomkinson [111], M.L. Newell [204, 205], D.J.S. Robinson [237, 238, 239, 240, 241, 243, 246, 247], E. Schenkmann [253], M.J. Tomkinson [271, 273, 275], D.I. Zaitsev [301, 302, 304, 307, 308, 310]; see also the survey L.S. Kazarin and L.A. Kurdachenko [132].

Now we formulate the necessary concepts in their more general form. In Chapter 3 we define the $\mathfrak{X}C$-center and the upper $\mathfrak{X}C$-hypercenter of a group and associated constructions. Proceeding in a similar way, we can translate the above concepts in those of the RG-center and the upper RG-hypercenter of modules over a group ring RG that is connected with some formation of groups.

Let $\mathfrak{X}$ be a class of groups. The factor C/B of a group G is said to be $\mathfrak{X}$-*central* (respectively, $\mathfrak{X}$-*eccentric*) if $G/C_G(C/B) \in \mathfrak{X}$ (respectively, $G/C_G(C/B) \notin \mathfrak{X}$). In general, the $\mathfrak{X}$-central and the $\mathfrak{X}$-eccentric factors of a group G can appear randomly in a composition series of G. Thus, discovering the cases in which all the $\mathfrak{X}$-central factors of a group or of a normal subgroup of it can be gathered in some place, while all the $\mathfrak{X}$-eccentric factors can be gathered in another place, appears to be a very interesting project. In this setting and for this purpose, R. Baer [8] introduced two important subgroups in a finite group G to rule out the $\mathfrak{X}$-centrality and the $\mathfrak{X}$-eccentrality with respect to G. These are the $\mathfrak{X}C$-*hypercenter* $\mathfrak{X}C_\infty(G)$ of G and *the* $\mathfrak{X}$-*hypereccenter* $E_\infty(G)$ of $\mathfrak{X}G$. The $\mathfrak{X}C$-hypercenter was defined in Chapter 3 (actually, in Chapter 3 it is denoted as the upper $\mathfrak{X}C$-hypercenter), while the $\mathfrak{X}$-hypereccenter is defined in the following way. Every G-chief factor of the normal subgroup $\mathfrak{X}E_\infty(G)$ is $\mathfrak{X}$-eccentric and $\mathfrak{X}E_\infty(G)$ is a maximal normal subgroup under this property. Clearly $\mathfrak{X}C_\infty(G) \cap \mathfrak{X}E_\infty(G) = \langle 1 \rangle$ always holds, but the decomposition

$$G = \mathfrak{X}C_\infty(G) \times \mathfrak{X}E_\infty(G)$$

usually fails. Baer himself achieved an important result in this direction, which we quote now (see R. Baer [8]).

Theorem. *Suppose that $\mathfrak{X}$ is a local formation of finite groups and A is a normal subgroup of a finite group G such that $Q = G/C_G(A)$ is $\mathfrak{X}C$-nilpotent. Then $A = (A \cap \mathfrak{X}C_\infty(G)) \times (A \cap \mathfrak{X}E_\infty(G)$.*

In passing, we mention that, under these assumptions, the stronger restriction $G \in \mathfrak{X}$ holds, since for a local formation $\mathfrak{X}$, the $\mathfrak{X}C$-nilpotency of G implies that $G \in \mathfrak{X}$ (see K. Doerk and T. O. Hawkes [56, Theorem IV.3.2]). We consider the question of the existence of Baer's decomposition for certain types of artinian modules. Our results concern infinite groups and artinian modules, all of which are associated in some way to a specific formation of groups. To proceed to the description, we need first to develop the concepts involved. As we did before for groups (see Chapter 3) we obtain a generalization of the RG-center and the upper RG-hypercenter with respect to certain formations of groups.

Let R be a ring, G a group, $\mathfrak{X}$ a class of groups and A an RG-module. As above, if $B \le C$ are RG-submodules of A, the factor C/B is said to be $\mathfrak{X}$-*central* (respectively, $\mathfrak{X}$-*eccentric*) if $G/C_G(C/B) \in \mathfrak{X}$ (respectively, $G/C_G(C/B) \notin \mathfrak{X}$).
To rule out these factors, we define

$$\mathfrak{X}C_{RG}(A) = \{a \in A \mid G/C_G(aRG) \in \mathfrak{X}\}.$$

Proposition 10.1. *Let A be an RG-module, where R is a ring and G is a group. If $\mathfrak{X}$ is a formation of groups, then $\mathfrak{X}C_{RG}(A)$ is an RG-submodule of A.*

Proof. If $a_1, a_2 \in \mathfrak{X}C_{RG}(A)$, then

$$G/(C_G(a_1RG) \cap C_G(a_2RG)) \le G/C_G(a_1RG) \times G/C_G(a_2RG),$$

and
$$C_G(a_1 RG) \cap C_G(a_2 RG) \leq C_G((a_1 - a_2)RG).$$
It follows that $G/C_G((a_1 - a_2)RG) \in \mathfrak{X}$; that is, $a_1 - a_2 \in \mathfrak{X}C_{RG}(A)$.

Now, suppose $a \in \mathfrak{X}C_{RG}(A)$ and $x \in RG$. Then $a_3 = ax \in aRG$ and so $a_3 RG \leq aRG$. Thus $C_G(a_3 RG) \geq C_G(aRG)$ and it follows that $G/C_G(a_3 RG) \in \mathfrak{X}$, whence $ax \in \mathfrak{X}C_{RG}(A)$. $\qquad\square$

The submodule $\mathfrak{X}C_{RG}(A)$ is called *the $\mathfrak{X}C$-RG-center of A* (in brief, the $\mathfrak{X}C$-center of A). Proceeding in a manner similar to the one we used for groups, we construct *the upper $\mathfrak{X}C$-RG-central series of the module A* as
$$\langle 0 \rangle = A_0 \leq A_1 \leq \cdots \leq A_\alpha \leq A_{\alpha+1} \leq \cdots \leq A_\gamma,$$
where $A_1 = \mathfrak{X}C_{RG}(A)$, $A_{\alpha+1}/A_\alpha = \mathfrak{X}C_{RG}(A/A_\alpha)$, for all ordinals $\alpha < \gamma$ and $\mathfrak{X}C_{RG}(A/A_\gamma) = \langle 0 \rangle$. The last term A_γ of this series is called *the upper $\mathfrak{X}C$-RG-hypercenter of A* (in short, the $\mathfrak{X}C$-hypercenter of A) and is denoted by $\mathfrak{X}C_{RG}^\infty(A)$; the terms A_α are called *the $\mathfrak{X}C$-RG-hypercenters of A*. If $A = A_\gamma$, then A is said to be *$\mathfrak{X}C$-RG-hypercentral*; if γ is finite, then A is called *$\mathfrak{X}C$-RG-nilpotent*.

We note that, if $\mathfrak{X} = \mathfrak{I}$ is the class of all identity groups, then $\mathfrak{X}C_{RG}(A) = \zeta_{RG}(A)$ is called *the RG-center of A* and $\mathfrak{X}C_{RG}^\infty(A) = \zeta_{RG}^\infty(A)$ is called *the upper RG-hypercenter of A*. If $\mathfrak{X} = \mathfrak{F}$ is the class of all finite groups, then $\mathfrak{X}C_{RG}(A) = FC_{RG}(A)$ is called *the FC-center of A* and $\mathfrak{X}C_{RG}^\infty(A) = FC_{RG}^\infty(A)$ is called *the upper FC-hypercenter of A*.

An RG-submodule C of A is said to be *$\mathfrak{X}$-RG-hypereccentric* if it has an ascending series
$$\langle 0 \rangle = C_0 \leq C_1 \leq \cdots C_\gamma \leq C_{\alpha+1} \leq \cdots \leq C_\gamma = C$$
of RG-submodules of A such that each factor $C_{\alpha+1}/C_\alpha$ is an $\mathfrak{X}$-eccentric simple RG-module, for every $\alpha < \gamma$.

We say that the RG-module A has *the Baer decomposition for the formation $\mathfrak{X}$* or A has *the Baer $\mathfrak{X}$-RG-decomposition* (in short, *the Baer $\mathfrak{X}$-decomposition*)), if the following equality holds:
$$A = \mathfrak{X}C_{RG}^\infty(A) \oplus \mathfrak{X}E_{RG}^\infty(A),$$
where $\mathfrak{X}E_{RG}^\infty(A)$ is the maximal $\mathfrak{X}$-RG-hypereccentric RG-submodule of A. Note that in this case, $\mathfrak{X}E_{RG}^\infty(A)$ contains every $\mathfrak{X}$-RG-hypereccentric RG-submodule and, in particular, it is unique. In fact, let B be an $\mathfrak{X}$-RG-hypereccentric RG-submodule of A and put $E = \mathfrak{X}E_{RG}^\infty(A)$. If $(B + E)/E$ is non-zero, it has a non-zero simple RG-submodule U/E. Since $(B+E)/E \cong B/(B \cap E)$, U/E is RG-isomorphic to some simple RG-factor of B, and it follows that $G/C_G(U/E) \notin \mathfrak{X}$. On the other hand, $(B + E)/E \leq A/E \cong \mathfrak{X}C_{RG}^\infty(A)$; that is. $G/C_G(U/E) \in \mathfrak{X}$. This contradiction shows that $B \leq E$. Hence $\mathfrak{X}E_{RG}^\infty(A)$ contains every $\mathfrak{X}$-RG-hypereccentric RG-submodule and, as we claimed, it is unique.

If $\mathfrak{X} = \mathfrak{I}$, the decomposition is simply called *the Z-decomposition,* whereas if $\mathfrak{X} = \mathfrak{F}$, we called it *the $\mathfrak{F}$-decomposition.* The first results on the existence of the Z-decomposition in infinite modules were obtained by B. Hartley and M.J. Tomkinson [111]. Later on, in the paper [301], D.I. Zaitsev proved that every artinian $\mathbb{Z}G$-module over a hypercentral group G has the Z-decomposition. The next natural step is to consider the formation $\mathfrak{F}$ of all finite groups. The existence of the $\mathfrak{F}$-decomposition in artinian modules over FC-hypercentral groups has been studied in the papers Z.Y. Duan [60, 62, 63], L.A. Kurdachenko, B.V. Petrenko and I.Ya. Subbotin [159], and D.I. Zaitsev [307, 308]. The study of the existence of the partial cases of the $\mathfrak{F}$-decomposition has played an important role in the study of the general case of the existence of the $\mathfrak{X}$-decomposition. Actually, its solution has allowed us to obtain solutions for many important formations $\mathfrak{X}$. To deal with this, it is convenient to split the complete study into two cases:

(1) $\mathfrak{F} \leq \mathfrak{X}$; and

(2) $\mathfrak{X}$ is a proper formation of finite groups.

A formation $\mathfrak{X}$ is said to be *overfinite* if it satisfies the following conditions:

(i) If $G \in \mathfrak{X}$, and H is a normal subgroup of G of finite index, then $H \in \mathfrak{X}$.

(ii) If G is a group, H is a normal subgroup of finite index of G, and $H \in \mathfrak{X}$, then $G \in \mathfrak{X}$.

(iii) $\mathfrak{I} \leq \mathfrak{X}$.

Clearly, an overfinite formation always contains $\mathcal{F}$. The most important examples of these formations are polycyclic groups, Chernikov groups, soluble minimax groups, soluble groups of finite special rank, and soluble groups of finite section rank.

Lemma 10.2 (L.A. Kurdachenko, B.V. Petrenko and I.Ya. Subbotin [159]). *Let $\mathfrak{X}$ be a formation of groups. Suppose that A is an RG-module, where R is a ring and G is a group. If B is an RG-submodule of A, then $\mathfrak{X}C_{RG}^{\infty}(B) \leq \mathfrak{X}C_{RG}^{\infty}(A)$.*

Proof. Let

$$\langle 0 \rangle = B_0 \leq B_1 \leq \cdots \leq B_\alpha \leq B_{\alpha+1} \leq \cdots \leq B_\delta,$$

and

$$\langle 0 \rangle = A_0 \leq A_1 \leq \cdots \leq A_\alpha \leq A_{\alpha+1} \leq \cdots \leq A_\gamma$$

be the upper $\mathfrak{X}C$-central series of B and A, respectively. It suffices to show that $B_\alpha \leq A_\alpha$, for every ordinal α, which will immediately imply the conclusion.

We proceed by transfinite induction. Clearly $B_1 \leq A_1$. Suppose that $\alpha > 1$. If α is a limit ordinal, by induction,

$$B_\alpha = \bigcup_{\beta < \alpha} B_\beta \leq \bigcup_{\beta < \alpha} A_\beta = A_\alpha,$$

and we finish in this case. Thus, suppose that $\alpha - 1$ exists. Then $B_{\alpha-1} \leq A_{\alpha-1}$. If $B_\alpha \leq A_{\alpha-1}$, then $B_\alpha \leq A_\alpha$ and we are done. Therefore, we may assume that A_{α_1} does not include B_α. Pick $c \in B_\alpha$, put $C = cRG$ and choose $d \in C$ and $g \in C_G((C + B_{\alpha-1})/B_{\alpha-1})$. Then there exists some $b_d \in B_{\alpha-1}$ such that $dg = d + b_d$. Therefore,

$$(d + A_{\alpha-1})g = dg + A_{\alpha-1} = d + b_d + A_{\alpha-1} = d + A_{\alpha-1}.$$

It follows that

$$C_G((C + B_{\alpha-1})/B_{\alpha-1}) \leq C_G((C + A_{\alpha-1})/A_{\alpha-1}),$$

and so

$$C_G((c + A_{\alpha-1})/A_{\alpha-1}) \in \mathfrak{X} \text{ and } c + A_{\alpha-1} \in \mathfrak{X}C_{RG}(A/A_{\alpha-1}) = A_\alpha/A_{\alpha-1}.$$

Hence, $B_\alpha \leq A_\alpha$, and we are done. $\qquad \square$

Lemma 10.3 ([159]). *Let $\mathfrak{X}$ be a formation of groups. Suppose that A is an RG-module, where R is a ring and G is a group. Let B and C be two RG-submodules of A such that $B \leq C$. If the non-zero RG-factors of both B and B/C are $\mathfrak{X}$-eccentric, then every non-zero factor of C is $\mathfrak{X}$-eccentric.*

Proof. Suppose that U and V are RG-submodules of C such that $U \geq V$, and $U \neq V$. If $V \geq B$, then there is nothing to prove, so suppose that $V + B \neq B$. If the factor $(U + B)/(V + B)$ is non-zero, then $G/C_G((U + B)/(V + B)) \notin \mathfrak{X}$. The inclusion $C_G(U/V) \leq C_G((U+B)/(V+B))$ implies that $G/C_G(U/V) \notin \mathfrak{X}$. Finally, let now $U+B = V+B$. Here $(V+B)/V$ is non-zero and $(V+B)/V \cong_{RG} B/(B\cap V)$, so that $G/C_G((V + B)/V) \notin \mathfrak{X}$. By the inclusion $C_G((V + B)/V) \geq C_G(U/V)$, we have $G/C_G(U/V) \notin \mathfrak{X}$. $\qquad \square$

Corollary 10.4 ([159]). *Let $\mathfrak{X}$ be a formation of groups. Suppose that A is an RG-module, where R is a ring and G is a group. Let*

$$\langle 0 \rangle = B_0 \leq B_1 \leq \cdots \leq B_\alpha \leq \cdots \leq B_\gamma$$

be an ascending series of RG-submodules of A satisfying the following condition:

(*) *If U and V are RG-submodules of A such that $B_\alpha \leq V \leq U \leq B_{\alpha+1}$ for some $\alpha < \gamma$, and $U \neq V$, then U/V is $\mathfrak{X}$-eccentric.*

Then every non-zero RG-factor of B_γ is $\mathfrak{X}$-eccentric.

Proof. We proceed by transfinite induction on α. If $\alpha = 1$, the assertion is trivial. Suppose that $\alpha > 1$. If $\alpha - 1$ exists, by Lemma 10.3, every non-zero RG-factor of

B_α is $\mathfrak{X}$-eccentric. Thus, suppose that α is a limit ordinal. Clearly,

$$U = \bigcup_{\beta < \alpha} (U \cap B_\beta), \; V = \bigcup_{\beta < \alpha} (V \cap B_\beta),$$

and

$$U/V = \bigcup_{\beta < \alpha} ((U \cap B_\beta) + V)/V).$$

Since $U < V$, there is an ordinal $\delta < \alpha$ such that $U \cap B_\delta \neq V \cap B_\delta$. By induction, $G/C_G((U \cap B_\delta)/(V \cap B_\delta) \notin \mathfrak{X}$. Since

$$C_G((U/V) \leq C_G((U \cap B_\delta)/(V \cap B_\delta)),$$

$G/G/C_G(U/V) \notin \mathfrak{X}$, as required. $\square$

Corollary 10.5 ([159]). *Let $\mathfrak{X}$ be a formation of groups. Suppose that A is an RG-module, where R is a ring and G is a group. Let $\{B_\lambda \mid \lambda \in \Lambda\}$ be a family of RG-submodules of A such that every non-zero RG-factor of B_λ is $\mathfrak{X}$-eccentric, for every $\lambda \in \Lambda$. If $B = \sum_{\lambda \in \Lambda} B_\lambda$, then every non-zero RG-factor of B is $\mathfrak{X}$-eccentric.*

Proof. Let γ be the least ordinal of cardinal $|\Lambda|$. Then

$$\{B_\lambda \mid \lambda \in \Lambda\} = \{B_\alpha \mid \alpha < \gamma\}.$$

If $\alpha < \gamma$, define $C_\alpha = \sum_{\beta < \alpha} B_\beta$. Then

$$C_{\alpha+1}/C_\alpha = (B_{\alpha+1} + C_\alpha)/C_\alpha \cong_{RG} B_{\alpha+1}/(B_{\alpha+1} \cap C_\alpha).$$

In particular, every non-zero RG-factor of $C_{\alpha+1}/C_\alpha$ is $\mathfrak{X}$-eccentric, and it suffices to apply Corollary 10.4. $\square$

Corollary 10.6 ([159]). *Let $\mathfrak{X}$ be a formation of groups. Suppose that A is an RG-module, where R is a ring and G is a group. Let $\{B_\lambda \mid \lambda \in \Lambda\}$ be a family of RG-submodules of A such that each member B_λ has the Baer $\mathfrak{X}$-RG-decomposition. Then $B = \sum_{\lambda \in \Lambda} B_\lambda$ has the Baer $\mathfrak{X}$-RG-decomposition.*

Proof. If $\lambda \in \Lambda$, by hypothesis, we have

$$B_\lambda = C_\lambda \oplus E_\lambda,$$

where $C_\lambda = \mathfrak{X}C_{RG}^\infty(B_\lambda)$ and $E_\lambda = \mathfrak{X}E_{RG}^\infty(B_\lambda)$. Define

$$C = \sum_{\lambda \in \Lambda} C_\lambda \text{ and } E = \sum_{\lambda \in \Lambda} E_\lambda.$$

By Lemma 10.2, $C \leq \mathfrak{X}C_{RG}^\infty(B)$. Further, every non-zero RG-factor of E is $\mathfrak{X}$-eccentric by Corollary 10.5. It follows that $C \cap E = \langle 0 \rangle$, and thus $B = C \oplus E$. Therefore $C = \mathfrak{X}C_{RG}^\infty(B)$ and $E = \mathfrak{X}E_{RG}^\infty(B)$, as required. $\square$

Corollary 10.7 ([159]). *Let $\mathfrak{X}$ be a formation of groups, R be a ring, and G a group. Then any RG-module A has a largest RG-submodule having the Baer $\mathfrak{X}$-RG-decomposition.*

Lemma 10.8 ([159]). *Let A be an RG-module, where R is a ring and G is a group. Suppose that H is a normal subgroup of G of finite index, and let B be an RH-submodule of A such that $A = BRG$. If $\mathfrak{X}$ is an overfinite formation of groups and B has the Baer $\mathfrak{X}$-RH-decomposition, then A has the Baer $\mathfrak{X}$-RG-decomposition.*

Proof. Let $\{g_1, \ldots, g_n\}$ be a transversal to H in G. Then

$$A = Bg_1 + \cdots + Bg_n.$$

If C and E are RH-submodules of B such that $C \geq E$, put $L = C_H(C/E)$. If $g \in G$, then $C_H(Cg/Eg) = g^{-1}Lg$. Therefore

$$H/C_H(Cg/Eg) = H/(g^{-1}Lg) \cong H/L.$$

Consequently, if the RH-factor C/E is $\mathfrak{X}$-central (respectively $\mathfrak{X}$-eccentric), then so is the RH-factor Cg/Eg. This means that the RH-module Bg has the Baer $\mathfrak{X}$-RH-decomposition.

Let $a \in \mathfrak{X}C_{RH}(A)$ and put $A_0 = aRH$ and $A_1 = aRG$. Then

$$A_1 = A_0g_1 + \cdots + A_0g_n.$$

If $U = C_H(A_0)$, then $H/U \in \mathfrak{X}$. It follows that $H/C_H(A_0g) \in \mathfrak{X}$, for every $g \in G$ and so $ag \in \mathfrak{X}C_{RH}(A)$. In particular, $\mathfrak{X}C_{RH}(A)$ is an RG-submodule of A.

Furthermore,

$$C_H(A_1) = g_1^{-1}Ug_1 \cap \cdots \cap g_n^{-1}Ug_n,$$

hence $H/C_H(A_1) \in \mathfrak{X}$. Since $\mathfrak{X}$ is overfinite, $G/C_G(A_1) \in \mathfrak{X}$, and thus $\mathfrak{X}C_{RH}(A) \leq \mathfrak{X}C_{RG}(A)$. As the converse inclusion is also valid, we get $\mathfrak{X}C_{RH}(A) = \mathfrak{X}C_{RG}(A)$. Applying transitive induction, we obtain that $\mathfrak{X}C_{RH}^{\infty}(A) = \mathfrak{X}C_{RG}^{\infty}(A)$.

Suppose that $B = B_1 \oplus B_2$, where $B_1 = \mathfrak{X}C_{RH}^{\infty}(B)$ and $B_2 = \mathfrak{X}E_{RH}^{\infty}(B)$. Then

$$B_3 = B_2g_1 + \cdots + B_2g_n$$

is an RG-submodule of A. By Corollary 10.5, $B_3 \leq \mathfrak{X}E_{RH}^{\infty}(A)$, whereas, by Lemma 10.2,

$$B_1g_1 + \cdots + B_1g_n \leq \mathfrak{X}C_{RG}^{\infty}(A)$$

and hence $\mathfrak{X}E_{RH}^{\infty}(A) = B_3$. In particular, $\mathfrak{X}E_{RH}^{\infty}(A)$ is an RG-submodule of A. Let U and V be RG-submodules of $\mathfrak{X}E_{RH}^{\infty}(A)$ such that $U \geq V$ but $U \neq V$. Then U/V is a non-zero RH-factor of $\mathfrak{X}E_{RH}^{\infty}(A)$ and so $H/C_H(U/V) \notin \mathfrak{X}$. Since $\mathfrak{X}$ is overfinite, $G/C_G(U/V) \notin \mathfrak{X}$. Thus, $\mathfrak{X}E_{RH}^{\infty}(A) = \mathfrak{X}E_{RG}^{\infty}(A)$, as required. $\square$

Lemma 10.9 ([159]). *Let A be a finitely generated DG-module, where D is a Dedekind domain and G is a policyclic-by-finite group. Suppose that A is monolithic with monolith M, and let $1 \neq g \in \zeta(G)$ such that $M(g-1) = \langle 0 \rangle$. Then there is some $m \in N$ such that $A(g-1)^m = \langle 0 \rangle$.*

Proof. Put $P =$ Ann $_D(M)$. First, we suppose that $P \neq \langle 0 \rangle$. Since M is a simple DG-submodule of A, $P \in \mathrm{Spec}(D)$. Let T be the D-periodic part of A. Since A is monolithic, T is the P-component of A. Since DG is a noetherian ring (see D. S. Passman [217, Theorem 10.2.7]), A is a noetherian DG-module by Lemma 1.1. According to L.A. Kurdachenko, J. Otal and I.Ya. Subbotin [157, Lemma 1.9], $L =$ Ann $_D(T) \neq \langle 0 \rangle$, and then there exists some $d \in \mathbb{N}$ such that $L = P^d$. There exists some D-submodule C such that $A = T \oplus C$ (see I. Kaplansky [123]). In particular, $AL \leq C$ and $AL \cap T = \langle 0 \rangle$. Since $M \leq T$, this means that $A = T$.

Put $T_1 = \Omega_{P,1}(T)$ and think of T_1 as an FG-module, where $F = D/P$ is a field. Put $R = F \langle x \rangle$, where $\langle x \rangle$ is an infinite cyclic group and define an action on x on T_1 by $ax = ag$, where $a \in T_1$. Then R is a principal ideal domain and T_1 is an RG-module. Since $M(g-1) = M(x-1) = \langle 0 \rangle$, the $(x-1)$-component of T_1 is non-zero, and as above, we obtain that T_1 is equal to its $(x-1)$-component. Since the RG-module T_1 is also noetherian, there must exist some $t \in \mathbb{N}$ such that $T_1(x-1)^t = T_1(g-1)^t = \langle 0 \rangle$. If $y \in P \setminus P^2$, then the mapping $\phi : a \mapsto ay$, $a \in \Omega_{P,2}(T) = T_2$ is a DG-endomorphism of T_2 and so Ker ϕ and Im ϕ are DG-submodules of T_2. Since Im $\phi = T_2 y \leq T_1$, (Im $\phi)(g-1)^t = \langle 0 \rangle$; that is, $T_2(g-1)^t \leq T_1$ and $T_2(g-1)^{2t} = \langle 0 \rangle$. A simple inductive argument shows that $A(g-1)^{dt} = \langle 0 \rangle$.

Suppose now that $P = \langle 0 \rangle$. Since M is a simple DG-submodule, M is D-divisible. Then we may consider M as a KG-module, where K is the field of fractions of D. Since A is monolithic, A is D-torsion-free. Let $B = A \otimes_D K$ so that B is also D-torsion-free. Clearly B is a monolithic KG-module with KG-monolith M. Put $S = K \langle y \rangle$, where $\langle y \rangle$ is an infinite cyclic group and define an action of y on B by $bx = bg$, $b \in B$. Obviously, B is a finitely generated KG-module, so that B is a finitely generated SG-module. Since S is also a principal ideal domain, as above, the ring SG is also noetherian (see D.S. Passman [127, Theorem 10.2.7]). Again, $M(x-1) = M(g-1) = \langle 0 \rangle$ and, in particular, Ann $_S(M) \neq \langle 0 \rangle$. In this case, we have already proved that there is some $m \in \mathbb{N}$ such that $B(g-1)^m = \langle 0 \rangle$. In particular, $A(g-1)^m = \langle 0 \rangle$. $\square$

Corollary 10.10 ([159]). *Let A be a finitely generated DG-module, where D is a Dedekind domain, G is a polycyclic-by-finite group, and $C_G(A) = \langle 1 \rangle$. If $1 \neq g \in \zeta(G)$ and $C_A(g) \neq \langle 0 \rangle$, then $A \neq A(g-1)$.*

Proof. Pick $0 \neq a \in C_A(g)$. Since $g \in \zeta(G)$, $C_A(g)$ is a DG-submodule. Let B_a be a maximal DG-submodule of A under $a \notin B_a$. Then A/B_a is a monolithic DG-module with monolith $(aDG+B_a)/B_a = M/B_a$. Then $(M/B_a)(g-1) = \langle 0 \rangle$, and it suffices to apply Lemma 10.9. $\square$

Lemma 10.11 ([159]). *Let A be a finitely generated DG-module, where D is a Dedekind domain, G is a locally polycyclic-by-finite group, and $C_G(A) = \langle 1 \rangle$. If $1 \neq g \in \zeta(G)$ and $C_A(g) \neq \langle 0 \rangle$, then $A \neq A(g-1)$.*

Proof. Let

$$A = a_1 DG + \cdots + a_n DG$$

and suppose that $A = A(g-1)$. Then there exist some $b_1, \ldots, b_n \in A$ such that $a_j = b_j(g-1)$, for every $1 \leq j \leq n$. Pick $0 \neq c \in C_A(g)$ and choose a finitely generated subgroup H of G such that $g \in H$ and $c, b_1, \ldots, b_n \in B = a_1 DH + \cdots + a_n DH$. If $b \in B$, then there are some $x_1, \ldots, x_n \in DH$ such that

$$b = a_1 x_1 + \cdots + a_n x_n = b_1(g-1)x_1 + \cdots + b_n(g-1)x_n$$
$$= (b_1 x_1 + \cdots + b_n x_n)(g-1).$$

In other words, $B = B(g-1)$. On the other hand, $c \in C_A(g) \cap B = C_B(g)$ and, in particular, $C_B(g) \neq \langle 0 \rangle$. By Corollary 10.10, $B \neq B(g-1)$, a contradiction. $\quad\square$

Theorem 10.12. *Let A be an artinian DG-module, where D is a ring and G is a locally soluble FC-hypercentral group. If $\mathfrak{X}$ is an overfinite formation of groups, then A has the Baer $\mathfrak{X}$-DG-decomposition.*

Proof. If $G \in \mathfrak{X}$, then $A = \mathfrak{X}C_{DG}^\infty(A)$. Thus we may assume that $G \notin \mathfrak{X}$.

Suppose that the result is false. Let $\mathcal{M}$ be the family of DG-submodules B of A such that B does not have the Baer $\mathfrak{X}$-DG-decomposition. Since $A \in \mathcal{M}$, $\mathcal{M} \neq \emptyset$. Since A is artinian, $\mathcal{M}$ has a minimal element C. By Corollary 10.7, C contains the largest DG-submodule M of A having the Baer $\mathfrak{X}$-DG-decomposition. By the choice of C, we deduce that M contains every proper DG-submodule of C. In particular, M is a maximal DG-submodule of C. Decompose $M = M_1 \oplus M_2$, where $M_1 = \mathfrak{X}C_{DG}^\infty(M)$ and $M_2 = \mathfrak{X}E_{DG}^\infty(M)$.

First, we suppose that $G/C_G(C/M) \notin \mathfrak{X}$. We may replace C by C/M_2 to assume that $M = \mathfrak{X}C_{DG}^\infty(M)$. We also may suppose that $C_G(C) = \langle 1 \rangle$. Put $S = Soc_{DG}(C)$ so that $S \leq M$. Since S is a direct sum of finitely many simple DG-submodules, $G/C_G(S) \in \mathfrak{X}$ and, in particular, $C_G(S) \neq \langle 1 \rangle$. By Lemma 3.15, $C_G(S) \cap FC(G) \neq \langle 1 \rangle$. Pick some $1 \neq x \in C_G(S) \cap FC(G)$. By Corollary 3.4, $\langle x \rangle^G$ is central-by-finite and the index $|G : C_G(\langle x \rangle^G)|$ is finite. Therefore either $\langle x \rangle^G$ includes a finite minimal G-invariant subgroup X or $\langle x \rangle^G$ includes a G-invariant finitely generated torsion-free abelian subgroup Y. We note that if X is finite, then X is also abelian since G is locally soluble. Define $Z = X$ in the first case and $Z = Y$ in the second, and let $H = C_G(Z)$. Then $Z \leq \zeta(H)$, and $|G : H|$ is finite. By Theorem 5.5, there exist some $n \in \mathbb{N}$ such that we may express

$$C/M = (B/M)g_1 \oplus \cdots \oplus (B/M)g_n,$$

where B/M is a simple DH-module and $g_1, \ldots, g_n \in G$.

We claim that $H/C_H(B/M) \notin \mathfrak{X}$. Otherwise, since

$$H/C_H((B/M)g) = H/(g^{-1}C_H(B/M)g) \cong H/C_H(B/M)$$

for every $g \in G$, we obtain $H/C_H(C/M) \in \mathfrak{X}$ and it follows that $G/C_G(C/M) \in \mathfrak{X}$, a contradiction. Thus our claim has just been proven. Since M does not include B, $BDG = C$. Thus B does have the Baer $\mathfrak{X}$-DH-decomposition. Otherwise, C would have the Baer $\mathfrak{X}$-DG-decomposition by Lemma 10.8.

Let $\mathcal{L}$ be the family of those DH-submodules Q of C such that Q does not have the Baer $\mathfrak{X}$-DH-decomposition. Since $B \in \mathcal{L}$, $\mathcal{L} \neq \varnothing$. By Theorem 5.2, C is an artinian DH-module and so $\mathcal{L}$ has a minimal element E. By Lemma 10.8, M cannot include E and so $B = E + M$. By Corollary 10.7, E contains the largest DH-submodule E_1 having the Baer $\mathfrak{X}$-DH-decomposition. By Lemma 10.8, $E_1 DG \leq M$ and, in particular, $E_1 \leq M$. Moreover, $E \cap M = E_1$ since B/M is a simple DH-module. Since

$$E/E_1 = E/(E \cap M) \cong_{DH} (E + M)/M = B/M,$$

we obtain that

$$H/C_H(E/E_1) \notin \mathfrak{X}.$$

Let $W =\text{Soc }_{DH}(C)$. Since C is an artinian DH-module,

$$W = L_1 \oplus \cdots \oplus L_s,$$

where $L_1,\ldots,L_s$ are simple DH-submodules. For every $1 \leq j \leq n$, since $X \leq \zeta(H)$, $L_j(\omega DX)$ is a DH-submodule of L_j, and then either $L_j(\omega DX) = \langle 0 \rangle$ or $L_j(\omega DX) = L_j$. Consequently,

$$W = C_W(X) \oplus W(\omega DX).$$

Since W is a DG-submodule of C and X is a normal subgroup of G, we have that $W(\omega DX)$ is a DG-submodule of W. We claim $W(\omega DX) = \langle 0 \rangle$. Otherwise, $W(\omega DX) \cap \text{Soc }_{DG}(A) \neq \langle 0 \rangle$. Since $X \leq C_G(\text{Soc }_{DG}(A))$,

$$W(\omega DX) \cap \text{Soc }_{DG}(A) \leq W(\omega DX) \cap C_W(X) = \langle 0 \rangle,$$

which is a contradiction. Thus we have just proven our claim; that is, $W(\omega DX) = \langle 0 \rangle$, so that $W \leq C_C(X)$. It follows that $E \cap C_C(X) \neq \langle 0 \rangle$. If $e \in E \setminus E_1$, then E_1 does not include eDH and so $eDH = E$. In particular, E is a finitely generated DH-submodule. Since G is FC-hypercentral, G is locally (polycyclic-by-finite) by Corollary 3.20. By Lemma 10.11, $E(g - 1) \neq E$, for every $g \in X$, and thus $E(g - 1) \leq E_1$. Since $M = \mathfrak{X}C_{DG}^\infty(M)$, $E_1 = \mathfrak{X}C_{DH}^\infty(E_1)$ by Lemma 10.8. The mapping

$$\theta : a + E_1 \mapsto a(g - 1) + E_1(g - 1), \ a \in E, \text{ and } g \in X,$$

is a DH-homomorphism. If $E(g-1) \neq E_1(g-1)$, then the inclusion $E(g-1) \leq E_1$ implies that

$$H/C_H(E(g - 1)/E_1(g - 1)) \in \mathfrak{X}.$$

Since $E(g - 1)/E_1(g - 1) \cong_{DH} E/E_1$, $C_H(E/E_1) = C_H(E(g - 1)/E_1(g - 1))$. Since $H/C_H(E/E_1) \notin \mathfrak{X}$, we get to a contradiction, which shows that $E(g - 1) = E_1(g-1)$, for every $g \in X$. In this case, it follows that $E = C_E(g)+E_1$, and then, by the choice of E, $E = C_E(g)$, because $C_E(g)$ is a DH-submodule. Since this is true for every $g \in X$, we deduce $E \leq C_C(X)$. In particular, M does not include $C_C(X)$.

The latter is a DG-submodule, therefore $C_C(X) = C$, i.e. $X \leq C_G(C) = \langle 1 \rangle$, a contradiction.

Now we suppose $G/C_G(C/M) \in \mathfrak{X}$. Proceeding as above, we may assume $M = \mathfrak{X}E^\infty_{DG}(M)$ and $C_G(C) = \langle 1 \rangle$. Since $C_G(C/M) \neq \langle 1 \rangle$, we may pick $1 \neq y \in C_G(C/M) \cap FC(G)$. Note that $Y = \langle y \rangle^G$ is central-by-finite and $|G : C_G(Y)|$ is finite. Put $R = C_G(Y)$. We may suppose again that Y is abelian; that is, $Y \leq \zeta(R)$. Let $\mathcal{Q}$ be the family of all DR-submodules Q of C of M does not include Q. Since $C \in \mathcal{Q}$, $\mathcal{Q} \neq \emptyset$. By Theorem 5.2, C is an artinian DR-submodule and so $\mathcal{Q}$ has a minimal element U. By Lemma 10.8, $M = \mathfrak{X}E^\infty_{DR}(M)$. For each $g \in Y$, $U(g-1)$ and $C_U(g)$ are DR-submodules, because $g \in \zeta(R)$ and $U(g-1) \cong_{DR} U/C_U(g)$. By the choice of Y, $U(g-1) \leq M$. If $C_U(g) \leq M$, then $U/C_U(g)$ has one non-zero $\mathfrak{X}$-central DR-factor. On the other hand, since $U(g-1) \leq M$, every non-zero DR-factor of $U(g-1)$ is $\mathfrak{X}$-eccentric. This contradiction shows that M does not include $C_U(g)$. By the choice of U, we obtain that $U = C_U(g)$. Since this holds for every $g \in Y$, $U \leq C_C(Y)$. In particular, M does not include $C_C(Y)$. Since Y is normal in G, $C_C(Y)$ is a DG-submodule. Then $C = C_C(Y)$, because M include every proper DG-submodule of C. Hence $Y \leq C_G(C) = \langle 1 \rangle$, and we come to a contradiction also in this case. $\qquad\square$

We note that the case $D = \mathbb{Z}$ of this result was considered by L.A. Kurdachenko, B.V. Petrenko and I.Ya. Subbotin [160]. Also, it is worth mentioning that, since every overfinite formation $\mathfrak{X}$ contains $\mathfrak{F}$, every FC-hypercentral group is likewise $\mathfrak{X}$-hypercentral.

Corollary 10.13 ([159]). *Let D be a Dedekind domain and G a locally soluble FC-hypercentral group. If A is an artinian DG-module, then A has the $\mathfrak{F}$-decomposition.*

Corollary 10.14 (Z.Y. Duan [60]). *Let G be a locally soluble group, having an ascending series of normal subgroups, every factor of which is finite or cyclic. If A is an artinian $\mathbb{Z}G$-module, then A has the $\mathfrak{F}$-decomposition.*

Corollary 10.15 (D.I. Zaitsev [308]). *Let G be a hyperfinite locally soluble group. If A is an artinian $\mathbb{Z}G$-module, then A has the $\mathfrak{F}$-decomposition.*

Corollary 10.16. *Let A be an artinian DG-module, where D is a Dedekind domain and G is a locally soluble FC-hypercentral group. Then A has the Baer $\mathfrak{X}$-DG-decomposition for the following formations $\mathfrak{X}$:*

(1) $\mathfrak{X} = \mathfrak{P}$, *the formation of all polycyclic groups.*

(2) $\mathfrak{X} = \mathfrak{C}$, *the formation of all Chernikov groups.*

(3) $\mathfrak{X} = \mathfrak{S}_2$, *the formation of all soluble minimax groups.*

(4) $\mathfrak{X} = \mathfrak{S}^\wedge$, *the formation of all soluble groups of finite special rank.*

(5) $\mathfrak{X} = \mathfrak{S}_0$, *the formation of all soluble groups of finite section rank.*

In connection with Theorem 10.12, the following question naturally arises: for what formations of finite groups does the Baer decomposition exist? Also note that we did not even complete the case of the formation $\mathfrak{I}$ of all identity groups. In the next results, we will first consider the existence of the $\mathbb{Z}$-RH-decomposition in artinian RG-modules when H is a hypercentral normal subgroup of G. As we shall see, these results will play an important role in finding the answers for both questions.

Lemma 10.17 (L.A. Kurdachenko, J. Otal and I.Ya. Subbotin [158]). *Let A be an RG-module, where R is a ring and G is a group. Suppose that Y and H are normal subgroups of G such that $H \leq Y$ and G/Y is finite. Suppose also that B/C is a chief RG-factor of A such that $\zeta_{RH}(B/C) = \langle 0 \rangle$. If U/V is a chief RY-factor such that $C \leq V \leq U \leq B$, then $\zeta_{RH}(U/V) = \langle 0 \rangle$.*

Proof. By Theorem 5.5, the RG-submodule B has an RY-submodule $E \geq C$ such that E/C is a simple RY-module and

$$B/C = (E/C)g_1 \oplus \cdots \oplus (E/C)g_n,$$

for some $g_1, \ldots, g_n$. If $C_H(E/C) = H$, then

$$C_H((E/C)g_i) = g_i^{-1}C_H(E/C)g_i = g_i^{-1}Hg_i = H$$

implies $H = C_H(B/C)$, a contradiction. Since every $(E/C)g_i$ is a simple RY-module and H is a normal subgroup of Y, $\zeta_{RH}((E/C)g_i)$ is an RY-submodule, hence, $\zeta_{RH}((E/C)g_i) = \langle 0 \rangle$ for every i, $1 \leq i \leq n$. Since the RY-module B/C is semisimple, the chief RY-factor U/V is RY-isomorphic with some $(E/C)g_i$. It follows that $\zeta_{RH}(U/V) = \zeta_{RH}((E/C)g_i) = \langle 0 \rangle$, as required. □

Lemma 10.18 ([158]). *Let A be an RG-module, where R is a ring and G is a group. Let $\mathfrak{X}$ be a formation of groups, and suppose that H is a normal subgroup of G such that A has an RH-submodule B satisfying the following conditions:*

(1) *Every non-zero RH-factor of B is $\mathfrak{X}$-eccentric.*

(2) $\mathfrak{X}C_{RG}^{\infty}(A/B) = A/B.$

Then B is an RG-submodule of A.

Proof. The proof of this lemma is rather obvious. □

Lemma 10.19 ([158]). *Let A be an artinian RG-module, where R is a ring and G is an FC-hypercentral group. Suppose that H is a normal hypercentral subgroup of G and A has an RH-submodule B satisfying the following conditions:*

(1) *If U/V is a non-zero RH-factor of B, then $C_H(U/V) \neq H$.*

(2) $C_H(A/B) = H.$

Then B is an RG-submodule of A, and there exists an RG-submodule M such that $A = B \oplus M$.

Proof. By Lemma 10.18, B is an RG-submodule. Put

$$\mathcal{M} = \{C \mid C \text{ is an } RG\text{-submodule such that } A = B + C\}.$$

Since $A \in \mathcal{M}$, $\mathcal{M} \neq \emptyset$. Since A is an artinian RG-module, $\mathcal{M}$ has a minimal element M. Clearly, we may assume that $C_G(M) = \langle 1 \rangle$.

Suppose that $M_1 = M \cap B \neq \langle 0 \rangle$. By Corollary 3.16, $\zeta(H) \cap FC(G) \neq \langle 1 \rangle$. Pick $1 \neq x \in \zeta(H) \cap FC(G)$, and define $X = \langle x \rangle^G$. Then the subgroup $Y = C_G(X)$ has finite index in G, and $H \leq Y$. In particular, $X \leq \zeta(Y)$. Now put

$$\mathcal{M}_1 = \{S \mid S \text{ is an } RY\text{-submodule such that } M = M_1 + S\}.$$

Again, $\mathcal{M}_1 \neq \emptyset$. By Theorem 5.2, A is an artinian RY-module so that $\mathcal{M}_1$ has a minimal element U. Put $U_1 = M_1 \cap U$. If $z \in X$, then $z \in \zeta(H)$ and so the mapping $\phi_z : u \mapsto u(z - 1)$, $u \in U$ is an RY-endomorphism; and, therefore, $U(z - 1)/U_1(z - 1)$ is an RY-epimorphic image of

$$U/U_1 = U/(M_1 \cap U) \cong_{RY} (U + M_1)/M_1$$
$$= M/M_1 = M/(M \cap B) \cong_{RY} (M + B)/B = A/B.$$

It follows that
$$H = C_H(U(z - 1)/U_1(z - 1)).$$

On the other hand, $A(z-1) \leq B$ by the condition (2). Since $U(z-1)$ and $U1(z-1)$ are RY-submodules, they are RH-submodules too. Thus, if $U(z - 1) \neq U_1(z - 1)$, by condition (1), $H \neq C_H(U(z-1)/U_1(z-1))$. This contradiction shows $U(z-1) = U_1(z-1)$. In this case, $U = U_1 + C_U(z)$, hence, $M = M_1 + U = M_1 + U_1 + C_U(z) = M_1 + C_U(z)$, which implies $C_U(z) \in \mathcal{M}_1$. By the choice of U, $U = C_U(z)$. Since this holds for every $z \in X$, $U = C_U(X)$ and, in particular, $U \leq C_M(X)$. Note that $C_M(X)$ is an RG-submodule since X is normal in G. Then

$$A = B + M = B + M_1 + U = B + C_M(X),$$

and, by the choice of M, $M = C_M(X)$. Hence, $X \leq C_G(M) = \langle 1 \rangle$, a contradiction. Therefore $M \cap B = \langle 0 \rangle$, hence, $A = B \oplus M$, as required. $\square$

Lemma 10.20 ([158]). *Let A an artinian DG-module, where D is a Dedekind domain and G is an FC-hypercentral group. Suppose that H is a normal hypercentral subgroup of G and A has a DG-submodule B satisfying the following conditions:*

(1) $B \leq \zeta_{DH}^\infty(A)$.

(2) A/B *is a simple DG-module.*

(3) $C_H(A/B) \neq H$.

Then there exists a DG-submodule M such that $A = B \oplus M$.

Proof. Given $a \in A \setminus B$, put $A_1 = aDG$. It suffices to show $A_1 = (A_1 \cap B) \oplus M$ for some DG-submodule M. For then

$$A = A_1 + B = ((A_1 \cap B) \oplus M) + B = B \oplus M,$$

as required. In other words, we may assume that A can be generated by any element $a \in A \setminus B$. Put

$$\mathcal{M} = \{C \mid C \text{ is a } DG\text{-submodule such that } A = B + C\}.$$

Since $A \in \mathcal{M}$, $\mathcal{M} \neq \emptyset$. Since A is an artinian DG-module, $\mathcal{M}$ has a minimal element M. Clearly, we may assume that $C_G(M) = \langle 1 \rangle$. Suppose that $M_1 = M \cap B \neq \langle 0 \rangle$. By Corollary 3.16, $\zeta(H) \cap FC(G) \neq \langle 1 \rangle$. Let $1 \neq x \in \zeta(H) \cap FC(G)$ and define $X = \langle x \rangle^G$. Then the subgroup $Y = C_G(X)$ has finite index in G and $H \leq Y$. In particular, $X \leq \zeta(Y)$. Since

$$M/M_1 = M/(M \cap B) \cong (M + B)/B = A/B$$

is a simple DG-module, by Theorem 5.5, M has a DY-submodule $U \geq M_1$ such that U/M_1 is a simple DY-module and

$$M/M_1 = (U/M_1)g_1 \oplus \cdots \oplus U/M_1)g_n,$$

for some $g_1, \ldots, g_n$. By Lemma 10.17, $C_H(U/M_1) \neq H$. Put

$$\mathcal{L} = \{S \mid S \text{ is a } DY\text{-submodule such that } U = M_1 + S\}.$$

Obviously, $\mathcal{L} \neq \emptyset$. By Theorem 5.2, A is an artinian DY-module, and then $\mathcal{L}$ has a minimal element Q. Put $Q_1 = M_1 \cap Q = M \cap Q$. Let R/S be a chief DY-factor of M_1. Since H is a normal subgroup of G, $\zeta_{DH}(R/S)$ is a DY-submodule of R/S. Since $M_1 \leq Z_{DH}^\infty(A)$, $\zeta_{DH}(R/S) \neq \langle 0 \rangle$. It follows that $C_H(R/S) = H$. Since

$$Q/Q_1 = Q/(M_1 \cap Q) \cong_{DY} (Q + M_1)/M_1 = U/M_1,$$

we deduce that Q/Q_1 is a simple DY-module. Then either $\zeta_{DH}(Q/Q_1) = Q/Q_1$ or $\zeta_{DH}(Q/Q_1) = \langle 0 \rangle$. If $H \leq C_Y(Q/Q_1)$, then $H \leq C_Y(U/M_1)$ too. But we have already proved that this is impossible. Thus $C_Y(Q/Q_1)$ does not include H. If $Q_1 = \langle 0 \rangle$, then we consider the DG-submodule QDG. In this case Q is a simple DY-submodule, so QDG is a semisimple DY-submodule; that is,

$$QDG = Qw_1 \oplus \cdots \oplus Qw_m,$$

for some $w_1, \ldots, w_m \in G$. If $QDG \cap B \neq \langle 0 \rangle$, then it has a simple DY-submodule V, and there exists an index $1 \leq j \leq m$ such that $V \cong_{DY} Qw_j$. In particular, $C_Y(V) = C_Y(Qw_j)$. However, $C_Y(Qw_j) = w_j^{-1}C_Y(Q)w_j$. Since $Q \cong_{DY} U/M_1$ and $C_H(U/M_1) \neq H$, we obtain that $C_H(Q) \neq H$ and

$$H = w_j^{-1}Hw_j \neq w_j^{-1}C_H(Q)w_j. = C_H(Qw_j) = C_H(V).$$

Since V is a DY-submodule and H is normal in Y, $C_V(H)$ is a DY-submodule. It follows that $C_V(H) = \langle 0 \rangle$. On the other hand, $B \leq \zeta^\infty_{DH}(A)$ and, in particular, $C_V(H) \neq \langle 0 \rangle$. This contradiction shows that $QDG \cap B = \langle 0 \rangle$. Since A/B is a simple DG-module, $QDG + B = A$. By the choice of M, $QDG = M$. Since we supposed $M \cap B \neq \langle 0 \rangle$, we may also assume that $Q_1 \neq \langle 0 \rangle$. Note that Q can be generated by every element $a \in Q \setminus M_1$. If $z \in X$, then $z \in \zeta(Y)$. By Lemma 10.11, $Q(z-1) \neq Q$. Since Q is a minimal element of $\mathcal{L}$, $Q(z-1) \notin \mathcal{L}$. This means that $Q(z-1) + M_1 \neq U$. It follows that $Q(z-1) \leq M_1$ because the DY-module U/M_1 is simple. Since $z \in \zeta(Y)$, the mapping $\phi_z : u \mapsto u(z-1)$, $u \in U$ is a DY-endomorphism. Thus the factor $Q(z-1)/Q_1(z-1)$ is a DY-epimorphic image of Q/Q_1. Since Q/Q_1 is a simple DY-module, then either

$$Q(z-1)/Q_1(z-1) \cong_{DY} Q/Q_1 \quad \text{or} \quad Q(z-1) = Q_1(z-1).$$

We have already proved that $C_Y(Q/Q_1)$ does not include H, so, in the first case, $C_Y(Q(z-1)/Q_1(z-1))$ does not either contains H. On the other hand, $Q(z-1) \leq M_1$ and we have already shown that the centralizer of every chief DY-factor of M_1 contains H. Then the first case is impossible, and so $Q(z-1) = Q_1(z-1)$. In this case $Q = Q_1 + C_Q(z)$. Since $z \in \zeta(Y)$, $C_Q(z)$ is a DY-submodule of Q, then

$$U = M_1 + Q = M_1 + Q_1 + C_Q(z) = M_1 + C_Q(z),$$

which yields $C_Q(z) \in \mathcal{L}$. By the choice of Q, $Q = C_Q(z)$. Since this holds for every $z \in X$, $Q = C_Q(X)$. In particular, $Q \leq C_M(X)$. Since $U \neq M_1$, we deduce that M_1 cannot include $C_M(X)$. Since M/M_1 is a simple DG-module and X is normal in G, $M_1 + C_M(X) = M$. Therefore

$$A = B + M = B + M_1 + C_M(X) = B + C_M(X)$$

and, by the choice of M, $M = C_M(X)$. Hence, $X \leq C_G(M) = \langle 1 \rangle$, a contradiction, which shows that $M \cap B = \langle 0 \rangle$ against our assumption. Hence, $A = B \oplus M$. $\square$

Theorem 10.21 ([158]). *Let A be an artinian DG-module, where D is a Dedekind domain and G is an FC-hypercentral group. If H is a normal hypercentral subgroup of G, then A has the Z-DH-decomposition.*

Proof. Suppose that the result is false. Let $\mathcal{M}$ be the family of all DG-submodules B of A that does not have the Z-DH-decomposition. Clearly $\mathcal{M} \neq \emptyset$. Since A is an artinian DG-module, $\mathcal{M}$ has a minimal element C. By Corollary 10.7, C contains the largest DG-submodule M having the Z-DH-decomposition. By the choice of C, M must include every proper DG-submodule. In particular, M is a maximal DG-submodule of C.

Let $M = M_1 \oplus M_2$, where $M_1 = \zeta^\infty_{DH}(M)$ and $M_2 = \zeta^*_{DH}(M)$. Since H is a normal subgroup of G, $\zeta_{DH}(C/M)$ is a DG-submodule of C/M. Then either

$\zeta_{DH}(C/M) = C/M$ or $\zeta_{DH}(C/M) = \langle 0 \rangle$. First, we suppose that $\zeta_{DH}(C/M) = C/M$ and consider the factor-module C/M_1. By Lemma 10.20, there exists a DG-submodule M_3/M_1 such that

$$C/M_1 = M/M_1 \oplus M_3/M_1.$$

In this case, $M_3 = \zeta_{DH}^{\infty}(C)$ and so $C = M_3 \oplus M_2$. Now suppose that $\zeta_{DH}(C/M) = \langle 0 \rangle$. By Lemma 10.20, there exists a DG-submodule M_4/M_2 such that

$$C/M_2 = M/M_2 \oplus M_4/M_2,$$

and we come to $M_4 = \zeta_{DH}^{*}(C)$. Thus $C = M_1 \oplus M_4$. $\qquad\square$

Corollary 10.22. *Let A be a DG-module, where D is a Dedekind domain and G is a hypercentral group. If A is artinian, then A has the Z-decomposition.*

Note that, for $D = \mathbb{Z}$ and $G = H$ the above result was proven by D. I. Zaitsev in [301]. Note that the result gives a positive answer for the question that we raised above involving the formation $\mathfrak{J}$.

Consider now another question, namely, the existence of the Baer decomposition for some natural formation of finite groups. Before dealing with it, we first notice that infinite groups behave badly with respect to some properties automatically satisfied by a finite group. For example, suppose that $\mathfrak{X}$ is a formation of finite groups and G is a group such that $G \in \mathbf{R}\mathfrak{X}$. If G is finite, then $G \in \mathfrak{X}$. However, if G is infinite, the situation can be totally different. To avoid this complication, we consider the following formation of finite groups.

A formation $\mathfrak{X}$ of finite groups is said to be *infinitely hereditary concerning a class of groups* $\mathfrak{H}$ if it satisfies the following condition:

(IH) Whenever an $\mathfrak{H}$-group G belongs to the class $\mathbf{R}\mathfrak{X}$, then every finite factor-group of G belongs to $\mathfrak{X}$.

It is worth mentioning that many formations of finite groups are infinitely hereditary concerning the class of FC-hypercentral groups; for example:

(1) $\mathfrak{A} \cap \mathfrak{F}$, the finite abelian groups;

(2) $\mathfrak{N}_c \cap \mathfrak{F}$, the finite nilpotent groups of class at most c;

(3) $\mathfrak{S}_d \cap \mathfrak{F}$, the finite soluble groups of derived length at most d;

(4) $\mathfrak{S} \cap \mathfrak{F}$, the finite soluble groups;

(5) $\mathfrak{B}(n) \cap \mathfrak{F}$, the finite groups of exponent dividing n;

among others. Moreover, all these five examples and

(6) $\mathfrak{N} \cap \mathfrak{F}$, the finite nilpotent groups;

(7) $\mathfrak{U} \cap \mathfrak{F}$, the finite supersoluble groups

are infinitely hereditary concerning both the classes of the FC-groups and hyper-finite groups.

Theorem 10.23 ([158]). *Let $\mathfrak{X}$ be a formation of finite groups, and let A be an artinian DG-module, where D is a Dedekind domain and G is an infinite locally soluble $\mathfrak{X}C$-hypercentral group. If $\mathfrak{X}$ is infinitely hereditary concerning the class of FC-hypercentral groups, then A has the Baer $\mathfrak{X}$-DG-decomposition.*

Proof. Since G is infinite, $G \notin \mathfrak{X}$. Suppose that A does not have the Baer $\mathfrak{X}$-DG-decomposition. Let $\mathcal{M}$ be the family of all DG-submodules B of A that does not have the $\mathfrak{X}$-DG-decomposition. Then $\mathcal{M} \neq \emptyset$. Since A is an artinian DG-module, $\mathcal{M}$ has a minimal element C. By Corollary 10.7, C includes the largest DG-submodule M having the $\mathfrak{X}$-DG-decomposition. By the choice of C, M contains every proper DG-submodule of C, and so M is a maximal DG-submodule of C.

Let
$$M = M_1 \oplus M_2,$$
where $M_1 = \mathfrak{X}C_{DG}^{\infty}(M)$ and $M_2 = \mathfrak{X}E_{DG}^{\infty}(M)$. Suppose that $G/C_G(C/M) \notin \mathfrak{X}$. Replacing C by C/M_2, we assume that $M = \mathfrak{X}C_{DG}^{\infty}(M)$. By Corollary 10.13,
$$C = FC_{DG}^{\infty}(C) \oplus \mathfrak{F}E_{DG}^{\infty}(C).$$
Since C is indecomposable and $M \leq F_{DG}^{\infty}(C)$, $C = F_{DG}^{\infty}(C)$. In particular, the factor-module C/M is finite.

Let $\mathcal{L}$ be the set of normal subgroups H of G such that $G/H \in \mathfrak{X}$, and let S be the intersection of the members of $\mathcal{L}$. In other words, S is the $\mathfrak{X}$-residual of G. If $S = \langle 1 \rangle$, then $G \in \mathbf{R}\mathfrak{X}$. Since $\mathfrak{X}$ is infinitely hereditary concerning the class of FC-hypercentral groups, the finite factor-group $G/C_G(C/M) \subset \mathfrak{X}$. This contradiction shows that this case cannot appear, and so $S \neq \langle 1 \rangle$. The group G has an ascending series of normal subgroups
$$\langle 1 \rangle = G_0 \leq G_1 \leq \cdots \leq G_\alpha \leq G_{\alpha+1} \leq \cdots \leq G_\gamma = G$$
such that $G/C_G(G_{\alpha+1}/G_\alpha) \in \mathfrak{X}$ for every $\alpha < \gamma$. It follows that
$$\bigcap_{\alpha < \gamma} C_G(G_{\alpha+1}/G_\alpha) \geq S.$$

By a result due to Sh. S. Kemhadze [135, Theorem 1], the Baer radical of G contains $\bigcap_{\alpha < \gamma} C_G(G_{\alpha+1}/G_\alpha)$. Note that the Baer radical of the group G is locally nilpotent (see D. J. S. Robinson [234, Theorem 2.31]), and a locally nilpotent FC-hypercentral group is hypercentral, so that, in particular, S is hypercentral. Since $G/S \in \mathbf{R}\mathfrak{X}$, $C_G(C/M) \not\geq S$. Further, M has an ascending series of DG-submodules
$$\langle 0 \rangle = U_0 \leq U_1 \leq \cdots U_\alpha \leq U_{\alpha+1} \leq \cdots \leq U_\gamma = M$$
such that $G/C_G(U_{\alpha+1}/U_\alpha) \in \mathfrak{X}$, for every $\alpha < \gamma$. It follows that
$$\bigcap_{\alpha < \gamma} C_G(U_{\alpha+1}/U_\alpha) \geq S,$$

and then M is DS-hypercentral. We have already remarked that $C_S(C/M) \neq S$. Therefore $M = \zeta_{DH}^\infty(C)$. By Theorem 10.21, $C = M \oplus E$, where $E = \mathfrak{J}E_{DG}^\infty(M)$. Since $E \cong_{DG} C/M$, $E = \mathfrak{X}E_{DG}^\infty(A)$, a contradiction.

Suppose now that $G/C_G(C/M) \in \mathfrak{X}$. In this case, we replace C by C/M_1 to assume $M = \mathfrak{X}E_{DG}^\infty(M)$. By Corollary 10.13,

$$C = FC_{DG}^\infty(C) \oplus \mathfrak{F}E_{DG}^\infty(C).$$

Since C is indecomposable and $G/C_G(C/M)$ is finite, we obtain $C = FC_{DG}^\infty(C)$. The DG-submodule M has an ascending series of DG-submodules

$$\langle 0 \rangle = U_0 \leq U_1 \leq \cdots \leq U_\alpha \leq U_{\alpha+1} \leq \cdots \leq U_\gamma = M$$

such that $U_{\alpha+1}/U_\alpha$ is a DG-chief factor and $G/C_G(U_{\alpha+1}/U_\alpha) \notin \mathfrak{X}$ for every $\alpha < \gamma$. It follows that $C_G(U_{\alpha+1}/U_\alpha)$ does not include S, and then $\zeta_{DS}(U_{\alpha+1}/U_\alpha) = \langle 0 \rangle$ for every $\alpha < \gamma$, which means that $M = \zeta_{DH}^*(C)$. By Theorem 10.21, $C = M \oplus E$, where $E = \zeta_{DH}^\infty(C)$. Since $E \cong_{DG} C/M$, $E = \mathfrak{X}C_{DG}^\infty(A)$, another contradiction. $\square$

Corollary 10.24 ([158]). *Let $\mathfrak{X}$ be a formation of finite groups and let A be an artinian DG-module, where D is a Dedekind domain and G is an infinite locally soluble $\mathfrak{X}C$-hypercentral group. Then A has the Baer $\mathfrak{X}$-DG-decomposition, provided $\mathfrak{X}$ is one of the following formations: $\mathfrak{A} \cap \mathfrak{F}$, $\mathfrak{N}_c \cap \mathfrak{F}$, $\mathfrak{S}_d \cap \mathfrak{F}$, $\mathfrak{S} \cap \mathfrak{F}$, and $\mathfrak{B}(n) \cap \mathfrak{F}$.*

Proof. It suffices to show that the formation of all finite soluble groups is infinitely hereditary concerning the class of FC-hypercentral groups. To do this, it will suffice to show that if an FC-hypercentral group G belongs to the class $\mathbf{R}(\mathfrak{S} \cap \mathfrak{F})$, then it is locally soluble. So, G will be hyperabelian, and then its finite factor-groups will be soluble. Let K a finitely generated subgroup of G. By Corollary 3.20, K is nilpotent-by-finite. In particular, K satisfies Max, and so K has a maximal normal soluble subgroup S. Assume that $K \neq S$. Then K has a subnormal subgroup L such that $L \geq S$ and L/S is a finite non-abelian simple group. Since $L \in \mathbf{R}(\mathfrak{S} \cap \mathfrak{F})$, L has a family $\{L_\lambda \mid \lambda \in \Lambda\}$ of normal subgroups such that any L/L_λ is finite soluble and

$$\bigcap_{\lambda \in \Lambda} L_\lambda = \langle 1 \rangle.$$

If we assume that some $L_\lambda S \neq L$, then $L/L_\lambda S$ must be a finite simple non-abelian group. This contradiction shows that $L_\lambda S = L$, for each $\lambda \in \Lambda$. Then

$$L/L_\lambda = L_\lambda S/L_\lambda \cong S/(S \cap L_\lambda).$$

It follows that each L/L_λ is a finite soluble group of derived length at most d, where d is the derived length of the soluble radical S. By Remak's theorem

$$L \hookrightarrow \prod_{\lambda \in \Lambda} L/L_\lambda,$$

and so L is a soluble subgroup of derived length at most d. In particular, $L = S$, a contradiction that shows $K = S$. Hence, K is soluble. □

Corollary 10.25 ([158]). *Let $\mathfrak{X}$ be a formation of finite groups, and let A be an artinian DG-module, where D is a Dedekind domain, and G is an infinite locally soluble $\mathfrak{X}C$-hypercentral group. If G is an FC-group, then A has the Baer $\mathfrak{X}$-DG-decomposition, provided $\mathfrak{X}$ is one of the following formations: $\mathfrak{A} \cap \mathfrak{F}$, $\mathfrak{N}_c \cap \mathfrak{F}$, $\mathfrak{N} \cap \mathfrak{F}$, $\mathfrak{S}_d \cap \mathfrak{F}$, $\mathfrak{S} \cap \mathfrak{F}$, $\mathfrak{B}(n) \cap \mathfrak{F}$, $\mathfrak{U} \cap \mathfrak{F}$.*

Corollary 10.26 ([158]). *Let $\mathfrak{X}$ be a formation of finite groups, and let A be an artinian DG-module, where D is a Dedekind domain, and G is an infinite locally soluble $\mathfrak{X}C$-hypercentral group. If G is a hyperfinite group, then A has the Baer $\mathfrak{X}$-DG-decomposition, provided $\mathfrak{X}$ is one of the following formations: $\mathfrak{A} \cap \mathfrak{F}$, $\mathfrak{N}_c \cap \mathfrak{F}$, $\mathfrak{N} \cap \mathfrak{F}$, $\mathfrak{S}_d \cap \mathfrak{F}$, $\mathfrak{S} \cap \mathfrak{F}$, $\mathfrak{B}(n) \cap \mathfrak{F}$, $\mathfrak{U} \cap \mathfrak{F}$.*

Corollary 10.27 ([158]). *Let $\mathfrak{X}$ be a formation of finite groups, and let A be an artinian DG-module, where D is a Dedekind domain, and G is an infinite $\mathfrak{X}C$-hypercentral group. If G is a soluble Chernikov group, then A has the Baer $\mathfrak{X}$-DG-decomposition.*

Proof. Let R be the $\mathfrak{X}$-residual of the group G. Since G/R is residually finite and G is Chernikov, G/R is finite. This means that any formation $\mathfrak{X}$ of finite groups is infinitely hereditary concerning the class of Chernikov groups, and it suffices to apply Theorem 10.23. □

Consider now another important type of modules with finiteness conditions, namely, noetherian modules. In this case, the situation is not so satisfactory as it is in the case of artinian modules. As a matter of fact, we can not ask about the existence of $\mathfrak{X}$-decompositions in noetherian modules, even in the case $\mathfrak{X} = \mathfrak{I}$, as the following elementary example shows. Let $U = \langle u \rangle \times \langle v \rangle$ be a free abelian group of rank 2, and let $\langle g \rangle$ be a cyclic group of order 3. We construct the semidirect product $G = U \leftthreetimes \langle g \rangle$, in which g acts on U by $u^g = v$ and $v^g = u^{-1}v^{-1}$. Then every non-identity G-invariant subgroup of U has finite index and, in particular, the $\mathbb{Z}\langle g \rangle$-module U is noetherian. However, U is directly indecomposable and has central G-chief factors and non-central G-chief factors.

Due to this, we may consider only a certain weakened version of the Baer $\mathfrak{X}$-RG-decomposition, as the following one. Given a DG-module A, then we define $\gamma^0_{DG}(A) = A\mathfrak{g}$, $\gamma^{\alpha+1}_{DG}(A) = (\gamma^\alpha_{DG}(A))\mathfrak{g}$ for every ordinal α and $\gamma^\lambda_{DG}(A) = \bigcap_{\mu < \lambda} \gamma^\mu_{DG}(A)$ for a limit ordinal λ, where $\mathfrak{g} = \omega(DG)$ is the augmentation ideal of DG. Thus we have just constructed *the lower DG-central series of the module* A,

$$A = \gamma^0_{DG}(A) \geq \gamma^1_{DG}(A) \geq \cdots \geq \gamma^\alpha_{DG}(A) \geq \gamma^{\alpha+1}_{DG}(A) \geq \cdots \geq \gamma^\gamma_{DG}(A).$$

D.J.S. Robinson employed the following form of the Z-decomposition for notherian modules [236]

Theorem 10.28. *Let G be a nilpotent group, R a ring and let A be a noetherian RG-module. Then there is some $k \geq 0$ such that*

$$\gamma_{RG}^k(A) \cap \zeta_{RG}(A) = \langle 0 \rangle.$$

In particular, $\gamma_{RG}^k(A)$ has no non-zero G-central factors. Moreover, $\gamma_{RG}^\omega(A) = \gamma_{RG}^{\omega+1}(A)$.

The interested reader can find the proof of this result in the book of D.J.S. Robinson [242, 15.3]. In the paper [236], Robinson has been able to construct an example showing that the discussed result cannot be extended to the case of hypercentral groups having an upper central series of length $\geq \omega + 1$. For hypercentral groups, D.I. Zaitsev [302] has been able to obtain the following even weaker form of the Z-decomposition for noetherian modules.

Theorem 10.29. *Let G be a hypercentral group and let A be a noetherian $\mathbb{Z}G$-module. Then A has a non-zero $\mathbb{Z}G$-central factor if and only if $A \neq \gamma_{DG}(A)$.*

In the paper [308] D. I. Zaitsev introduced the following weak form of the $\mathfrak{F}$-decomposition for noetherian modules:

Theorem 10.30. *Let G be a hyperfinite group and A a noetherian $\mathbb{Z}G$-module. Then A has a non-zero finite $\mathbb{Z}G$-factor if and only if A has a proper $\mathbb{Z}G$-submodule having finite index.*

Finally, Z.Y. Duan [61] proved that *every noetherian $\mathbb{Z}G$-module over a hyperfinite locally soluble group G has the $\mathfrak{F}$-decomposition.*

Chapter 11

On the countability of artinian modules over FC-hypercentral groups

One of the first important problems of the theory of artinian modules is the evaluation of their socular height. We initially discussed this question in Chapter 4, studying the conditions under which this socle height is at most ω, where ω is the first infinite ordinal. We showed that this is valid for artinian R-modules over a commutative ring R (Theorem 4.12). For artinian modules over a more general group ring, this question can be reformulated in the following way.

Let G be a group, D a Dedekind domain. *What can be said about a group G if the socular height of an arbitrary artinian DG-module (D is a Dedekind domain) is at most ω? Or more generally, what can be said about a countable group G if an arbitrary artinian DG-module (D is a Dedekind domain) has a countable set of generators as a D-module?*

In his paper [105], B. Hartley has constructed several impressive examples of uncountable artinian monolithic $\mathbb{Z}G$-modules for different types of groups G including some nilpotent groups (see Theorem 4.14). All these examples are uniserial modules, so, in particular, they are monolithic. We note that in all these examples the factor-group $G/C_G(\mu_{\mathbb{Z}G}(A))$ is not abelian-by-finite. Therefore the question about the study of nilpotent groups G such that $G/C_G(\mu_{\mathbb{Z}G}(A))$ is abelian-by-finite arises naturally. In Chapter 9 we realized that abelian-by-finite groups play a special role in the theory of artinian modules over group rings, and this will be supported by the results stated below. We consider the problem of countability for a fairly wide generalization of nilpotent groups, namely, for FC-hypercentral groups. These results have been obtained by L.A. Kurdachenko, N.N. Semko and I.Ya. Subbotin [164].

We begin with the following theorem that has many applications.

Theorem 11.1. *Let G be an FC-hypercentral group, D a Dedekind domain, A an artinian monolithic DG-module, $H = C_G(\mu_{DG}(A))$. Then A is a DH-hypercentral module.*

Proof. Since A is an artinian DG-module, A has an ascending series of DG-submodules

$$\langle 0 \rangle = A_0 \leq A_1 \leq \cdots \leq A_\alpha \leq A_{\alpha+1} \leq \cdots \leq A_\gamma = A,$$

all factors of which are simple DG-modules. In particular, $M = A_1$ is the monolith of A.

Suppose that the result is false. Then there must exist an ordinal β such that $C_H(A_{\beta+1}/A_\beta) \neq H$ but the upper DH-hypercenter of $A_{\beta+1}$ contains A (and hence, the upper DH-hypercenter of $A_{\beta+1}$ coincides with A_β). Put $C = A_{\beta+1}$, and $B = A_\beta$. There is no loss if we assume that $C_G(C) = \langle 1 \rangle$. Let G_1 be a normal subgroup having finite index in G. By the Hartley–Zaitsev theorem (Theorem 5.5),

$$C/B = \bigoplus_{x \in S}(L/B)x,$$

where L/B is a simple DG_1-submodule, and S is a finite subset. Suppose that $C_H((L/B)g) = H$ for some element $g \in S$. If $x \in S$, then we have

$$C_H((L/B)x) = C_H(((L/B)g)g^{-1}x) = (g^{-1}x)^{-1}C_H((L/B)g)g^{-1}x$$
$$= (g^{-1}x)^{-1}Hg^{-1}x = H,$$

which implies the equation $H = C_H(C/B)$. This is impossible. Consequently, $C_H((L/B)g) \neq H$ for every $g \in S$.

We claim that $C_H(B) = \langle 1 \rangle$. Otherwise, by Lemma 3.15, $FC(G) \cap C_H(B) \neq \langle 1 \rangle$. Let $1 \neq h \in FC(G) \cap C_H(B)$, and put $H_1 = \langle h \rangle^G$, and $G_1 = C_G(H_1)$. Then G_1 is a normal subgroup of G having finite index. As we have seen above

$$C/B = \bigoplus_{x \in S}(L/B)x,$$

where L/B is a simple DG_1-submodule, and S is a finite subset of G. By the choice of h, the mapping $\eta : c \mapsto c(h-1)$, $c \in C$ is a DG_1-endomorphism, so that $\operatorname{Ker} \eta = C_C(h)$, and $\operatorname{Im} \eta = C(h-1)$ are DG_1-submodules. The equation $C_G(C) = \langle 1 \rangle$ shows that $\operatorname{Ker} \eta \neq C$. Since $B \leq \operatorname{Ker} \eta$, $\operatorname{Im} \eta$ is a semisimple DG_1-submodule of finite composition length. In particular, $\operatorname{Im} \eta$ contains some simple DG_1-submodule R. It happens that R must be isomorphic to some DG_1-submodules $(L/B)x$, $x \in S$, and by the argument given above, $C_H(R) \neq H$. The finiteness of the index $|G : G_1|$ implies that

$$RDG = Rx_1 \oplus \cdots \oplus Rx_m$$

for some finite subset $\{x_1, \ldots, x_m\} \subseteq G$. In particular, RDG is a semisimple DG_1-submodule of finite composition length. If we suppose that $RDG \cap B \neq \langle 0 \rangle$, then this intersection contains a simple DG_1-submodule Q. As above, Q Is a

DG_1-submodule isomorphic to some Rx_j, and so $C_H(Q) = C_H(Rx_j) \neq H$. On the other hand, B is DH-hypercentral; therefore, the inclusion $Q \leq B$ together with the DG_1-simplicity of Q imply the equation $C_H(Q) = H$, a contradiction. This contradiction yields that $RDG \cap B = \langle 0 \rangle$. In particuar, $RDG \cap \mu_{DG}(A) = \langle 0 \rangle$, which is impossible. This final contradiction establishes our claim; that is, $C_H(B) = \langle 1 \rangle$.

Choose in H an arbitrary finitely generated subgroup K. Since K acts trivially on all factors of the ascending series of B, K is residually nilpotent (see P. Hall and B. Hartley [99, Theorem A2]). It follows that the set of all elements of H having finite order is a (characteristic) locally nilpotent subgroup of H (see, for example, [235, Theorem 6.14]). In particular, $FC(H)$ is a locally soluble subgroup. Let $\mathfrak{M}$ be the set of all DG-submodule M such that B does not contain M. Since A is artinian, $\mathfrak{M}$ has a minimal element E. We note that $E \cap B \neq \langle 0 \rangle$. Otherwise, $E \cap B = \langle 0 \rangle$, and then the inclusion $\mu_{DG}(A) \leq B$ leads to a contradiction, because A is DG-monolithic. Hence $E \cap B \neq \langle 0 \rangle$. Since C/B is a simple DG-module, $B + C = E$, so that

$$C/B = (B + E)/B \cong_{DG} E/(E \cap B)$$

is a simple DG-module, and $C_H(E/(E \cap B)) = C_H(C/B) \neq H$. There is no loss if we assume that $C_G(E) = \langle 1 \rangle$. Let $1 \neq x \in FC(G) \cap H$, and put $X = \langle x \rangle^G$. We noted above that $FC(H)$ is locally soluble. Then X includes a non-identity G-invariant abelian subgroup Y. By the choice of X, and Y, $Z = C_G(Y)$ has finite index in G, and $Y \leq \zeta(Z)$. By Theorem 5.5, we have

$$E/(E \cap B) = \bigoplus_{x \in S}(U/(E \cap B))x,$$

where $U/(E \cap B)$ is a simple DZ-submodule, and S is some finite subset of G. Let now $\mathfrak{U}$ be the set of all DZ-submodules M of U such that $E \cap B$ does not contain M. By Theorem 5.2, U is an artinian DZ-module, and then $\mathfrak{U}$ has a minimal element V. We have now $U = V + (B \cap E)$, because $U/(B \cap E)$ is a simple DZ-module, so that

$$U/(B \cap E) = (V + (B \cap E))/(B \cap E) \cong_{DZ} V/(VB)$$

is also a simple DZ-module, and $C_H(U/(B \cap E)) = C_H(V/(V \cap B))$. As we demonstrated above $C_H(V/(V \cap B)) \neq H$. If we assume that $V \cap B = \langle 0 \rangle$; then, using the arguments given above, we come to a contradiction. This contradiction shows that $W = V \cap B \neq \langle 0 \rangle$. Since W is a DH-hypercentral submodule, $C_W(y) \neq \langle 0 \rangle$ for each element $y \in Y$, and thus $C_V(y) \neq \langle 0 \rangle$. The submodule W cannot contain any vDZ for every element $v \in V \setminus W$. The choice of V yields that $vDZ = V$, and, in particular, V is a cyclic DZ-submodule. By Corollary 3.20, every finitely generated subgroup of an FC-hypercentral group is nilpotent-by-finite. Thus Lemma 10.11 implies that $V(y - 1) \neq V$. By the choice of V, W contains every proper

DZ-submodule of V, so that, $V(y-1) \leq W$. Since $Y \leq \zeta(Z)$, the mapping $\nu : v \mapsto v(y-1)$, $v \in V$ is a DZ-homomorphism. It follows that $V(y-1)/W(y-1)$ is a DZ-homomorphic image of V/W. If $V(y-1)/W(y-1)$ is non-zero, then by the simplicity of V/W, we have $V/W \cong_{DZ} V(y-1)/W(y-1)$. In particular,

$$C_H(V(y-1)/W(y-1)) = C_H(V/W) \neq H.$$

On the other hand, W is a DZ-hypercentral submodule. Therefore every one of its simple DZ-factors is H-central. Then $C_H(V(y-1)/W(y-1)) = H$. This contradiction shows that $V(y-1) = W(y-1)$. It follows that for each element $v \in V$, there exists an element $w \in W$ such that $v(y-1) = w(y-1)$; that is, $(v-w)(y-1) = 0$. Thus, we see that W cannot contain $C_V(y)$. Clearly, $C_V(y)$ is a DZ-submodule, and, by the choice of B, we have $C_V(y) = V$. This holds for each element $y \in Y$, therefore $C_V(Y) \geq V$. Moreover, $C_V(Y)$ is a DG-submodule, because Y is a normal subgroup of G, and B cannot contain $C_V(Y)$. By the choice of E, we obtain that $C_V(Y) = E$. In other words, $Y \leq C_G(E) = \langle 1 \rangle$. This final contradiction proves the theorem.   □

Lemma 11.2. *Let G be an abelian-by-finite group, D a Dedekind domain, and A an artinian DG-module. If B is a finitely generated DG-submodule of A, then it has finite DG-composition series. In particular, the socular height of A is at most ω.*

Proof. There is no loss if we assume that B is a cyclic DG-submodule; that is, $B = bDG$, for some element $b \in A$. Let R be a normal abelian subgroup of G having finite index in G. The DR-module A is artinian by Theorem 5.2. Put $C = bDR$. Thus $C \cong_{DR} DR/\text{Ann }_{DR}(b)$. Since R is abelian, the group ring DR is commutative so that $J = \text{Ann }_{DR}(b)$ is a two-sided ideal of DR, and we may consider the factor-ring DR/J, which is artinian. But an artinian commutative ring is also noetherian (see R. Sharp [263, Theorem 8.44]). So the DR-submodule C is both artinian and noetherian; that is, C has finite DR-composition series. Since R has finite index in G,

$$B = CDG = Cx_1 \oplus \cdots \oplus Cx_m$$

for some finite subset $\{x_1, \ldots, x_m\} \subseteq G$. In particular, B has finite DR-composition length. Then B has finite composition length as DG-module. It follows that every finitely generated DG-submodule of A lies in some term of the upper socular series of A that corresponds to a finite ordinal. This means that the socular height of A is at most ω.   □

Lemma 11.3. *Let G be a countable FC-hypercentral group, F a field, and A an artinian monolithic FG-module. Suppose that $\dim_F A$ is uncountable, but every proper FG-submodule of A has countable dimension, and $C_G(A) = \langle 1 \rangle$. Then $C_G(\mu_{FG}(A))$ has no non-identity periodic G-invariant subgroups.*

Proof. Put $H = C_G(\mu_{FG}(A))$. Proceeding as we did in the proof of Theorem 11.1, we notice that the set T of all elements of H having finite order is a (characteristic) subgroup of H. Suppose the contrary. Let $T \neq \langle 1 \rangle$. Since G is an FC-hypercentral group, $T \cap FC(G) \neq \langle 1 \rangle$ by Lemma 3.15. Let $1 \neq x \in FC(G) \cap T$, and put $X = \langle x \rangle^G$. Then X is finite by Corollary 3.4. Furthermore, X has a non-identity G-invariant abelian subgroup Y, because T is locally nilpotent. Then $Z = C_G(Y)$ has finite index in G, and $Y \leq \zeta(Z)$. There is no loss if we assume that Y is a q-subgroup, for some prime q. By Theorem 11.1, A is an FH-hypercentral module, and hence A is likewise FY-hypercentral. It follows that the natural semidirect product $A \rtimes Y$ is a hypercentral group.

Suppose first that char $F = 0$. In a hypercentral group, all elements of finite order form a (characteristic) subgroup, that is, in this case, $B \rtimes Y = B \times Y$, and $YC_G(B) = \langle 1 \rangle$, a contradiction. Let now char $F = p > 0$. If we suppose that $q \neq p$, then, by elementary properties of hypercentral groups, we obtain again that $A \rtimes Y = A \times Y$, which leads to the contradiction $Y \leq C_G(A) = \langle 1 \rangle$. Suppose that $q = p$. Let $\mathfrak{M}$ be the set of all FZ-submodules U having uncountable F-dimension. Since A is an artinian FZ-module by Theorem 5.2, $\mathfrak{M}$ has a minimal element C. Thus $\dim_F C$ is uncountable but every proper FZ-submodule of C has countable dimension over F. The inclusion $Y \leq \zeta(Z)$ implies that the mapping $\nu : c \mapsto c(y-1)$, $c \in C$ is an FZ-homomorphism for every element $y \in Y$. It follows that Ker $\nu = C_C(y)$, and Im $\nu = C(y-1)$ are FZ-submodules of C, and $C(y-1) \cong_{FZ} C/C_C(y)$. If $C(y-1)$ is a proper FZ-submodule of C, then it has countable dimension. By the isomorphism $C(y-1) \cong_{FZ} C/C_C(y)$, the submodule $C_C(y)$ must have uncountable dimension. By the choice of C, $C = C_C(y)$. Since this holds for each element $y \in Y$, $C_A(Y) \geq C$. Since Y is a normal subgroup of G, $C_A(Y)$ is an FG-submodule. The inclusion $C_A(Y) \geq C$ shows that $C_A(Y)$ has uncountable dimension too. Under the conditions of this lemma, this implies that $C_A(Y) = A$. In other words, $Y \leq C_G(A) = \langle 1 \rangle$. This contradiction shows that $C(y-1) = C$.

Consider now the natural semidirect product $C \rtimes \langle y \rangle$, $y \in Y$. Since char $F = p$, the additive group of C is an elementary abelian p-subgroup. The finiteness of $\langle y \rangle$ implies that $E = C \rtimes \langle y \rangle$ is nilpotent (see, for example, [235, Lemma 6.35]). Clearly, $[E, E] = C(y-1) = C$, and, in particular, $E/[E, E]$ is finite. This implies that E is finite too (see, for example, [234, Theorem 2.2]). This final contradiction proves that $T = \langle 1 \rangle$. $\qquad \square$

Proposition 11.4. *Let D be a Dedekind domain, P a maximal ideal of D, $y \in P \backslash P^2$, and A a P-module over the ring D. Then A is D-divisible if and only if $A = Ay$.*

Proof. If A is divisible, then clearly $A = Ay$. Conversely, suppose that $A = Ay$. Let $a \in A$, and $0 \neq x \in D$. If $x \in P$, then $xD + P = D$. Since A is a P-module, Ann $_D(a) = P^s$ for some $s \in \mathbb{N}$. By Lemma 6.7, $xD + P^s = D$, and so $1 = xu + y$ where $u \in D$, and $y \in P^s$. We have now

$$a = a \cdot 1 = a(xu + y) = axu + ay = aux = (au)x,$$

and thus A is D-divisible. Suppose that $x \in P$. Since $A = Ay$, $A = Ay^n$ for every $n \in \mathbb{N}$. By Theorem 6.3, Lemma 6.7 and Lemma 6.8, $xD = P^tQ$, where $P+Q = D$. It follows that $x \in P^t \setminus P^{t+1}$, and so $x = x_1 z$, where $x_1 \in P^t$, and $z \in Q$. Moreover, $z \in P$. From the equation $A = Ay^t$ we obtain that $a = a_1 y^t$, for some element $a_1 \in A$. Since Ann $_D(a) = P^s$, we have $P^{s+t} \leq$ Ann $_D(a_1)$. Proposition 6.13 yields that $P^t = Dy^t + P^{t+s}$, so $x_1 = y^t u + v$, $u \in D$, and $v \in P^{t+s}$. From $x \notin P^{t+1}$ it follows that $u \notin P$. Since P is a prime ideal, $u \notin P$, $z \notin P$, and we can conclude that $uz \notin P$. Hence, $uzP + P = D$. This implies that $uzD + P^{t+s} = D$, so $1 = uzx_2 + x_3$, where $x_2 \in D$, and $x_3 \in P^{t+s}$. Then

$$a_1 = a_1 \cdot 1 = a_1(uzx_2 + x_3) = (a_1x_2)uz + a_1x_3 = a_2uz,$$

where $a_2 = a_1x_2$. Thus Ann $_D(a_1) =$ Ann $_D(a_2)$; that is, $P^{t+s} \leq$ Ann $_D(a_2)$. It follows that

$$a = a_1y^t = (a_2uz)y^t = a_2uzy^t + a_2vz = a_2(uzy^t + vz) = a_2x,$$

and hence $A = Ax$.   □

Lemma 11.5. *Let G be a countable FC-hypercentral group, F a field, and A an artinian monolithic FG-module. Suppose that $\dim_F A$ is uncountable, but every proper FG-submodule of A has countable dimension over F. If $C_G(A) = \langle 1 \rangle$, then $C_G(\mu_{FG}(A)) = \langle 1 \rangle$.*

Proof. Put $H = C_G(\mu_{FG}(A))$. Suppose that the result is false, that is $H \neq \langle 1 \rangle$. Since G is FC-hypercentral, $H \cap FC(G) \neq \langle 1 \rangle$ by Lemma 3.15. All elements of an FC-group having finite order form a (characteristic) subgroup (see Corollary 3.4). By Lemma 11.3, H has no non-identity periodic G-invariant subgroups, so that $H \cap FC(G)$ is torsion-free. But a torsion-free FC-group is abelian (see again Corollary 3.4), so $H \cap FC(G)$ is an abelian torsion-free subgroup.

Let $1 \neq x \in FC(G) \cap H$, and put $X = \langle x \rangle^G$. The subgroup X is finitely generated torsion-free abelian, and its centralizer $Z = C_G(X)$ has finite index in G, and $X \leq \zeta(Z)$. Let $\mathfrak{M}$ be the set of all FZ-submodules having uncountable F-dimension. Since A is an artinian FZ-module by Theorem 5.2, $\mathfrak{M}$ has a minimal element C. Thus, $\dim_F C$ has uncountable F-dimension, but every proper FZ-submodule of C has countable dimension over F. The inclusion $Y \leq \zeta(Z)$ implies that the mapping $\nu : c \mapsto c(y-1)$, $c \in C$ is an FZ-homomorphism for all elements $y \in Y$. It follows that Ker $\nu = C_C(y)$, and Im $\nu = C(y-1)$ are FZ-submodules of C, and $C(y-1) \cong_{FZ} C/C_C(y)$. If $C(y-1)$ is a proper FZ-submodule of C, then it has countable dimension. By the isomorphism $C(y-1) \cong_{FZ} C/C_C(y)$, the submodule $C_C(y)$ must have uncountable dimension. By the choice of C this gives that $C = C_C(y)$. Since this holds for every element $y \in Y$, $C_A(Y) \geq C$. Since Y is a normal subgroup of G, $C_A(Y)$ is an FG-submodule. The inclusion $C_A(Y) \geq C$ shows that $C_A(Y)$ has uncountable dimension too. By hypothesis, this implies that $C_A(Y) = A$. In other words, $Y \leq C_G(A) = \langle 1 \rangle$. This contradiction shows that $C(y-1) = C$. By Theorem 11.1, A is an FH-hypercentral module, so that A is FX-hypercentral. It follows that the natural semidirect product $A \rtimes X$ is a hypercentral group. Let $D = F\langle x \rangle$, where x is a non-identity element of X. Since x has infinite order, D is a principal ideal domain, in particular, D is a Dedekind domain. Obviously, the ideal $P = (x - 1)F\langle x \rangle$ is prime. Since $C \rtimes \langle x \rangle$ is a hypercentral group, the D-module C coincides with its P-component. The equation $C(x - 1) = C$, together with Proposition 11.4 implies that C is a D-divisible module. By Theorem 7.25,

$$C = \bigoplus_{\lambda \in \Lambda} A_\lambda,$$

where A_λ is a Prüfer P-module for every $\lambda \in \Lambda$. Since $\dim_F C$ is uncountable, then the set of indexes Λ cannot be finite. It follows that $\dim_F C = \dim_F \Omega_{P,1}(C)$, and, in particular, $\dim_F \Omega_{P,1}(C)$ is uncountable. The inclusion $X \leq \zeta(Z)$ shows that $\Omega_{P,1}(C)$ is an FZ-submodule of C. By the choice of C, every proper FZ-submodule of C has countable dimension. This contradiction gives the required result. $\qquad\square$

Theorem 11.6. *Let G be a countable FC-hypercentral group, F a field, and A an artinian FG-module. If the factor-group $G/C_G(\mathrm{Soc}\ _{FG}(A))$ is abelian-by-finite, then $\dim_F A$ is countable. In particular, if the field F is countable, then A is likewise countable.*

Proof. Consider first the case when the module A is monolithic. In this case $\mathrm{Soc}\ _{FG}(A) = \mu_{FG}(A)$. Put $H = C_G(\mu_{FG}(A))$. Suppose again that $\dim_F A$ is uncountable. Let $\mathfrak{M}$ be the set of all FG-submodules having uncountable F-dimension. Since A is an artinian FG-module, $\mathfrak{M}$ has a minimal element B. In other words, $\dim_F B$ is uncountable, but every proper FG-submodule of B has countable dimension over F. By Lemma 11.5, $H \leq C_G(B)$. Therefore, we can think of B as an $F(G/H)$-module. The factor-group G/H is abelian-by-finite, and now Lemma 11.2 yields that the socle height of B is at most ω. Since A is artinian, every factor of the upper socle series is decomposed into a direct sum of finitely many simple submodules. Thus B has an ascending series of submodules

$$\langle 0 \rangle = B_0 \leq B_1 \leq \cdots \leq B_n \leq B_{n+1} \leq \cdots \leq B_\rho$$

with simple factors, and $\rho \leq \omega$. Each factor of this series is a cyclic FG-module, so that it is an epimorphic image of the FG-module FG. Since G is countable, $\dim_F FG$ is countable. Thus, every factor of this series has finite or countable dimension over F. Hence $\dim_F B$ is countable, and we come to a contradiction.

Consider now the general case. Suppose that

$$\mathrm{Soc}\ _{FG}(A) = M_1 \oplus \cdots \oplus M_k,$$

where $M_1, \ldots, M_k$ are simple FG-modules. For each index j, $1 \leq j \leq k$, let U_j be a maximal FG-submodule under the following properties:

$$M_1, \ldots, M_{j-1}, M_{j+1}, \ldots, M_k \leq U_j,$$
$$U_j \cap M_j = \langle 0 \rangle.$$

Then the factor-module A/U_j is monolithic, and with the monolith $(M_j+U_j)/U_j$. Clearly, $H \leq C_G((M_j + U_j)/U_j)$, so that $G/C_G((M_j + U_j)/U_j)$ is abelian-by-finite. Applying the above paragraph, we obtain that every A/U_j has countable dimension over F. But

$$U_1 \cap \cdots \cap U_k \cap \mathrm{Soc}\ _{FG}(A) = \langle 0 \rangle .$$

Since A is artinian, this means that

$$U_1 \cap \cdots \cap U_k = \langle 0 \rangle .$$

Application of Remak's theorem gives the embedding

$$A \hookrightarrow A/U_1 \oplus \cdots \oplus A/U_k,$$

which shows that $\dim_F A$ is countable.   □

Lemma 11.7. *Let R be an integral domain, F the field of fraction of R, G a group, and A an artinian RG-module. If A is R-torsion-free, then $E = A \otimes_R F$ is an artinian FG-module.*

Proof. Let $\{E_n \mid n \in \mathbb{N}\}$ be a descending chain of FG-submodules of E. Then $\{A \cap E_n \mid n \in \mathbb{N}\}$ is a descending chain of RG-submodules of A. Since A is artinian, there exists some $k \in \mathbb{N}$ such that $A \cap E_k = A \cap E_{k+n}$ for every $n \in \mathbb{N}$. The R-module $E_k/(A \cap E_k)$ is periodic, so that $E_k/(A \cap E_{k+n})$ is likewise periodic for every $n \in \mathbb{N}$. This means that the R-divisible envelope of every $A \cap E_{k+n}$ coincides with E_k. On the other hand, that envelope is E_{k+n}. Hence $E_k = E_{k+n}$ for every $n \in \mathbb{N}$. This proves that E is an artinian FG-module.   □

Theorem 11.8. *Let G be a countable FC-hypercentral group, D a Dedekind domain and A an artinian DG-module. If $G/C_G(\mathrm{Soc}_{DG}(A))$ is abelian-by-finite, then $|A| \leq |D|\aleph_0$. In particular, if D is countable, then A is countable.*

Proof. If D is a field, then our assertion follows from Theorem 11.6. Therefore we may suppose that D is not a field. In particular, this implies that D is infinite. Let T be the D-periodic part of A. By Corollary 6.25,

$$T = \bigoplus_{Q \in \pi} A_Q,$$

where $\pi = \mathrm{Ass}_D(A)$, and A_Q is the Q-component of A. Note also that A_Q is a DG-submodule of A. Since A is artinian, the set π is finite. Put $B_Q = \Omega_{Q,1}(A_Q)$. Then we may think of B_Q as an $F_Q G$-module, where $F_Q = D/Q$ is a field. By Theorem 11.6, B_Q has countable dimension over F_Q. Let $y \in Q \setminus Q^2$, and consider the ideal $yD + Q^n$. By Proposition 6.13,

$$yD + Q^n = Q^k$$

for some $1 \le k \le n$. Since $y \notin Q^2$, $k = 1$. In other words,

$$yD + Q^n = Q$$

for every $n \in \mathbb{N}$. Put $C_Q = \Omega_{Q,2}(A_Q)$. The mapping $\phi : c \mapsto cy$, $c \in C_Q$ is a D-endomorphism such hat $\operatorname{Ker} \phi = B_Q$, and $\operatorname{Im} \phi = (C_Q)y = (C_Q)Q \le B_Q$, so that C_Q/B_Q is isomorphic to some D-submodule of B_Q. It follows that

$$|C_Q| \le |F_Q|\aleph_0 \le |D|\aleph_0 = |D|.$$

Similar arguments give that $|\Omega_{Q,n}(A_Q)| \le |D|$ for every $n \in \mathbb{N}$, and, consequently, $|A_Q|| \le |D|$. Since π is finite, $|T| \le |D|$.

The factor-module A/T is D-torsion-free. By Lemma 11.7, $E = (A/T) \otimes_R F$ is an artinian FG-module, where F is the field of fractions of the integral domain D. Since D is infinite, $|F| = |D|$. By Theorem 11.6,

$$|E| \le |F|\aleph_0 \le |D|\aleph_0 = |D|,$$

and hence $|A| \le ||D|$. $\qquad\square$

With the help of some additional restrictions on G, one can obtain conditions under which the socle height of any artinian DG-module A is at most ω.

Let G be a group. From now on, we denote by $P(G)$ *the maximal normal periodic subgroup of G.*

Lemma 11.9. *Let G be an FC-hypercentral group with $P(G) = \langle 1 \rangle$. If G has a finite 0-rank, then G has a normal nilpotent torsion-free subgroup of finite index.*

Proof. The elements of $A_1 = FC(G)$ having finite order form a characteristic subgroup (see Corollary 3.4), so that the equation $P(G) = \langle 1 \rangle$ implies that A_1 is torsion-free. Since it has a finite 0-rank, it follows that A_1 has a finitely generated subgroup B_1 such that A_1/B_1 is periodic. Put $C_1 = (B_1)^G$. The inclusion $B_1 \le FC(G)$ implies that C_1 is a finitely generated subgroup, and the index $|G : C_G(C_1)|$ is finite. Put $E_1/C = P(G/C_1) \cap C_G(C_1)/C_1$ so that $C_1 \le \zeta(E_1)$, and E_1/C_1 is periodic. By Corollary 3.3, it follows that $[E_1, E_1]$ is also periodic. In particular the elements of E_1 having finite orders form a characteristic subgroup. It turns out that E_1 is an abelian torsion-free subgroup. Since the operation of taking roots in an abelian torsion-free group is unique, $C_G(C_1) = C_G(E_1)$ because E_1/C_1 is periodic, and hence the index $|G : C_G(E_1)|$ is finite. Put $Z_1 = C_G(E_1)$, and $A_2/E_1 = FC(G/E_1) \cap Z_1/E_1$. As above, it can be shown that A_2/E_1 is an abelian torsion-free group. Iterating the previous arguments, we find a normal subgroup E_2 such that $E_1 \le E_2$, E_2/E_1 is abelian and torsion-free, and $Z_2 = Z_1 \cap C_G(E_2/E_1)$ has finite index in G. Moreover, if G/E_2 is infinite, then $A_3/E_2 = FC(G/E_2) \cap Z_2/E_2$ is torsion-free abelian. Proceeding by ordinary induction, we continue the process of construction of subgroups E_n. Since the 0-rank of G is finite, this process cannot be infinite; that is, after finitely many steps we come to a nilpotent torsion-free subgroup E_n having finite index. $\qquad\square$

Lemma 11.10. *Let D be a Dedekind domain with infinite set Spec (D), G be a locally (polycyclic-by-finite) group of finite 0-rank, and A an artinian DG-module. Then A is D-periodic. If A_P is the P-component of A, $P \in Ass_D(A)$ and $p = char(D/P)$, then the additive group of A_P is a p-group whenever $p > 0$, and the additive group of A_P is torsion-free whenever $p = 0$.*

Proof. Since A is artinian, A has an ascending series of DG-submodules $\{A_\alpha \mid \alpha < \gamma\}$, every factor $A_{\alpha+1}/A_\alpha$ of which is a simple DG-module. According to L.A. Kurdachenko, J. Otal and I.Ya. Subbotin [157, Corollary 1.16], Ann $_D(A_{\alpha+1}/A_\alpha) \in$ Spec (D) for every index $\alpha < \gamma$. It is not hard to see now that A is D-periodic.

Let $P \in$ Ass $_D(A)$, and suppose that $p = $ char (D/P). If $p > 0$, then the additive group of D/P is an elementary abelian p-group. By Corollary 6.17, $P/P^2 \cong_D D/P$, and so the orders of the elements of D/P^2 divide p^2. By the same logic, we obtain that the orders of the elements of D/P^n divide p^n. If $a \in A_P$, then Ann $_D(a) = P^n$ for some $n \in \mathbb{N}$, so that $aD \cong_D D/P^n$. In particular, the order of a divides p^n. This means that the additive group of A_P is a p-group. If $p = 0$, then the additive group of D/P is torsion-free. Using again the isomorphism $P/P^2 \cong_D D/P$, we obtain that the additive group of D/P^2 is torsion-free. And similarly we obtain that the additive group of D/P^n is torsion-free for each $n \in \mathbb{N}$. If $a \in A_P$, then Ann $_D(a) = P^n$ for some $n \in \mathbb{N}$, so that $aD \cong_D D/P^n$. In particular, it follows that the order of a is infinite. □

Corollary 11.11. *Let D be a Dedekind domain with infinite set Spec(D), G be a locally (polycyclic-by-finite) group of finite 0-rank, and A an artinian monolithic DG-module. Then $P = Ann(\mu_{DG}(A)) \neq \langle 0 \rangle$ and A is a P-module. Furthermore, if $p = $ char $D/P > 0$, then the additive group of A_P is a p-group; and if $p = 0$, then the additive group of A is torsion-free.*

Proof. Indeed, by Lemma 11.10, A is D-periodic. By Corollary 6.25,

$$A = \bigoplus_{Q \in \pi} A_Q,$$

where $\pi = $Ass $_D(A)$. Since every primary component A_Q is a DG-submodule, the inclusion $\mu_{DG}(A) \leq A_P$ implies that $A = A_P$. □

Lemma 11.12. *Let G be an FC-hypercentral group, D a Dedekind domain, A a monolithic D-periodic DG-module, $H = C_G(\mu_{DG}(A))$, $C_G(A) = \langle 1 \rangle$, and $p = $ char$(D/Ann_D(\mu_{DG}(A)))$. Suppose also that A is DH-hypercentral. If $p > 0$, and the group G has finite section p-rank, then H has a normal nilpotent torsion-free subgroup of finite index. If $p = 0$, and G has finite 0-rank, then $P(H) = \langle 1 \rangle$, and H has a normal nilpotent torsion-free subgroup of finite index.*

Proof. Using Lemma 2.10, it is not hard to see that if an FC-hypercentral group has finite section p-rank for some prime p, then it has also finite 0-rank. In other words, in each of these cases, the 0-rank of the group is finite. Let L be an arbitrary finitely generated subgroup of H. Since L acts trivially on the factors of an ascending series of A, then L is residually nilpotent (see P. Hall and B. Hartley [99, Theorem A2]. It follows that the set T of all elements of H having finite order form a (characteristic) locally nilpotent subgroup of H (see, for example, [235, Theorem 6.14]).

Suppose that $p > 0$, and let P be the Sylow p-subgroup of T, and Q the Sylow p'-subgroup of T. Since every abelian subgroup of P is Chernikov, P is likewise Chernikov (see for example, [234, Theorem 3.32]). Let D be the divisible part of P. Proceeding as we did in the proof of Corollary 11.11, we obtain that the additive group of A is a p-group. The natural semidirect product $A \rtimes T$ is a hypercentral group, because A is FH-hypercentral. But, then Theorem 3.10 shows that $D \times Q \leq C_H(A)$, that is the periodic part of H is finite. Applying Lemma 11.9 to H/T we obtain that H/T is nilpotent-by-finite and almost torsion-free. According to D.I. Zaitsev [291, Lemma 3], we obtain that H is nilpotent-by-finite and almost torsion-free. Let now $p = 0$. Proceeding as we did in the proof of Corollory 11.11, we obtain that the additive group of A is torsion-free. Since the natural semidirect product $A \rtimes T$ is a hypercentral group, and all elements of hypercentral group having finite orders form a subgroup, then $A \rtimes T = A \times T$, that is, $T \leq C_G(A) = \langle 1 \rangle$. Applying Lemma 11.9 to H, we obtain that H/T is nilpotent-by-finite and almost torsion-free. $\square$

Corollary 11.13. *Let G be an FC-hypercentral group, D a Dedekind domain with infinite set $Spec(D)$, A an artinian monolithic DG-module, $H = C_G(\mu_{DG}(A))$, $C_G(A) = \langle 1 \rangle$, and $p = \mathrm{char}(D/Ann\,_D(\mu_{DG}(A))$. If $p > 0$, and the group G has finite section p-rank, then H has a normal nilpotent torsion-free subgroup of finite index. If $p = 0$, and G has finite 0-rank, then $P(H) = \langle 1 \rangle$, and H has a normal nilpotent torsion-free subgroup of finite index.*

Proof. By Lemma 11.10, A is D-periodic. Theorem 11.1 yields that A is a DH-hypercentral module. Thus all conditions of Lemma 11.12 are satisfied. $\square$

Lemma 11.14. *Let G be a nilpotent torsion-free group of finite 0-rank, and let p be a prime. If H is a subgroup satisfying $r_0(H) = r_0(G)$, then G has a subgroup L with the following properties:*

(1) *$H \leq L$ and the index $|L : H|$ is finite.*

(2) *There exists a subnormal series*

$$L = L_0 \lhd L_1 \lhd \cdots \lhd L_n = G,$$

all factors of which are periodic p-divisible groups.

Proof. Since G is nilpotent, H is subnormal with defect at most k, where k is the nilpotency class of G. Let

$$H = H_0 \lhd H_1 \lhd \cdots \lhd H_n = G$$

be a subnormal series. We proceed by induction on n. If H is normal in G, then the equation $r_0(H) = r_0(G)$ yields that G/H is periodic. Since G is a nilpotent torsion-free group of finite 0-rank, then G has finite special rank. It follows that the Sylow p-subgroup P/H of G/H is Chernikov (see, for example, [235, Corollary 1 to Theorem 6.36]). Denote by D/H the divisible part of P/H. Note that the center of a nilpotent periodic group contains every divisible subgroup (see, for example, [234, Lemma 3.13]), so that $P/H = (D/H)(L/H)$ for some finite G-invariant subgroup L/H. Clearly, G/L is p-divisible.

Let now $n > 0$, and suppose that our statement has been shown for subnormal subgroups whose defect is less than n. In particular, there is a subgroup $L_1 \geq H_1$ such that the index $|L_1 : H_1|$ is finite, and L_1 is a member of a finite subnormal series of G with p-divisible factors. There is no loss if we suppose that H_1 is normal in L_1. The family $\{H^g \mid g \in L_1\}$ is finite, that is,

$$\{H^g \mid g \in L_1\} = \{g_1^{-1}Hg_1, \ldots, g_s^{-1}Hg_s\}.$$

From $H_1/H^g = H_1^g/H^g = H_1/H$, we obtain that H_1/H^g is periodic for each element $g \in L_1$. By Remak's Theorem,

$$H_1/K \hookrightarrow H_1/(g_1^{-1}Hg_1) \times \cdots \times H_1/(g_s^{-1}Hg_s),$$

where $K = g_1^{-1}Hg_1 \cap \cdots \cap g_s^{-1}Hg_s$ is an L_1-invariant subgroup. This embedding shows that L_1/K is also periodic. Let σ be the set of all prime divisors of the number $|L_1 : H_1|$, and put $\pi = \sigma \cup \{p\}$. We have

$$L_1/K = U/K \times V/K,$$

where U/K is the Sylow π-subgroup, and V/K is the Sylow π'-subgroup. By the choice of π, we obtain the inclusion $V/K \leq H_1/K$, which proves that H is V-invariant. Since π is finite, U/K is a Chernikov group. Let D/K be the divisible part of U/K. Since the center of a nilpotent periodic group includes every divisible subgroup (see, for example, [234, Lemma 3.13]), $U/K = (W/K)(D/K)$ for some finite L_1-invariant subgroup W/K. Put $U_1/K = U/K \cap H/K$, and $V_1/K = V/K \cap H/K$. We have already noted that V_1/K is normal in V/K. Since $(U/K)/(W/K)$ is abelian, $(U_1/K)(W/K)$ is normal in U/K. It follows that $L/K = (U_1/K)(W/K)(V_1/K)$ is normal in L_1/K. By the choice of L_1/L, we obtain that the factor-group L_1/L is p-divisible so that L is a member of a finite subnormal series of G with p-divisible factors. The finiteness of W/K, and the equation $L/K = (H/K)(W/K)$ imply that $|L : H|$ is finite; that is, conditions (1) and (2) hold. $\square$

Corollary 11.15. *Let G be a nilpotent torsion-free group of finite 0-rank, and let p be a prime. Then G has a finitely generated subgroup L with the following properties:*

(1) $r_0(L) = r_0(G)$.

(2) *There exists a subnormal series*

$$L = L_0 \lhd L_1 \lhd \cdots \lhd L_n = G,$$

all factors of which are periodic p-divisible groups.

Proof. Let

$$\langle 1 \rangle = Z_0 \leq Z_1 \leq \cdots \leq Z_n = G$$

be the upper central series of G. For every $1 \leq j \leq n$, we choose elements

$$g_{j,1}, \ldots, g_{j,k(j)} \in Z_j \setminus Z_{j-1}$$

such that $Z_j / \langle g_{j,1}, \ldots, g_{j,k(j)} \rangle Z_{j-1}$ is periodic, that is

$$r_0(\langle g_{j,1}, \ldots, g_{j,k(j)} \rangle Z_{j-1}/Z_{j-1}) = r_0(Z_j/Z_{j-1}).$$

Put now $H = \langle g_{j1}, \ldots, g_{jk(j)} | 1 \leq j \leq n \rangle$. The subgroup H is finitely generated, and $r_0(H) = r_0(G)$. It suffices to apply Lemma 11.14 to H. □

Lemma 11.16. *Let G be an FC-hypercentral group, F a field of characteristic p, A a monolithic FG-module, $H = C_G(\mu_{FG}(A))$, and $C_G(A) = \langle 1 \rangle$. Suppose also that if $p = 0$, then G has finite 0-rank, and if $p > 0$, then G has finite section p-rank. If A is an FH-hypercentral module, then the length of the upper FH-hypercentral series of A is at most ω.*

Proof. Corollary 11.13 yields that H has a normal nilpotent subgroup L having finite index in H. Let $|H/L| = k$. Then $H^k \leq L$, and, in particular, H^k is torsion-free nilpotent. Clearly, H^k is also G-invariant. Since the orders of the elements of H/H^k are bounded, and H has finite special rank, it follows that H/H^k is finite. In order to avoid additional notation we can assume that L is a G-invariant subgroup of H. By Corollary 11.15, L has a finitely generated subgroup K such that $r_0(L) = r_0(K)$, and there exists a finite subnormal series

$$K = K_0 \triangleleft K_1 \triangleleft \cdots \triangleleft K_n = L,$$

whose factors are periodic, and p-divisible. Let a be an arbitrary element of A. The natural semidirect product $A \leftthreetimes H$ is a hypercentral group, because A is an FH-hypercentral module. Then the subgroup $\langle a, K \rangle$ is finitely generated and nilpotent. It follows that a is contained in some FK-hypercenter of A corresponding to a finite ordinal m. It follows that the length of the upper FK-central series of A is at most ω. Put $C = \zeta_{FK}(A)$. Since K is normal in K_1, C is naturally an FK_1-submodule.

Suppose first that $p > 0$. Since K_1/K is p-divisible, Lemma 3.10 yields the inclusion $C \leq C_A(K_1)$; this proves that the FK_1-center of the module A coincides with its FK-center. Similarly, we prove that every FK_1-hypercenter of the module A coincides with its FK-hypercenter having the same number. Using the p-divisibility of the factors $K_2/K_1, \ldots, K_n/K_{n-1}$, and ordinary induction, we obtain at last that every FL-hypercenter of the module A coincides with its FK-hypercenter having the same number. Since C is FH-hypercentral, from the finiteness of H/L, and reasoning as above, we can deduce that $H/C_H(C)$ is a finite nilpotent group. It follows that $E = C \leftthreetimes H$ is nilpotent (see, for example, [235,

Lemma 6.34]). In other words, C has an FH-central series of finite length. The same is valid for each factor of the upper FL-central series, which means that the length of the FH-central series of the module A is at most ω.

Now let $p = 0$. In this case, the additive group of the module A is torsion-free. Since the natural semidirect product $A \leftthreetimes K_1$ is a hypercentral group, and the elements of a hypercentral group having finite order form a subgroup, $A \leftthreetimes K_1 = A \times K_1$, that is $C \leq C_A(K_1)$, and hence the FK_1-center of the module A coincides with its FK-center. As above, it follows that every FL-hypercenter of the module A coincides with its FK-hypercenter having the same number. Since C is FH-hypercentral, we obtain again that $H/C_H(C)$ is a finite nilpotent group. It turns out that $C \leq C_A(H)$, and hence the FH-center of A coincides with its FK-center. Therefore, every FH-hypercenter of the module A coincides with its FK-hypercenter having the same number. Thus, in this case, the length of the FH-central series of the module A is at most ω. $\qquad\square$

Lemma 11.17. *Let G be a group, D be a Dedekind domain, and A a simple DG-module. If $Q = G/C_G(A)$ has a normal abelian subgroup R of finite index, then the periodic part T of R has finite special rank.*

Proof. By Theorem 5.5,

$$A = \bigoplus_{x \in S} Cx,$$

where C is a simple DR-submodule, and S is a finite set. Obviously, Cx is a simple DR-submodule; therefore $t(R/C_R(Cx))$ is a locally cyclic p'-group, where $p = \mathrm{char}$ $(D/\mathrm{Ann}\,_D(A))$ for every $x \in S$ (see L.A. Kurdachenko, J. Otal and I.Ya. Subbotin [157, Corollaries 2.4, 2.5]). The equation

$$\bigcap_{x \in S} C_R(Cx) = C_R(A) = \langle 1 \rangle,$$

and Remak's Theorem together imply the embedding

$$R \hookrightarrow \mathrm{Dr}_{x \in S} R/C_R(Cx),$$

which shows that $t(R)$ has finite special rank. $\qquad\square$

Lemma 11.18. *Let G be a group of finite special rank. If $[G, G]$ is finite, then $G/\zeta(G)$ is also finite.*

Proof. We first consider the case, when G is a periodic group. Put $K = [G, G]$, and $C = C_G(K)$ so that $K \cap C \leq \zeta(C)$. Thus C is nilpotent. Put $\pi = \Pi(K)$, and let P be the Sylow π-subgroup of C, and Q the Sylow π'-subgroup of C. Clearly, Q is G-invariant. Since $Q \cap K = \langle 1 \rangle$, $Q \leq \zeta(G)$. Denote by D the divisible part of P. Since G is an FC-group, $D \leq \zeta(G)$. By the finiteness of the set π, we obtain that P is Chernikov so that P/D is finite. Then C/DQ is finite. Since K is finite, G/C is finite. Then G/DQ is finite, and it follows that $G/\zeta(G)$ is finite.

Consider now the general case. Let U be a maximal torsion-free subgroup of $\zeta(G)$. The factor-group G/U is periodic (see D. J.S. Robinson [234, Theorem 4.32]). Applying the result shown in the above paragraph, $Z/U = \zeta(G/U)$ has finite index in G/U. Let $z \in Z$, and $g \in G$ so that $[g, z] \in U$, and then $[g, z] \in U \cap [G, G] = \langle 1 \rangle$. It follows that $Z = \zeta(G)$, and hence $G/\zeta(G)$ is finite as required. $\qquad \square$

Proposition 11.19. *Let G be an FC-hypercentral group, F a field of characteristic p, and A an artinian monolithic FG-module. Suppose further that if $p > 0$, then G has finite section p-rank, and if $p = 0$, then G has finite 0-rank. If the factor-group $G/C_G(\mu_{FG}(A))$ is abelian-by-finite, then every finitely generated FG-submodule of A has finite FG-composition series. In particular, the socular height of A is at most ω.*

Proof. Put $H = C_G(\mu_{FG}(A))$. By Corollary 11.13, H has a normal torsion-free subgroup L having finite index. We proceed by induction on $r_0(L)$. If $r_0(L) = 0$, then H is finite. By Lemma 11.17, and Lemma 11.18, G is abelian-by-finite, and it suffices to apply Lemma 11.2. Let now $r_0(L) > 0$. In this case $\zeta(L) \neq \langle 1 \rangle$, which, by Corollary 3.16 gives that $\zeta(L) \cap FC(G) \neq \langle 1 \rangle$. Pick $1 \neq x \in FC(G) \cap \zeta(L)$, and put $X = \langle x \rangle^G$ and $Y = C_G(X)$. Then X is a finitely generated abelian torsion-free subgroup, Y is a normal subgroup of finite index, and $X \leq \zeta(Y)$. The FY-module A is artinian by Theorem 5.2. Theorem 11.1 yields that A is FH-hypercentral. By Theorem 11.6, the length of the upper FH-hypercentral series of A is at most ω. In particular, every element b of A is contained in some FH-hypercenter of A with a finite number $m(b)$. Thus, the FY-submodule $B = bFY$ is FH-nilpotent. Then B is $F\langle g \rangle$-nilpotent for each element $g \in H$ so that $B(g - 1)^{m(b)} = \langle 0 \rangle$. Since $x \in \zeta(Y)$, the mapping $a \mapsto a(x - 1)^n$, $a \in B$, is an FY-endomorphism for every $n \in \mathbb{N}$. Put $B_0 = B$, and $B_{n+1} = B_n(x - 1)$, $n > 1$. Then B_n is a finitely generated FY-submodule for every $n \in \mathbb{N}$.

Suppose that A is a monolithic FY-module. There is some $t \leq m(b)$ such that $B(g-1)^{t+1} = \langle 0 \rangle$ but $B(g-1)^t \neq \langle 0 \rangle$. Then $x \in C_L(B_t)$, and hence $r_0(L/C_L(B_t) < r_0(L)$. It follows that B_t has a finite FY-composition series. As we have already noted above the mapping $\eta : a \mapsto a(x - 1)$, $a \in B_{t-1}$ is an FY-endomorphism, and Im $\eta = B_t(x - 1) = B_t$, and Ker $\eta = B_t$ so that $B_{t-1}/B_t \cong_{FY} B_t$. It follows that B_{t-1}/B_t, and hence B_{t-1} have finite FY-composition series. Proceeding by ordinary induction, after finitely many steps we obtain that $B_0 = B$ has finite FY-composition series. By the finiteness of G/Y, the FG-submodule BFG has finite FY-composition series, thus BFG has also finite FG-composition series.

Consider now the general case. By Theorem 5.5,

$$M = \mu_{FG}(A) = \bigoplus_{x \in S} Cx,$$

where C is a simple FY-submodule, and S is a finite subset. For every $x \in S$, we denote by U_x a maximal submodule among the FG-submodules that contains $\bigoplus_{x \neq v \in S} Cv$ and meets Cx trivially. Then the factor-module A/U_x is FY-monolithic with the monolith $(Cx + Ux)/Ux$. By the result shown in the above paragraph, $(B + Ux)/Ux$ has finite FY-composition series for every $x \in S$. Furthermore,

$$\bigcap_{x \in S} Ux$$

is an FG-submodule such that

$$\left(\bigcap_{x \in S} Ux\right) \cap M = \langle 0 \rangle.$$

Since A is a monolithic FG-module,

$$\bigcap_{x \in S} Ux = \langle 0 \rangle.$$

Application of Remak's Theorem gives the embedding

$$A \hookrightarrow \bigoplus_{x \in S} A/Ux.$$

In this setting

$$B \hookrightarrow \bigoplus x \in S(B + Ux)/Ux,$$

so that B has finite FY-composition series. Finally, by the finiteness of G/Y, the FG-submodule BFG has finite FY-composition series; thus BFG has also finite FG-composition series. $\qquad\square$

Theorem 11.20. *Let G be an FC-hypercentral group, D a Dedekind domain with infinite set $\mathrm{Spec}(D)$, A an artinian monolithic DG-module, and p the characteristic of $D/\mathrm{Ann}_D(\mu_{DG}(A))$. Suppose further that if $p > 0$, then G has finite section p-rank, and if $p = 0$, then G has finite 0-rank. If the factor-group $G/C_G(\mu_{FG}(A))$ is abelian-by-finite, then every finitely generated DG-submodule of A has finite DG-composition series. In particular, the socular height of A is at most ω.*

Proof. Put $H = C_G(\mu_{DG}(A))$. By Lemma 11.10, A is a P-module, where $P = \mathrm{Ann}_D(\mu_{DG}(A))$. Pick $b \in A$, and put $B = bDG$. Then $\mathrm{Ann}_D(b) = P^n$ for some $n \in \mathbb{N}$, that is, $B = \Omega_{P,n}(B)$. We proceed by induction on n. If $n = 1$, then $b \in \Omega_{P,1}(A)$. Think of $\Omega_{P,1}(A)$ as an FG-module, where $F = D/P$ is a field. By Proposition 11.19, B has finite DG-composition series. Let now $n > 1$. Pick some $y \in P \setminus P^2$. By Proposition 6.13, $yD + P^m = P$ for every $m \in \mathbb{N}$. The mapping $\phi : c \mapsto cy^m$, $c \in B$ is a DG-endomorphism, and hence By^m is a finitely generated DG-submodule for every $m \in \mathbb{N}$. In particular, By^{n-1} is a finitely generated DG-submodule. Since $y \in \mathrm{Ann}_D(By^{n-1})$, the equation $P = yD + P^n$ implies that $\mathrm{Ann}_D(By^{n-1}) = P$. By the induction hypothesis, By^{n-1} has finite DG-composition series. Furthermore, $\mathrm{Im}\,\phi = (By^{n-2})y = By^{n-1}$ and $\mathrm{Ker}\,\phi = By^{n-1}$, so that

$$By^{n-2}/By^{n-1} \cong By^{n-1}.$$

It follows that By^{n-2}/By^{n-1}, and hence By^{n-2} has finite DG-composition series. Proceeding by ordinary induction, after finitely many steps we obtain that B has finite DG-composition series. □

Corollary 11.21. *Let G be an FC-hypercentral group of finite section rank, D a Dedekind domain with infinite set $\mathrm{Spec}(D)$, and A an artinian DG-module. If the factor-group $G/C_G(\mathrm{Soc}\ _{FG}(A))$ is abelian-by-finite, then every finitely generated DG-submodule of A has finite DG-composition series. In particular, the socular height of A is at most ω.*

Proof. As in some of the above proofs, the general case can be reduced to the case of a monolithic module, and for the latter it suffices to apply Theorem 11.20. □

Lemma 11.22. *Let F be a field of characteristic p, G a periodic group, and A a monolithic FG-module. Suppose further that the following conditions hold:*

(1) *G has a normal abelian subgroup U of finite index.*

(2) *If $p > 0$, then U is a p'-subgroup.*

Then A has finite FG-composition series.

Proof. Let M be the FG-monolith of A. Applying Theorem 7.28, we have that

$$A = M \oplus B$$

for some FU-submodule B. By Theorem 5.5,

$$M = \bigoplus_{x \in S} Cx,$$

where C is a simple FU-submodule, and S is a finite subset of G. In particular, A/B has finite FU-composition series. This remains true for every factor-module A/Bg, $g \in G$. Since G/U is finite, the set $\{Bg \mid g \in G\}$ is finite. Let

$$\{Bg \mid g \in G\} = \{Bg_1, \ldots, Bg_m\}.$$

Since the intersection
$$C = Bg_1 \cap \cdots \cap Bg_m$$

is an FG-submodule, by its construction, $C \cap M = \langle 0 \rangle$. It follows that $C = \langle 0 \rangle$. By Remak's Theorem

$$A \hookrightarrow A/Bg_1 \oplus \cdots \oplus A/Bg_m,$$

which shows that A has finite FU-composition length. Then A has finite FG-composition length. □

Proposition 11.23. *Let G be an FC-hypercentral group, F a field of characteristic p, A a monolithic FG-module, and $H = C_G(\mu_{FG}(A))$. Suppose further that G satisfies the following conditions:*

(1) *A is FH-nilpotent.*

(2) *If $p > 0$, then G has finite section p-rank.*

(3) *If $p = 0$, then G has finite 0-rank.*

If G/H is abelian-by-finite, then A has finite FG-composition series.

Proof. By Lemma 11.12, H contains a normal torsion-free subgroup L of finite index. We proceed by induction on $r_0(L)$. If $r_0(L) = 0$, then H is finite. Moreover, if $p > 0$, then H is a p-subgroup. By Lemma 11.17 and Lemma 11.18, G is periodic, and abelian-by-finite. Let U/H be a normal abelian subgroup of G/H having finite index. Put $M = \mu_{FG}(A)$. By Theorem 5.5,

$$M = \bigoplus_{x \in S} Cx,$$

where C is a simple FU-submodule, and S is a finite subset of G. Clearly, Cx is a simple FU-submodule for every $x \in S$, and therefore $U/C_U(Cx))$ is a locally cyclic p'-group (see L.A. Kurdachenko, J. Otal and I.Ya. Subbotin [158, Corollaries 2.4 and 2.5]). The equation

$$\bigcap_{x \in S} C_U(Cx) = C_U(A) = H$$

and Remak's Theorem allow us to deduce the embedding

$$U/H \hookrightarrow \mathrm{Dr}\ _{x \in S} U/C_U(Cx),$$

which establishes that U/H is a p'-group. By Lemma 11.17 and Lemma 11.18, G is abelian-by-finite. Now it is not hard to obtain that G has a normal abelian p'-subgroup V of finite index. In this case, it suffices to apply Lemma 11.22.

Let now $r_0(L) > 0$. In this case, it suffices to repeat almost word for word the arguments given in the proof of Proposition 11.19. □

Corollary 11.24. *Let G be an FC-hypercentral group, F a field of characteristic p, A a monolithic FG-module, and $H = C_G(\mu_{FG}(A))$. Suppose further that G satisfies the following conditions:*

(1) *A is FH-nilpotent.*

(2) *$\dim_F(\mu_{FG}(A))$ is finite.*

(3) *If $p > 0$, then G has a finite section p-rank.*

(4) *If $p = 0$, then G has finite 0-rank.*

If G/H is periodic, then A has finite FG-composition series.

Proof. Put $M = \mu_{FG}(A)$. The factor-group G/H is nilpotent-by-finite (see L.A. Kurdachenko, J. Otal and I.Ya. Subbotin [157, Corollary 6.13]). Let U be a normal subgroup of G having finite index such that U/H is nilpotent. By Theorem 5.5,

$$M = \bigoplus_{x \in S} Cx,$$

where C is a simple FU-submodule, and S is a finite subset of G. Clearly, Cx is a simple FU-submodule for every $x \in S$; therefore $U/C_U(Cx))$ is abelian-by-finite (see B.A.F. Wehrfritz [280, Lemma 3.5]). The equation

$$\bigcap_{x \in S} C_U(Cx) = C_U(A) = H$$

and Remak's Theorem give the embedding

$$U/H \hookrightarrow \mathrm{Dr}_{x \in S} U/C_U(Cx),$$

which establishes that U/H is abelian-by-finite. Now it suffices to apply Proposition 11.23. □

Corollary 11.25. *Let F be a locally finite field of characteristic p, G an FC-hypercentral group of finite section p-rank, A a monolithic FG-module, and $H = C_G(\mu_{FG}(A))$. If A is FH-nilpotent, and G/H is abelian-by-finite, then A has finite FG-composition series.*

Proof. We recall that the finiteness of the section p-rank implies the finiteness of the 0-rank. Also note that an FC-hypercentral group is locally polycyclic-by-finite. Let U be a normal abelian subgroup having finite index in G. Put $M = \mu_{FG}(A)$. By Theorem 5.5,

$$M = \bigoplus_{x \in S} Cx,$$

where C is a simple FU-submodule, and S is a finite subset of G. Clearly, Cx is a simple FU-submodule for every $x \in S$, and therefore $U/C_U(Cx))$ is a periodic p'-group (see L.A. Kurdachenko, J. Otal and I.Ya. Subbotin [157, Corollary 1.21]). The equation

$$\bigcap_{x \in S} C_U(Cx) = C_U(A) = H$$

and Remak's Theorem give the embedding

$$U/H \hookrightarrow \mathrm{Dr}_{x \in S} U/C_U(Cx),$$

which establishes that U/H is periodic. This implies that G/H is likewise periodic. Now it suffices to apply Proposition 11.23. □

Corollaries 11.24 and 11.25 at once give

Corollary 11.26. *Let F be a locally finite field of characteristic p, G an FC-hypercentral group of finite section $p-$rank, A a monolithic FG-module, and $H = C_G(\mu_{FG}(A))$. If A is FH-nilpotent, and $\dim_F(\mu_{FG}(A))$ is finite, then A has finite FG-composition series.*

Corollary 11.27. *Let G be an FC-hypercentral group, F a field of characteristic p, A a monolithic FG-module, and $H = C_G(\mu_{FG}(A))$. Suppose further that G satisfies the following conditions:*

(1) A is FH-hypercentral.

(2) If $p > 0$, then G has finite section p-rank.

(3) If $p = 0$, then G has finite 0-rank.

If G/H is periodic abelian-by-finite, then each finitely generated FG-submodule of A has finite FG-composition series. In particular, the socular height of A is finite.

Proof. Indeed, by Theorem 11.6, the length of the upper FH-central series of A is at most ω. If $a \in A$, then a belongs to a term of the upper FH-central series having a finite number. This means that aFG is FH-nilpotent, and it suffices to apply Proposition 11.23. □

Corollary 11.28. *Let G be an FC-hypercentral group, F a field of characteristic p, A a monolithic FG-module, and $H = C_G(\mu_{FG}(A))$. Suppose further that G satisfies the following conditions:*

(1) A is FH-hypercentral.

(2) $\dim_F(\mu_{FG}(A))$ is finite.

(3) If $p > 0$, then G has a finite section p-rank.

(4) If $p = 0$, then G has finite 0-rank.

If G/H is periodic, then every finitely generated FG-submodule of A has finite FG-composition series. In particular, the socular height of A is at most ω.

Corollary 11.29. *Let F be a locally finite field of characteristic p, G an FC-hypercentral group of finite section p-rank, A a monolithic FG-module, and $H = C_G(\mu_{FG}(A))$. If A is FH-hypercentral, and G/H is abelian-by-finite, then every finitely generated FG-submodule of A has finite FG-composition series. In particular, the socular height of A is at most ω.*

Corollary 11.30. *Let F be a locally finite field of characteristic p, G an FC-hypercentral group of finite section p-rank, A a monolithic FG-module, and $H = C_G(\mu_{FG}(A))$. If A is FH-hypercentral, and $\dim_F(\mu_{FG}(A))$ is finite, then every finitely generated FG-submodule of A has finite FG-composition series. In particular, the socular height of A is at most ω.*

Corollary 11.31. *Let G be an FC-hypercentral group, F a field of characteristic p, A a monolithic FG-module, and $H = C_G(\mu_{FG}(A))$. Suppose further that G satisfies the following conditions:*

(1) *A is FH-hypercentral.*

(2) *If $p > 0$, then G has finite section p-rank.*

(3) *If $p = 0$, then G has finite 0-rank.*

If G/H is periodic abelian-by-finite, then every term of the upper socular series of A has finite FG-composition series. In particular, A is an artinian FG-module.

Proof. In fact, by Corollary 11.27, the length of the upper socle series of A is at most ω. Let

$$\langle 0 \rangle = S_0 \leq S_1 \leq \cdots \leq S_n \leq S_{n+1} \leq \cdots$$

be the upper socular series of A. Since A is monolithic, S_1 is the FG-monolith of A. We have

$$S_2/S_1 = \bigoplus_{\lambda \in \Lambda} B_\lambda/S_1,$$

where B_λ/S_1 is a simple FG-submodule for every $\lambda \in \Lambda$. By (1), $B_\lambda(h-1) \leq S_1$ for every $h \in H$, and every index $\lambda \in \Lambda$, so that $S_2(h-1) \leq S_1$ for every $h \in H$. This means that S_2 is FH-nilpotent. By Proposition 11.23, S_2 has finite FG-composition series. Using similar arguments, and applying ordinary induction, we obtain that S_n has finite FG-composition series for every $n \in \mathbb{N}$. Now it is not hard to show that A is an artinian FG-module. ⊔

Corollary 11.32. *Let G be an FC-hypercentral group, F a field of characteristic p, A a monolithic FG-module, and $H = C_G(\mu_{FG}(A))$. Suppose further that G satisfies the following conditions:*

(1) *A is FH-hypercentral.*

(2) *$\dim_F(\mu_{FG}(A))$ is finite.*

(3) *If $p > 0$, then G has finite section p-rank.*

(4) *If $p = 0$, then G has finite 0-rank.*

If G/H is periodic, then every term of the upper socular series of A has finite FG-composition series. In particular, A is an artinian FG-module.

Corollary 11.33. *Let F be a locally finite field of characteristic p, G an FC-hypercentral group of finite section p-rank, A a monolithic FG-module, and $H = C_G(\mu_{FG}(A))$. If A is FH-hypercentral, and G/H is abelian-by-finite, then every term of the upper socular series of A has finite FG-composition series. In particular, A is an artinian FG-module.*

Corollary 11.34. *Let F be a locally finite field of characteristic p, G an FC-hypercentral group of finite section p-rank, A a monolithic FG-module, and $H = C_G(\mu_{FG}(A))$. If A is FH-hypercentral, and $\dim_F(\mu_{FG}(A))$ is finite, then every term of the upper socular series of A has finite FG-composition series. In particular, A is an artinian FG-module.*

Theorem 11.35. *Let G be an FC-hypercentral group, D a Dedekind domain with infinite set $\mathrm{Spec}(D)$, A a monolithic DG-module, $H = C_G(\mu_{DG}(A))$, and p the characteristic of $D/\mathrm{Ann}_D(\mu_{DG}(A))$. Suppose further that G satisfies the following conditions:*

(1) *A is DH-hypercentral and D-periodic.*

(2) *If $p > 0$, then G has finite section p-rank.*

(3) *If $p = 0$, then G has finite 0-rank.*

If G/H is periodic and abelian-by-finite, then A satisfies the following properties:

  (i) *Each finitely generated DG-submodule of A has finite DG-composition series*

 (ii) *The socular height of A is at most ω.*

(iii) *A is an artinian DG-module.*

Proof. Since A is monolithic and D-periodic, we have that A is a P-module, where $P = \mathrm{Ann}_D(\mu_{DG}(A))$. Repeating almost word for word the arguments of the proof of Theorem 11.20, we obtain (i). The statement (ii) is a direct consequence of (i). Finally, by Corollary 11.31, $\Omega_{P,1}(A)$ is artinian. Therefore, it suffices to apply Proposition 1.8. $\qquad\square$

Theorem 11.36. *Let G be an FC-hypercentral group of finite section rank, D a Dedekind Z-domain, A a monolithic DG-module, and $H = C_G(\mu_{DG}(A))$. Suppose further that the following conditions hold:*

(a) *A is DH-hypercentral.*

(b) *A is D-periodic.*

(c) *G/H is abelian-by-finite.*

Then A satisfies the following properties:

  (i) *Every finitely generated DG-submodule of A has finite DG-composition series.*

 (ii) *The socular height of A is at most ω.*

(iii) *A is an artinian DG-module.*

Chapter 12

Artinian modules over periodic abelian groups

From now on, in this and other chapters, our goal is the consideration of artinian modules over a group ring DG, where D is a Dedekind domain, and G is an abelian group of finite special rank. The first step here is the study of artinian modules over periodic groups. B. Hartley and D. McDougall [110] considered artinian modules over a ring having the form $\mathbb{Z}G$, where G is an abelian Chernikov group. Their results could be extended to the case of a ring of the form DG, where G is a periodic abelian group of finite rank, and D is a Dedekind domain. As we shall see in this investigation, the previously obtained results about the existence of complements to submodules play a crucial role here.

We start with some general results.

Lemma 12.1 (L.A. Kurdachenko [150]). *Let R be a commutative ring. Let A be an artinian monolithic R-module with monolith M, and put $L = \mathrm{Ann}_R(M)$. Then L annihilates every factor of any ascending composition series of A.*

Proof. Assume the contrary. Then there exist two submodules $B \geq C$ of A such that B/C is simple, and L annihilates all the factors of any ascending series of C but $\mathrm{Ann}_R(B/C)$ does not include L. Since B/C is simple, there exists some $b \in B$ such that $B/C = (b + C)R$. Put $B_1 = bR$. Then $B_1 \cong_R R/\mathrm{Ann}_R(b)$. Since R is commutative, $\mathrm{Ann}_R(b)$ is a two-sided ideal of R. Hence $R/\mathrm{Ann}_R(b)$ is an artinian ring, and then it is noetherian (see R. Sharp [263, Theorem 8.44]). Therefore B_1 has at least a finite composition series. If $x \in L$, by the Fitting lemma (Proposition 1.5), there exists a decomposition $B_1 = U \oplus V$, where x induces on U an automorphism, and $Vx^n = \langle 1 \rangle$ for some $n \in \mathbb{N}$. It follows that $M \leq V$, and so $U = \langle 1 \rangle$. Hence, $B_1x^n = \langle 1 \rangle$. In particular, $Bx \leq C$, and since this holds for every $x \in L$, $L \leq \mathrm{Ann}_R(B/C)$, a contradiction. $\qquad\square$

Corollary 12.2 ([150]). *Let A be an artinian monolithic RG-module, where R is a commutative ring, and G is an abelian group. Suppose that M is the RG-monolith of A. If $H = C_G(M)$, then A is an RH-hypercentral module.*

Proof. Indeed, if $x \in H$, then $x-1 \in \mathrm{Ann}_{RG}(M)$. By Lemma 12.1, $x-1$ annihilates all the factors of every ascending composition series of A. $\qquad\square$

Let A be an R-module, and pick $y \in R$. We say that A *is a y-module* (over R) if A is an L-module, where $L = Ry$. In other words, A is a y-module if and only if every element of A is annihilated by some element of the form y^t, where $t \in \mathbb{N}$.

A Dedekind domain D is said to be *a Dedekind Z-domain* if D satisfies the following conditions:

(1) Spec (D) is infinite.

(2) If $\langle 0 \rangle \neq P \in$ Spec (D), then the field D/P is locally finite.

Clearly, $\mathbb{Z}$ is a Dedekind Z-domain. Another important example for its applications in the theory of groups is the group algebra of an infinite cyclic group over a finite field.

Theorem 12.3 ([150]). *Let R be a commutative ring, G an abelian group, and A an artinian RG-module. If $S =$Soc $_{RG}(A)$ and $H = C_G(S)$, then*

(1) *A is RH-hypercentral. In particular, A is an $(x-1)$-module, for every $x \in H$.*

(2) *If R is a Dedekind Z-domain, and $r_0(G)$ is finite, then G/H is a periodic group of finite special rank.*

Proof. We have
$$S = M_1 \oplus \cdots \oplus M_n,$$
where $M_1, \ldots, M_n$ are some simple RG-submodules of A. Choose submodules $B_1, \ldots, B_n$ of A such that, for every $1 \leq j \leq n$, every B_j is maximal under

(i) $\bigoplus_{k \neq j} M_k \leq B_j$;

(ii) $M_j \cap B_j = \langle 1 \rangle$.

Then every A/B_j is a monolithic RG-module with monolith $(M_j + B_j)/B_j$. Let $H_j = C_G((M_j + B_j)/B_j)$. By Corollary 12.2, A/B_j is RH_j-hypercentral. Let $x \in B_1 \cap \cdots \cap B_n$. Then $a(x-1) \in B_j$, for every $a \in M_j$. But $a(x-1) \in M_j$, so that $a(x-1) = 0$, and $x \in C_G(M_j)$. Thus $H = H_1 \cap \cdots \cap H_n$.

Put
$$E = A/B_1 \oplus \cdots \oplus A/B_n.$$
Clearly, E is an RH-hypercentral module. The mapping
$$\phi : a \mapsto (a + B_1, \ldots, a + B_n), a \in A$$
is an RG-homomorphism, hence Im ϕ is RH-hypercentral. Since
$$\mathrm{Ker}\ \phi \cap S = (B_1 \cap \cdots \cap B_n) \cap S = \langle 0 \rangle,$$
$A \cong_{RG}$ Im ϕ, and thus A is RH-hypercentral.

If $x \in H$, then the natural semidirect product $A \rtimes \langle a, x \rangle$ is a hypercentral group. It follows that $\langle a, x \rangle$ is a nilpotent subgroup for every $a \in A$, so $a(x-1)^s = 0$, for some $s \in \mathbb{N}$. This means that A is an $(x-1)$-module for every $x \in H$.

To finish, we note that if R is a Dedekind $\mathbb{Z}$-domain, and $r_0(G)$ is finite, then every G/H_j is a locally cyclic periodic group (see L. A. Kurdachenko, J. Otal and I. Ya. Subbotin [157, Theorem 3.2]). By Remak's theorem

$$G/H \hookrightarrow G/H_1 \cap \cdots \cap G/H_n,$$

therefore G/H is a periodic group of finite special rank, as required. $\qquad\square$

Lemma 12.4 ([150]). *Let A be an artinian DG-module, where D is a Dedekind $\mathbb{Z}$-domain, and G is an abelian group of finite 0-rank. Then A is a D-periodic module. Moreover, if $P \in \mathrm{Ass}_D(A)$, and A_P is the P-component of A, then the additive group of A_p is a p-group where $p = \mathrm{char}(D/P)$. In particular, A is also $\mathbb{Z}$-periodic.*

Proof. Since A is artinian, A has an ascending series of DG-submodules

$$\langle 0 \rangle = A_0 \leq A_1 \leq \cdots \leq A_\alpha \leq A_{\alpha+1} \leq \cdots \leq A_\gamma$$

such that every factor $A_{\alpha+1}/A_\alpha$ of this series is a simple DG-module, $\alpha < \gamma$. Acoording to L.A. Kurdachenko, J. Otal and I.Ya. Subbotin [157, Corollary 1.16], $\mathrm{Ann}_D(A_{\alpha+1}/A_\alpha) \in \mathrm{Spec}\,(D)$, for every $\alpha < \gamma$. This means that the additive group of A is D-periodic.

Put $P \in \mathrm{Ass}_D(A)$, and let A_P be the P-component of A. Since D/P is a locally finite field, the additive group of D/P is a p-group, where $p = \mathrm{char}\,(D/P)$. By Corollary 6.17, $P/P^2 \cong_D D/P$. It follows that the order of each element of D/P^2 divides p^2. Similarly, we obtain that the order of each element of D/P^n divides p^n. If $a \in A_P$, then there is some $n \in \mathbb{N}$ such that $\mathrm{Ann}_D(a) = P^n$, and so $aD \cong_D D/P^n$. In particular, the additive order of a divides p^n. Thus A_P is a p-group. Since A_P is a DG-submodule, $\mathrm{Ass}_D(A)$ is finite. By Corollary 6.25,

$$A = \bigoplus_{P \in \pi} A_P,$$

where $\pi = \mathrm{Ass}_D(A)$, then it follows that A is also $\mathbb{Z}$-periodic. $\qquad\square$

Proposition 12.5 ([150]). *Let D be a Dedekind domain, and let G be an abelian group of finite special rank. If $P \in \mathrm{Spec}\,D(A)$, let A be an artinian P-module over DG, and put $H = C_G(\mathrm{Soc}_{DG}(A))$. If H is periodic, then G has a subgroup Q of finite index such that $H \cap Q \leq C_G(A)$.*

Proof. Put $p = \mathrm{char}\,(D/P)$. If $p = 0$, then the additive group of A is torsion-free. By Theorem 12.3, A is a DH-hypercentral module. Put $A_1 = \zeta_{DH}(A)$, and $A_2/A_1 = \zeta_{DH}(A/A_1)$. Then H acts trivially on the factors of the series

$$\langle 0 \rangle = A_0 \leq A_1 \leq A_2.$$

According to O.H. Kegel and B.A.F. Wehrfritz [134, 1.C.3], $H/C_H(A_2)$ is isomorphic to some subgroup of $\mathrm{Hom}_{\mathbb{Z}}(A_2/A_1, A_1)$. Since A_1 is $\mathbb{Z}$-torsion-free, $\mathrm{Hom}_{\mathbb{Z}}(A_2/A_1, A_1)$ is $\mathbb{Z}$-torsion-free. As H is periodic, $H = C_H(A_2)$, and hence $A_2 = A_1$. Then $A = A_1$, and so $H = C_H(A)$.

Suppose now that $p > 0$. By Lemma 12.4, the additive group of A is a p-group. For every $n \in \mathbb{N}$, put

$$A_n = \Omega_{p,n}(A) = \{a \in A \mid p^n a = 0\}.$$

We note that $H = U \times V \times W$, where U is a finite p-subgroup, V is a divisible p-subgroup, and W is a p'-subgroup. Put $B_1 = \zeta_{DVW}(A_n)$, and $B_2/B_1 = \zeta_{DVW}(A/A_n)$. Then VW acts trivially on the factors of the series

$$\langle 0 \rangle = B_0 \leq B_1 \leq B_2,$$

and then $VW/C_{VW}(B_2)$ can be embedded in $\mathrm{Hom}_{\mathbb{Z}}(B_2/B_1, B_1)$. Since the order of any element of B_2 divides p^n, the order of an element of $\mathrm{Hom}_{\mathbb{Z}}(B_2/B_1, B_1)$ divides p^n. But VW has no non-identity bounded p-factor-groups, and so $C_{VW}(B_2) = VW$, that is $B_2 = B_1$, and $VW \leq C_G(A_n)$. Since this holds for any $n \in \mathbb{N}$, $VW \leq C_G(A)$. Put

$$\mathrm{Soc}_{DG}(A) = M_1 \oplus \cdots \oplus M_s,$$

where $M_1, \ldots, M_s$ are simple DG-submodules. The periodic part of every factor $G/C_G(M_j)$ is a locally cyclic p'-subgroup (see, for example, L.A. Kurdachenko, J. Otal and I.Ya. Subbotin [157, Corollary 2.2]). Since

$$H = C_G(M_1) \cap \cdots \cap C_G(M_s),$$

by Remak's theorem

$$G/H \hookrightarrow G/C_G(M1) \times \cdots \times G/C_G(M_s).$$

In particular, G/H is a p'-group. This means that H/VW is the finite Sylow p-subgroup of G/VW. Therefore, there exists some subgroup Q such that $G/VW = H/VW \times Q/VW$. Hence, the index $|G : Q|$ is finite, and $H \cap Q = VW \leq C_G(A)$. $\qquad \square$

If G is a group, and H is a normal subgroup of finite index of G, then an artinian DH-module A is artinian as a DG-module. Conversely, an artinian DG-module A is likewise an artinian DH-module by Theorem 5.2. This shows that the study of artinian modules over a group ring can be carried out in subgroups of finite index of the group in consideration. In particular, if G is a periodic abelian group of finite rank, by Proposition 12.5, we may assume that $C_G(\mathrm{Soc}_{DG}(A)) = C_G(A)$, and in this case, $G/C_G(A)$ is a p'-group, where $p = \mathrm{char}\,(D/P)$. We will focus now on this case. The results we state below are slight variations on some results due to B. Hartley and D. McDougall [110]. Before dealing with them, we need to establish the following property of the injective envelopes.

Lemma 12.6 (B. Hartley and D. McDougall [110]). *Let G be a periodic central-by-finite group, D a Dedekind domain, and $P \in \mathrm{Spec}(D)$. Suppose that A is an artinian P-module over DG, and let E be the DG-injective envelope of A. If $\mathrm{char}(D/P) = p > 0$, suppose further that $p \notin \Pi(G)$. Then*

(1) *E is a P-module, and $\Omega_{P,1}(E) = \Omega_{P,1}(A)$.*

(2) *A is DG-injective if and only if A is D-divisible.*

Proof. By Corollary 7.30, $\Omega_{P,1}(A)$ is a direct sum of finitely many simple DG-submodules. Since E is a maximal essential extension of A, E is also a maximal essential extension of $\Omega_{P,1}(A)$. In particular, $\mathrm{Ass}_D(E) = \{P\}$. According to D.W. Sharpe and P. Vamos [265, Proposition 2.23], E is a direct sum of injective envelopes of direct summands of $\Omega_{P,1}(A)$, and so we may assume that $\Omega_{P,1}(A)$ is a simple DG-module. If $\Omega_{P,1}(A) \neq \Omega_{P,1}(E)$, by Corollary 7.30, there exists a DG-submodule C such that $\Omega_{P,1}(E) = \Omega_{P,1}(A) \oplus C$, a contradiction. Thus $\Omega_{P,1}(E) = \Omega_{P,1}(A)$.

Let T be the D-periodic part of E. Then $\mathrm{Ass}_D(T) = \{P\}$. By Proposition 7.7, E is D-divisible, and so is T by Lemma 7.18. Moreover, T has a D-complement in E. Thus T is a DG-monolithic submodule with monolith $\Omega_{P,1}(A)$. By Theorem 7.28, T has a DG-complement in E, and because E is an essential extension of A, we have a contradiction. Hence, $E = T$, and, in particular, E is a P-module.

If A is injective, by Proposition 7.7, A is D-divisible. Conversely, if A is D-divisible, and E is a DG-injective envelope of A, then E is likewise D-divisible. Since $\Omega_{P,1}(A) = \Omega_{P,1}(E)$, we have $E = A$, and all the assertions of the result have been proven. $\square$

Lemma 12.7. *Let D be a Dedekind domain, and let $P \in \mathrm{Spec}(D)$. Suppose that A is a P-module over D such that $\Omega_{P,n}(A) = A$ for some $n \in \mathbb{N}$. If B is a D-submodule of A such that*

$$A/B = \bigoplus_{\lambda \in \Lambda}(a_\lambda + B)D,$$

where $\mathrm{Ann}_D(a_\lambda + B) = P^n$ for every $\lambda \in \Lambda$, then there exists some D-submodule C such that $A = B \oplus C$.

Proof. Pick $y \in P \setminus P^2$. By Proposition 6.13, $P = Dy + P^n$. Put

$$C = \bigoplus_{\lambda \in \Lambda} a_\lambda D$$

so that $A = B + C$. If $b \in B \cap C$, then there are a finite subset $\Lambda_1 \subseteq \Lambda$, and some elements $x_\lambda \in D$, $\lambda \in \Lambda_1$ such that

$$b = \sum_{\lambda \in \Lambda_1} a_\lambda y^{k_\lambda} x_\lambda.$$

Thus

$$\sum_{\lambda \in \Lambda_1} (a_\lambda + B)y^{k_\lambda} x_\lambda = B$$

and it follows that $k_\lambda = n$ for every $\lambda \in \Lambda_1$. But in this case $b = 0$. Hence, $B \cap C = \langle 0 \rangle$; that is, $A = B \oplus C$, as required.   □

Let G be a periodic abelian group, D a Dedekind domain, $P \in \mathrm{Spec}\,(D)$, and $F = D/P$. We recall that the simple FG-modules over periodic abelian group G were already described by B. Hartley and D. McDougall (see, for example, [157, Theorem 2.3]). Suppose that $\{A_\lambda \mid \lambda \in \Lambda\}$ is a complete set of representatives of the isomorphism types of simple FG-modules. We consider each A_λ as a DG-module, and define E_λ to be the DG-injective envelope of A_λ. We write

$$\Omega_{P,\infty}(E_\lambda) = \bigcup_{n \in \mathbb{N}} \Omega_{P,n}(E_\lambda).$$

Up to an isomorphism, the submodule $\Omega_{P,n}(E_\lambda)$ is determined by λ and n. If char $F = p > 0$, by Lemma 12.6, $A_\lambda = \Omega_{P,1}(E_\lambda)$, and $E_\lambda = \Omega_{P,\infty}(E_\lambda)$. Therefore

$$\Omega_{P,n+1}(E_\lambda)/\Omega_{P,n}(E_\lambda) \cong \Omega_{P,1}(E_\lambda) = A_\lambda$$

is a simple DG-module. It follows that the $\Omega_{P,n}(E_\lambda)$, where $n \in \mathbb{N}$ or $n = \infty$ are the only submodules of E_λ.

Theorem 12.8 ([110]). *Let G be a periodic abelian group of finite special rank, D a Dedekind domain, $P \in \mathrm{Spec}(D)$, A a P-module over DG, and if $\mathrm{char}(D/P) = p > 0$, then $p \notin \Pi(G)$. Then*

(1) *A module A is artinian if, and only if*

$$A = A_1 \oplus \cdots \oplus A_n,$$

   where each direct summand is isomorphic to certain $\Omega_{P,r}(E_\lambda)$, for suitable $\lambda \in \Lambda$, and $r \in \mathbb{N}$ or $r = \infty$.

(2) *If*

$$A = B_1 \oplus \cdots \oplus B_k$$

   is another direct decomposition of A of such kind, then $n = k$, and there are an automorphism α of the module A, and a permutation $\sigma \in S_n$ such that $B_j\alpha = A_{j\sigma}$, for every $1 \leq j \leq n$.

Proof. By Proposition 1.8, every DG-module $\Omega_{P,n}(E_\lambda)$ is artinian, for $n \in \mathbb{N}$ or $n = \infty$, and a direct sum of finitely many of such modules is artinian by Lemma 1.2.

Conversely, suppose that A is an artinian P-module over DG. Suppose that A cannot be expressed as a direct sum. Then among the submodules of A that cannot be expressed in such a way, we choose a minimal one. Replacing A by this, we may assume that every proper submodule of A admits such a decomposition but A does not. In particular, A is indecomposable.

Let V be a maximal D-divisible submodule of A. Clearly, V is a DG-submodule. By Lemma 12.6, V is DG-injective. Then V has a DG-complement in A, and then $V = A$. By Corollary 7.30, $\Omega_{P,1}(E_\lambda)$ is a direct sum of finitely many simple DG-submodules. According to D.W. Sharpe, and P. Vamos [268, Proposition 2.23], A is a direct sum of the injective envelopes of these simple submodules. This contradiction shows that $V = \langle 1 \rangle$. Pick $y \in P \setminus P^2$. By [157, Proposition 5.2], $A \neq Ay$. Furthermore, by [157, Proposition 5.3] yields that $AP^s = Ay^s$, for every $s \in \mathbb{N}$. Note that the chain

$$A \geq Ay \geq Ay^2 \geq \cdots \geq Ay^t \geq \cdots$$

breaks after finitely many steps. In other words, there is some $m \in \mathbb{N}$ such that $Ay^m = Ay^{m+1}$. According to L.A. Kurdachenko, J. Otal and I.Ya. Subbotin [157, Proposition 5.2], Ay^m is D-divisible. It follows that $Ay^m = \langle 1 \rangle$. We may assume that $Ay^{m-1} \neq \langle 1 \rangle$. In particular, Ay^{m-1} has a non-zero simple DG-submodule U. There is an isomorphism of U into some A_λ, and this may be extended to a homomorphism ϕ of A into the injective module $\Omega_{P,\infty}(E_\lambda)$. We have

$$(A\phi)y^{m-1} = (Ay^{m-1})\phi \geq U\phi = A_\lambda,$$

so that

$$\Omega_{P,n}(A\phi) = A\phi, \text{ and } \Omega_{P,n-1}(A\phi) \neq A\phi.$$

It follows that $A\phi = \Omega_{P,m}(E_\lambda)$. Put $W = \mathrm{Ker} \ \phi$ so that

$$A/W \cong \mathrm{Im} \ \phi = \Omega_{P,m}(E_\lambda)$$

is a direct sum of isomorphism copies D/P^m. By Lemma 12.7, W has a D-complement. Since $W \neq A$, W is a direct sum of finitely many monolithic submodules of types $\Omega_{P,1}(E_\lambda)$. By Corollary 7.29, W has a DG-complement, a contradiction. Assertion (2) follows from Theorem 1.7. $\qquad\square$

Suppose now that D is a Dedekind Z-domain, G is a periodic abelian group of finite special rank, and A is an artinian DG-module. By Lemma 12.4, A is D-periodic. By Corollary 6.25

$$A = \bigoplus_{P \in \pi} A_P,$$

where A_P is the P-component of A, and $\pi = \mathrm{Ass}_D(A)$ is finite. Thus the general case reduces to the case of a P-module.

If $\mathrm{char}(D/P) = p > 0$, put

$$\mathrm{Soc}_{DG}(A) = M_1 \oplus \cdots \oplus M_n,$$

where $M_1, \ldots, M_n$ are simple,, and define $H = C_G(\mathrm{Soc}_{DG}(A))$. Then

$$H = C_G(M_1) \cap \cdots \cap C_G(M_n)$$

and every $G/C_G(M_j)$ is a locally cyclic p'-group (see, for example, L.A. Kur-dachenko, J. Otal and I.Ya. Subbotin [157, Theorem 2.3]). Suppose that $C_G(A) = \langle 1 \rangle$. By Proposition 12.5, the Sylow p-subgroup of G is finite or, that is to say, the Sylow p'-subgroup Q of G has finite index. By Theorem 5.2, A is an artinian DQ-module. In other words, we may reduce this case to the case in which A is an artinian P-module over DG, where $\mathrm{char}(D/P) \notin \Pi(G)$. Note that Theorem 12.8 gives a description of such modules.

Chapter 13

Nearly injective modules

In this chapter we continue our study of complementability of modules over group rings. Starting from modules over rings with the form FG, where F is a field, we move toward modules over rings with the form RG, where R is a certain commutative ring. In particular, we move toward modules over rings with the form $\mathbb{Z}G$. If A is a simple $\mathbb{Z}G$-module, in many cases we have $pA = \langle 0 \rangle$ for some prime p (the reader can find such cases in L.A. Kurdachenko, J. Otal and I.Ya. Subbotin [157, Chapter 1]). Let E be a $\mathbb{Z}G$-injective envelope of A. Inside E we consider the $\mathbb{Z}$-injective envelope U of A. In connection with the results of the previous chapter, the following question naturally arises: *In which cases is E equal to U?* This question was already considered by B. Hartley in [107]. This chapter is devoted to Hartley's results.

Let R be a commutative ring whose underlying additive group is torsion-free. Suppose that p is a prime, and A is an R-module such that $pA = \langle 0 \rangle$. Since $\mathbb{Z} \cdot 1 \cong \mathbb{Z}$, the elements of $\mathbb{Z} \cdot 1$ are not zero divisors in R; so the R–injective envelope E of A is $\mathbb{Z}$-divisible. Let T be the $\mathbb{Z}$-periodic part of E, which is clearly a divisible p-subgroup. Actually, T is an R-submodule of E. Consider $\Omega_1(T)$ as an R/pR-module. Of course, this submodule is an essential extension of A. Let U and V be (R/pR)-modules such that $U \leq V$, and let $\phi : U \longrightarrow \Omega_1(T)$ be an (R/pR)-homomorphism. Think of U and V as R-modules. Since E is R-injective, ϕ can be extended to an R-homomorphism $\phi_1 : V \longrightarrow E$. Since $p(\operatorname{Im} \phi_1) = \langle 0 \rangle$, $\operatorname{Im} \phi_1 \leq \Omega_1(T)$. Thus, ϕ_1 is in fact an (R/pR)-homomorphism. This shows that $\Omega_1(T)$ is (R/pR)-injective, and hence $\Omega_1(T)$ contains an (R/pR)-injective envelope B of A. By Proposition 7.2, there exists some (R/pR)-submodule C such that $\Omega_1(T) = B \oplus C$. But E is an essential extension of A, so that $C = \langle 0 \rangle$. Therefore $\Omega_1(T)$ becomes an (R/pR)-injective envelope of A. Consequently,

$$\Omega_1(T) = A \ \text{ if and only if } \ A \text{ is } (R/pR)\text{-injective.}$$

Let R be a commutative ring whose underlying additive group is torsion-free. Suppose that p is a prime, and A is an R-module such that $pA = \langle 0 \rangle$. Following B.

Hartley, the R-module A is said to be *nearly R-injective* if the injective envelope $\mathrm{IE}_R(A)$ of A is a p-group, and $\Omega_1(\mathrm{IE}_R(A)) = A$. Roughly speaking, this definition means that A is nearly injective if its R-injective envelope is as small as possible. In fact, this envelope is a $\mathbb{Z}$-essential extension of A.

Lemma 13.1. *Let R be a commutative ring whose underlying additive group is torsion-free, and suppose that p is a prime, and A is an R-module such that $pA = \langle 0 \rangle$. Suppose further that we have a decomposition*

$$A = A_1 \oplus \cdots \oplus A_n$$

of A as a direct sum of some R-submodules $A_1, \ldots, A_n$. Then A is nearly injective if and only if $A_1, \ldots, A_n$ are nearly injective.

Proof. Indeed, according to D.W. Sharpe and P. Vamos [265, Proposition 2.23] the injective envelope of a finite direct sum of modules is isomorphic to the direct sum of the injective envelopes of the summands. $\qquad\square$

Suppose now that A is a simple nearly injective R-module such that $pA = \langle 0 \rangle$, for a certain prime p. Put $E = \mathrm{IE}_R(A)$. If $j \in \mathbb{N}$, then the mapping $\lambda_j : a \mapsto p^j a$, $a \in \Omega_{j+1}(E)$ is an R-homomorphism such that $\mathrm{Im}\,\lambda_j = \Omega_1(E)$, and $\mathrm{Ker}\,\lambda_j = \Omega_j(E)$. Then $\Omega_{j+1}(E)/\Omega_j(E) \cong \Omega_1(E)$, and, in particular, $\Omega_{j+1}(E)/\Omega_j(E)$ is a simple R-module. Therefore the family of all R-submodules of E is exactly

$$\{\langle 0 \rangle, \Omega_1(E), \Omega_2(E), \ldots, \Omega_j(E), \ldots\}$$

and $E = \bigcup_{j \in \mathbb{N}} \Omega_j(E)$.

We set up the framework in which we are interested. Let p be a prime, and let $A_1, \ldots, A_n$ a finite set of simple nearly injective R-modules such that $pA_1 = \cdots = pA_n = \langle 0 \rangle$. If $B_1 \leq \mathrm{IE}_R(A_1), \ldots, B_n \leq \mathrm{IE}_R(A_n)$ are (not necessarily proper) R-submodules, then an R-submodule B having the form

$$B = B_1 \oplus \cdots \oplus B_n$$

is said to be an $\mathcal{H}$-*module*. A useful property of such modules is given by the next result.

Lemma 13.2 (B. Hartley [107]). *Let R be a commutative ring whose underlying additive group is torsion-free, and let U be an R-module. Suppose that B is an R-submodule of U such that there is a subgroup C of U satisfying $U = B \oplus C$. If B is an $\mathcal{H}$-submodule, then there is an R-submodule D of U such that $U = B \oplus D$.*

Proof. As indicated above, we suppose

$$B = B_1 \oplus \cdots \oplus B_n,$$

where B_j is an R-submodule of $\mathrm{IE}_R(A_j)$, and A_j is a simple nearly injective R-module such that $pA_j = \langle 0 \rangle$ (p a prime). We proceed by induction on n.

Suppose that $n = 1$. Then either there exists some $j \in \mathbb{N}$ such that $B = \Omega_j(E)$ or $B = E$, where $E = \text{IE}_R(A)$, and A is a simple nearly injective R-module. If $B = E$, then B is injective, and so B is a direct summand of U (see Proposition 7.2). Suppose that $B = \Omega_j(E)$, for some $j \in \mathbb{N}$. Since $U = B \oplus C$, $p^j U \cap B = \langle 0 \rangle$. Let D be an R-submodule of U including $p^j U$, and maximal under $B \cap D = \langle 0 \rangle$. If $\nu : U \longrightarrow U/D$ is the canonical mapping, then $U\nu$ is an essential extension of $B\nu$, and so of $A\nu$. Thus $A\nu = \Omega_1(U\nu)$. Since $p^j(U\nu) = \langle 0 \rangle$, $B\nu$ is a direct sum of cyclic groups of order p^j, and therefore $B\nu$ is additively a direct summand of $U\nu$ (see I. Kaplansky [124, Theorem 7]). Since $\Omega_1(U\nu) = \Omega_1(B\nu)$, $U_\nu = B_\nu$. In other words, $U = B \oplus D$, as required.

If $n > 1$, a straightforward induction yields the general case. For

$$U/B_1 = B/B_1 \oplus (C + B_1)/B_1$$

and so, by induction, we may assume that $U = B + V$, for some R-submodule V such that $V \cap B = B_1$. Since

$$U = B_1 \oplus \cdots \oplus B_n \oplus C = B_1 \oplus W,$$

where

$$W = B_2 \oplus \cdots \oplus B_n \oplus C,$$

we have $V = B_1 \oplus (W \cap V)$. Applying the case $n = 1$, there exists an R-submodule D of V such that $V = B_1 \oplus D$, and it readily follows that $A = B \oplus D$. $\qquad \square$

An abstract description of $\mathcal{H}$-modules is given in the next result.

Theorem 13.3 ([107]). *Let R be a commutative ring whose additive group is torsion-free, and let A be an R-module. Suppose that*

$$A = A_1 \oplus \cdots \oplus A_n,$$

where $A_1, \ldots, A_n$ are simple nearly injective R-submodules. If S is an essential extension of A, then:

(1) *S is an artinian R-module;*

(2) *S is an $\mathcal{H}$-module with indecomposable direct summands;*

(3) *if $S = B_1 \oplus \cdots \oplus B_n = C_1 \oplus \cdots \oplus C_r$ are two direct decompositions of S, and all the submodules $B_1, \ldots, B_n$, $C_1, \ldots, C_r$ are indecomposable, then $n = r$ and there exist a permutation $\sigma \in S_n$ and an automorphism ϕ of A such that $C_j \phi = B_{j\sigma}$ for every $1 \le j \le n$.*

Proof. (1) By the definition of an injective envelope, the identity map on A can be extended to an embedding of S in $E = \text{IE}_R(A)$. By Lemma 13.1, A is nearly injective, and we have $\Omega_1(E) = \Omega_1(A) = \Omega_1(S)$. By Lemma 1.2, $\Omega_1(A)$ satisfies the minimal condition on R-submodules, and, by Corollary 1.9, so does S.

(2) Suppose the contrary. By (1), S has an R-submodule X, every proper submodule of which is an $\mathcal{H}$-module. Then X is an essential extension of $X \cap A$, which has the same form of A. Therefore, we may replace S by X and assume that every proper submodule of S is an $\mathcal{H}$-module. In this case, S is clearly indecomposable. Let D be a maximal divisible subgroup of S, and suppose that $D \neq \langle 0 \rangle$. Actually, D is an R-submodule, and we claim $D = S$. Otherwise, by Theorem 7.16, there exists some subgroup L such that $S = D \oplus L$, and then by Lemma 13.2, D has an R-complement, which is impossible. Thus $D = S$ as claimed, and so S itself is divisible. Then it follows that the identity map on A can be extended to an embedding

$$S \longrightarrow \mathrm{IE}_R(A) = \mathrm{IE}_R(A_1) \oplus \cdots \oplus \mathrm{IE}_R(A_n),$$

and, since the image is divisible, and contains A, this embedding is in fact an isomorphism. Thus, the assumption $D \neq \langle 0 \rangle$ produces a contradiction, and so S has no non-zero divisible subgroups. Since S is artinian, the descending chain

$$S \geq pS \geq \cdots \geq p^k S \geq \cdots$$

has to break after finitely many steps, and, since the lower term of it is divisible, it has to be zero. Therefore, there exists some $m > 0$ such that $p^m S = \langle 0 \rangle$. We choose m as small as possible. Actually $m \geq 1$, and so $p^{m-1} S \neq \langle 0 \rangle$. By (1), $p^{m-1} S$ has a minimal R-submodule M, and by the definition of injectivity, the identity mapping on M extends to an R-homomorphism $\phi : S \longrightarrow \mathrm{IE}_R(M)$. Since $p^{m-1}(S\phi) = (p^{m-1}S)\phi \neq \langle 0 \rangle$, $S\phi = \Omega_m(\mathrm{IE}_R(M))$. Since $\mathrm{Ker}\,\phi \neq S$, $\mathrm{Ker}\,\phi$ is an $\mathcal{H}$-module. Furthermore, $p^m S = \langle 0 \rangle$, and $S/\mathrm{Ker}\,\phi$ is a direct sum of cyclic groups of order p^m precisely. Therefore $\mathrm{Ker}\,\phi$ is additively a direct summand of S (see I. Kaplansky [124, Theorem 7]). By Lemma 13.2, $\mathrm{Ker}\,\phi$ has an R-complement. It follows that $\mathrm{Ker}\,\phi = \langle 0 \rangle$, and so $S \cong_R \Omega_m(\mathrm{IE}_R(M))$, a final contradiction.

(3) Since S is artinian, the submodules $B_1, \ldots, B_n, C_1, \ldots, C_r$ are monolithic, and it suffices to apply Theorem 1.7. $\square$

Theorem 13.3 and Lemma 13.2 together give the next result.

Theorem 13.4 ([107]). *Let R be a commutative ring whose additive group is torsion-free, and suppose that*

$$A = A_1 \oplus \cdots \oplus A_n$$

is a nearly injective R-module such that $pA = \langle 0 \rangle$ for some prime p, and $A_1, \ldots, A_n$ are simple. Let U be an R-module including an R-submodule S which is an essential extension of A. If $U = S \oplus C$ for some subgroup C, then $U = S \oplus D$ for some R-submodule D.

Suppose that G is a locally finite p'-group, and let A be a simple $\mathbb{F}_p G$-module, which we view as a $\mathbb{Z}G$-module in the obvious way. We now consider the following question: *when is A nearly injective?*

Lemma 13.5 ([107]). *Suppose that G is a finite group of order n, and let A be a $\mathbb{Z}G$-module whose additive group is divisible, and has no elements of order dividing n. Then A is $\mathbb{Z}G$-injective.*

Proof. Denote by E the $\mathbb{Z}G$-injective envelope of A. By Theorem 7.16, there exists some subgroup L such that $E = A \oplus L$. By Corollary 5.11, there exists a $\mathbb{Z}G$-submodule U such that $E = A \oplus U$. Since E is an essential extension of A, $E = A$, and, in particular, A is $\mathbb{Z}G$-injective. $\qquad\square$

Theorem 13.6 ([107]). *Let G be a locally finite p'-group, and let A be a simple $\mathbb{F}_p G$-module. If $G/C_G(A)$ is abelian-by-finite, then A is nearly $\mathbb{Z}G$-injective.*

Proof. Put $E = \mathrm{IE}_{\mathbb{Z}G}(A)$, and let T be the $\mathbb{Z}$-periodic part of E. By Lemma 9.5 and Lemma 9.4, A is $\mathbb{F}_p G$-injective; the arguments given at the beginning of this section yield that $\Omega_1(T) = A$. We are going to show that $T = E$ by proving that T itself is injective. Let ν be a $\mathbb{Z}G$-endomorphism of A so that ν can be extended to E since the latter is injective. Since T is a fully invariant subgroup of E, this extension determines an endomorphism of T. Thus, if $D = \mathrm{End}_{\mathbb{Z}G}(A)$, and $Y = \mathrm{End}_{\mathbb{Z}G}(T)$, the above restriction determines an epimorphism from Y to D. By Lemma 9.5, A is finite dimensional over D. In particular, A satisfies the minimal condition on Y-submodules. By Theorem 1.7 so does T.

Let ϕ be an arbitrary $\mathbb{Z}G$-homomorphism from a right ideal L of $\mathbb{Z}G$ into T. It will suffice to show that ϕ can be extended to the whole of $\mathbb{Z}G$. In this case, by Baer's criterion (Theorem 7.1), T is $\mathbb{Z}G$-injective; consequently, there must exist an element $v \in T$ such that $x\phi = vx$ for all $x \in L$. Suppose no such v exists. Given a finite subgroup H of G, we put

$$S(H) = \{v \in T \mid x\phi = vx \text{ for all } x \in L \cap \mathbb{Z}H\}.$$

By Lemma 13.5, the restriction of ϕ on $L \cap \mathbb{Z}H$ can be extended to $\mathbb{Z}H$, and so $\mathfrak{S}(H) \neq \emptyset$. Let $\mathfrak{M}$ be the family of all finite subgroups of G. Then our assumption implies that

$$\bigcap_{H \in \mathfrak{M}} S(H) = \emptyset.$$

Clearly, the annihilator

$$\mathrm{Ann}_T(L \cap \mathbb{Z}H) = \{x \in T \mid x(L \cap \mathbb{Z}H) = 0\}$$

is a Y-submodule of T; since T satisfies Min-Y, there exists a finite subgroup H of G such that $\mathrm{Ann}_T(L \cap \mathbb{Z}H)$ is minimal among all these annihilators. Notice that $S(H) = v + \mathrm{Ann}_T(L \cap \mathbb{Z}H)$, for every $v \in \mathfrak{S}(H)$. Since

$$\bigcap_{H \in \mathfrak{M}} S(H) = \emptyset,$$

there exists a finite subgroup K of H such that $S(K) \cap S(H) \subseteq S(H)$. Let $F = \langle K, H \rangle$. Then F is finite, and

$$\emptyset \neq S(F) \subseteq S(K) \cap S(H) \subset S(H).$$

Let $v \in S(F)$. Then $v + \mathrm{Ann}_T(L \cap \mathbb{Z}F) \subset v + \mathrm{Ann}_T(L \cap \mathbb{Z}H)$, whence $\mathrm{Ann}_T(L \cap \mathbb{Z}F) < \mathrm{Ann}_T(L \cap \mathbb{Z}H)$, a contradiction. $\qquad\square$

Using these arguments, Theorem 13.4, and the straightforward fact that a direct sum of finitely many nearly injective modules annihilated by the same prime is nearly injective (see Lemma 13.1), we obtain

Theorem 13.7 ([107])**.** *Let G be a periodic abelian-by-finite p'-group, and let U be a $\mathbb{Z}G$-module that includes a submodule S. Suppose that S is an essential extension of a submodule A, which is a direct sum of finitely many simple submodules, and satisfies $pA = \langle 0 \rangle$. If $U = S \oplus C$ for some subgroup C, then there exists a $\mathbb{Z}G$-submodule D such that $U = S \oplus D$.*

For the case of a ring $\mathbb{Z}$, this theorem is a generalization of Corollary 7.29.

On the other hand, it seems natural to ask whether other restrictions on the size of the $\mathbb{Z}G$-injective envelope E of a simple $\mathbb{F}_pG$-module A might be obtained for suitable groups G. For example, when can we be sure that E is $\mathbb{Z}$-periodic? This seems to be a rather difficult question. In the paper [107], B. Hartley constructed an example of a 2-group G whose commutator subgroup is finite, and a simple $\mathbb{F}_pG$-module A, where p is an odd prime, such that $\mathrm{IE}_{\mathbb{Z}G}(A)$ is not $\mathbb{Z}$-periodic. This construction is far from the scope of this book, and therefore we will not give more details here about it.

Our previous results, and other results that we do not mention here, show that the theory of artinian modules over periodic abelian-by-finite groups is very fruitful. The examples of uncountable artinian groups constructed by B. Hartley [105] show that it is unlikely to expect significant progress in the theory of artinian DG-modules over such groups.

Chapter 14

Artinian modules over abelian groups of finite section rank

The description of artinian modules over non-periodic groups is quite different from that of the periodic case. If $|G/G^p| \geq p^2$, then regarding representation theory, the problem of description of A is "wild" (S.A. Kruglyak [140]). We can enlighten this rough idea with the following analogy. The problem of the description of abelian torsion-free groups of finite rank is also "wild" (A.V. Yakovlev [288]), but the $\mathbb{Z}$-injective envelopes of these modules are known to be direct sums of copies of the additive group of the rational numbers. Therefore, the following two questions arise: to describe the injective envelopes of artinian modules, and to describe artinian modules over a group G such that $|G/G^p| \leq p$. Both problems were solved in the paper by L.A. Kurdachenko [150]. The results of this article formed the base for this chapter. The paper [150] considered the artinian modules over abelian groups of finite special rank. Here we will consider a generalized situation.

Let R be a ring, and suppose that A is an artinian R-module. Then

$$\mathrm{Soc}_R(A) = M_1 \oplus \cdots \oplus M_n,$$

where $M_1, \ldots, M_n$ are minimal R-submodules of A. If $1 \leq j \leq n$, let U_j be an R-submodule of A maximal under

$$M_j \cap U_j = \langle 1 \rangle, \text{ and } M_1 \oplus \cdots \oplus M_{j-1} \oplus M_{j+1} \oplus M_n \leq U_j.$$

Then A/U_j is a monolithic R-module with R-monolith $(M_j + U_j)/U_j$. Put $U = U_1 \cap \cdots \cap U_n$. Then $U \cap \mathrm{Soc}_R(A) = \langle 0 \rangle$, and so $U = \langle 0 \rangle$. By Remak's theorem,

$$A \hookrightarrow A/U_1 \oplus \cdots \oplus A/U_n$$

and, by Lemma 1.2, the latter is an artinian R-module. This shows that the case of monolithic artinian R-modules is fundamental for our interests.

Let D be a Dedekind $\mathbb{Z}$-domain, and let G be an abelian group of finite 0-rank. Suppose that A is a monolithic artinian DG-module with DG-monolith

M. Then $\mathrm{Ann}_D(M) = P \in \mathrm{Spec}(D)$ (see L.A. Kurdachenko, J. Otal and I.Ya. Subbotin [157, Theorem 1.15]), and D/P is a locally finite field. Put $p = \mathrm{char}\, D/P$, and $H = C_G(M)$ so that G/H is a periodic locally cyclic p'-group (see L.A. Kurdachenko, J. Otal and I.Ya. Subbotin [157, Theorem 2.3]). Suppose that G has finite p-rank, and choose a finitely generated torsion-free subgroup L of H such that G/L is periodic. Let S/L be the Sylow p-subgroup of G/L. Thus S/L is a Chernikov subgroup, and, moreover, $S \leq H$. Let U/L be the Sylow p'-subgroup of G/L. Then $H \leq U$, and

$$G/L = U/L \times V/L \times W/L,$$

where V/L is a divisible Chernikov p-subgroup, and W/L is a finite p-subgroup. Put $S = UV$. As a consequence of the Theorem of Wilson (Chapter 5), the study of artinian modules over group rings can be reduced to subgroups of finite index, so that A is an artinian DS-module. In other words, we can further assume that $H = C_G(M)$ has a finitely generated torsion-free subgroup L such that H/L has no subgroups of index p.

Further reductions can be derived from the following result.

Lemma 14.1 (L.A. Kurdachenko [144]). *Let G be a hypercentral group, and let H be a normal subgroup of G such that G/H has no subgroups of index p, (p a prime). Let T be a G-invariant p-subgroup of H. Then every H-invariant subgroup of T is G-invariant.*

Proof. Let L be an H-invariant subgroup of T. We may assume that $L \neq \langle 1 \rangle$. Since H is hypercentral, $L \cap \zeta(H) \neq \langle 1 \rangle$. Note that $L \cap \zeta(H) \leq T \cap \zeta(H) = U$. Clearly, U is a G-invariant subgroup, and $C_G(U) \geq H$. In particular, $G/C_G(U)$ contains no subgroups of index p. By Lemma 3.10, $U \leq \zeta(G)$, and, in particular, $L \cap \zeta(H) \leq \zeta(G)$. It suffices to apply transfinite induction and similar arguments to obtain that L is G-invariant, as required. $\qquad\square$

Lemma 14.2. *Let D be a Dedekind Z-domain, and let G be an abelian group of finite section rank. Let A be a monolithic artinian DG-module with monolith M. Suppose that $H = C_G(M)$ includes a finitely generated torsion-free subgroup L such that H/L contains no subgroups of index p, where $p = \mathrm{char}(D/\mathrm{Ann}_D(M))$. Then G has a subgroup $K \geq L$ such that K/L is a periodic p'-group of special rank 1, and every DK-submodule of A is likewise a DG-submodule. In particular, the DK-module A is artinian.*

Proof. Note that G/H is a periodic p'-group of special rank 1 (see L.A. Kurdachenko, J. Otal and I.Ya. Subbotin [157, Theorem 2.3]). Therefore, we have a decomposition

$$G/L = S/L \times K/L \times Q/L,$$

where S/L is the Sylow p-subgroup of G/L, $Q \leq H$, K/L has finite special rank 1, $\Pi(K/L) = \Pi(G/H)$, and $KH = G$. Let $P = \mathrm{Ann}_D(M)$. By Lemma 12.4, A is a

D-periodic module, and therefore a P-module. Also, the additive group of A is a p-group. Since H/L contains no subgroups of index p, S/L is a divisible Chernikov group. By Corollary 12.2, the DH-module A is DH-hypercentral. In particular, the natural semidirect product $A \rtimes H$ is a hypercentral group. Let B be a DK-submodule of A. Since SQ/L contains no subgroups of index p, every L-invariant subgroup of A is SQ-invariant. In particular, B is a $D(SQ)$-submodule of A. Since $KSQ = G$, B is a DG-submodule. □

This allows us to assume that G has a finitely generated torsion-free subgroup $L \leq C_G(M)$ such that G/L is a p'-group of finite special rank. In particular, we may reduce the study to the case in which G has finite special rank.

Lemma 14.3 (L.A. Kurdachenko [150]). *Let D be a Dedekind $\mathbb{Z}$-domain, and let G be an abelian group of finite section rank. If A is an artinian DG-module, B a finitely generated DG-submodule of A, and $H = C_G(Soc_{DG}(A))$, then there exists some $n \in \mathbb{N}$ such that the nth term of the DH-central series of A contains B. In particular, B is DH-nilpotent.*

Proof. By Lemma 12.4, the additive group of A is periodic. Since A is artinian, $\mathrm{Ass}_{\mathbb{Z}}(A) = \Pi(A)$ is finite. There exists a finitely generated subgroup X of H such that H/X is periodic, and p-divisible for every $p \in \Pi(A)$. Suppose that $b_1, \ldots, b_k \in B$ satisfy

$$B = b_1 DG + \cdots + b_k DG.$$

Put

$$B_1 = b_1 \mathbb{Z} + \cdots + b_k \mathbb{Z} = \langle b_1, \ldots, b_k \rangle.$$

By Theorem 12.3, A is DH-hypercentral, and so is B_1. In particular, B_1 is $\mathbb{Z}X$-hypercentral. Then the natural semidirect product $B_1 \rtimes X$ is a finitely generated hypercentral group; hence, it is nilpotent. Its periodic part is finite, and contains B_1. It follows that B_1 is included in the n_1^{th}-$\mathbb{Z}X$-hypercenter of A for some $n_1 \in \mathbb{N}$. Then the nth-DX-hypercenter of A contains $B_2 = B_1 D$. Since each term of the upper DX-central series of A is a DG-submodule, it follows that there exists some $n_2 \in \mathbb{N}$ such that the n_2^{th}-DX-hypercenter of A contains $B = B_2 DG$.

To finish we are going to show that the upper DX-central series, and the upper DH-central series of A coincide. It suffices to prove that $A_1 = \zeta_{DX}(A) \leq \zeta_{DH}(A)$ or that $\zeta_{DH}(A)$ contains every finitely generated DG-submodule C of A_1. Suppose that the latter is false, that is, $C_1 = \zeta_{DH}(C) \neq C$. Put $C_2/C_1 = \zeta_{DH}(C/C_1)$. Then $C_2 \neq C_1$ by our assumption. Every element of $H/C_H(C_2)$ acts trivially on the factors of series

$$\langle 0 \rangle = C_0 \leq C_1 \leq C_2.$$

According to O.H. Kegel and B.A.F. Wehrfritz [134, 1.C.3], $H/C_H(C_2)$ can be embedded in $\mathrm{Hom}_{\mathbb{Z}}(C_2/C_1, C_1)$. Since C is finitely generated, the orders of the elements of C are bounded. Since $C \leq A_1$ the choice of X, implies that $H/C_H(C_2)$ is a periodic p-divisible group for every $p \in \Pi(A)$. This contradiction shows that $\zeta_{DH}(A) = \zeta_{DH}(A)$, as required. □

Corollary 14.4 ([150]). *If the conditions of Lemma 14.3 hold, then the DH-hyper-central length of A is at most ω.*

Corollary 14.5. *If the conditions of Lemma 14.3 hold, then the DG-socular height of A is at most ω.*

Proof. Pick $b \in B$, and put $B = bDG$. By Lemma 14.3, B is DH-nilpotent, and there exists some $n \in \mathbb{N}$ such that $B = \Omega_{P,n}(B)$. We claim that B has a finite composition DG-series, and proceed by induction on n. Suppose that $n = 1$. Let

$$\langle 0 \rangle = B_0 \leq B_1 \leq \cdots \leq B_c = B$$

be the upper DH-central series of B. If $1 \leq j \leq c$, each term B_j is clearly a DG-submodule of B. Since $C_G(B_{j+1}/B_j) \geq H$, $G/C_G(B_{j+1}/B_j)$ is a periodic p'-group. By Corollary 7.30, B_{j+1}/B_j is a direct sum of finitely many simple DG-submodules, and so B_{j+1}/B_j has a finite DG-composition series. Since this holds for every j, B has a finite DG-composition series. Suppose now that $n > 1$. Put $E = BP$. Then E is DH-nilpotent, and $E = \Omega_{P,n-1}(E)$. By induction, E has a finite DG-composition series. This is also true for B/E, hence B has a finite DG-composition series, as claimed. $\square$

Lemma 14.6 ([150]). *Let R be a commutative ring, G an abelian group, and A an RG-module whose additive group is a p-group, for some prime p. If H is a subgroup of G such that A is RH-hypercentral, and B is an RG-submodule of A, then $H/C_H(B)$ has no non-identity p-divisible subgroups. In particular, if G/H is a periodic p'-group, and $G/C_G(B)$ is periodic, then $G/C_G(B)$ has no non-identity divisible p-subgroups.*

Proof. Let $P/C_H(B)$ be a p-divisible subgroup of $H/C_H(B)$. For every $n \in \mathbb{N}$, put $B_n = \Omega_{P,n}(B)$, $C_1 = \zeta_{RP}(B_n)$, and $C_2/C_1 = \zeta_{RP}(B_n/C_1)$. Then $P/C_P(C_2)$ acts trivially on the factors of the series

$$\langle 0 \rangle = C_0 \leq C_1 \leq C_2.$$

According to O.H. Kegel and B.A.F. Wehrfritz [134, 1.C.3], $P/C_P(C_2)$ can be embedded in $\mathrm{Hom}_{\mathbb{Z}}(C_2/C_1, C_1)$. Since the order of each element of C_2 divides p^n, the order of each element of $\mathrm{Hom}_{\mathbb{Z}}(C_2/C_1, C_1)$ divides p^n. But $P/C_P(C_2)$ has no bounded p-factor-groups so that $P = C_P(C_2)$, and hence $C_2 = C_1$. It follows that $P \leq C_G(B_n)$. Since this holds for all $n \in \mathbb{N}$, $P \leq C_G(B)$. $\square$

Before we continue further considerations, we need to make the following remark. Let R be a ring, A be an R-module. Put $L = \mathrm{Ann}_R(A)$. We think of A as a K-module, where $K = R/L$. Then we may look at the consideration of injective envelopes from two points of view. First we consider a K-injective envelope E of A, and think of E as an R-module such that $L = \mathrm{Ann}_R(E)$. This seems to be very clever, because L has no influence on the structure of A. Alternatively, we could consider a full injective R-envelope I. In this case $\mathrm{Ann}_R(I)$ could not include L

itself, and therefore the structure of I depends on some elements of the ideal L. This case was explicitly considered above, so that we will look at the first one. Actually, this was the way chosen by B. Hartley and D. McDougall (see Chapter 12). Given a Chernikov abelian group G and an artinian $\mathbb{Z}G$-module A, Hartley and McDougall were able to reduce their study to the case in which the additive group of A is a p-group, where p is a prime. Then $G/C_G(A)$ has a finite Sylow p-subgroup, and, by the theorem of Wilson, it can be supposed that $G/C_G(A)$ is a p'-group; this supposition enables us to consider injective envelopes. Below we will follow the mentioned approach of Hartley and McDougall.

Theorem 14.7. *Let D be a Dedekind Z-domain, and let G be an abelian group of finite section rank. If A is an artinian DG-module, $I = \mathrm{Ann}_{DG}(A)$, $Q = DG/I$. If E is a Q-injective envelope of A, then E is an artinian DG-module.*

Proof. Suppose that $\mathrm{Soc}_{DG}(A) = M_1 \oplus \cdots \oplus M_t$, where $M_1, \ldots, M_t$ are simple DG-submodules. According to D.W. Sharpe and P. Vamos [265, Proposition 2.23],

$$E = E_1 \oplus \cdots \oplus E_t,$$

where each E_j is the DG-injective envelope of M_j. Hence, without loss of generality, it can be assumed that $\mathrm{Soc}_{DG}(A) = M_1$ is a simple DG-submodule, i.e. A is a monolithic module with the monolith $M = M_1$. Put $L = \mathrm{Soc}_{DG}(M)$, and $H = C_G(M)$. In particular DG/L is a field. By Lemma 12.4, A is D-periodic and then A is a P-module, where $P = \mathrm{Ann}_D(M) \in \mathrm{Spec}(D)$. Since A is an artinian DG-module, the upper socular series of A reaches A and every factor of this series can be decomposed into a direct sum of finitely many simple modules. This and Corollary 14.5 shows that A has an ascending series of submodules

$$\langle 0 \rangle = A_0 \leq A_1 = M \leq A_2 \leq \cdots \leq A_n \leq A_{n+1} \leq \cdots \leq A_\omega = A,$$

every factor of which is a simple DG-module. Suppose first that A has a finite composition series. Then A is finitely generated, say $A = a_1 DG + \cdots + a_m DG$. Since each summand $a_j DG \cong DG/\mathrm{Ann}_{DG}(a_j)$, A has a finite composition series, and, in particular, A is noetherian. Since

$$\mathrm{Ann}_{DG}(A) = \mathrm{Ann}_{DG}(a_1) \cap \cdots \cap \mathrm{Ann}_{DG}(a_m),$$

Remak's theorem gives the embedding

$$DG/\mathrm{Ann}_{DG}(A) \hookrightarrow DG/\mathrm{Ann}_{DG}(a_1) \oplus \cdots \oplus DG/\mathrm{Ann}_{DG}(a_m).$$

By Lemma 1.1, the factor-ring $Q = DG/\mathrm{Ann}_{DG}(A)$ is noetherian. By a result due to E. Matlis [183], $\mathrm{IE}_Q(A)$ is likewise an artinian DG-module.

We suppose now that the ascending DG-composition series of A are infinite. By Lemma 12.1, L annihilates every factor of any ascending composition series of A. It follows that $L^n \leq \mathrm{Ann}_{DG}(A_n)$, for each $n \in \mathbb{N}$. It turns out that $\mathrm{Ann}_{DG}(A) = \bigcap_{n \in \mathbb{N}} L^n$. Let

$$R = \varprojlim \{DG/L^n \mid n \in \mathbb{N}\}.$$

By Remak's theorem, $Q = DG/\mathrm{Ann}DG(A) \hookrightarrow R$. According to R. Sharp [264, Theorem 3.2], R is a noetherian local ring, and A is an artinian R-module. Further, $\mathrm{IE}_R(A)$ is again an artinian R-module (E. Matlis [183]). We claim that $\mathrm{IE}_R(A)$ is in fact a Q-injective module. For, suppose that V is a Q-module such that $\mathrm{IE}_R(A) \leq V$. Put $V_1 = L \otimes_Q R$. Then V_1 contains $\mathrm{IE}_R(A) \otimes_Q R = \mathrm{IE}_R(A)$. Since $\mathrm{IE}_R(A)$ is R-injective, there exists an R-submodule U such that $V_1 = \mathrm{IE}_R(A) \oplus U$. But $\mathrm{IE}_R(A) \leq V \leq V_1$, and so $V = \mathrm{IE}_R(A) \oplus (U \cap L)$, and hence $U \cap L$ is a Q-submodule. Thus $\mathrm{IE}_R(A)$ is a direct summand of every Q-module including $\mathrm{IE}_R(A)$. By Proposition 7.2, this means that $\mathrm{IE}_R(A)$ is an injective Q-module, and our claim has been proven. It follows that $E = \mathrm{IE}_Q(A) \leq \mathrm{IE}_R(A)$. Since $\mathrm{IE}_R(A)$ is an R-essential extension of M, $\mathrm{IE}_R(A)$ is an artinian monolithic R-module with the monolith M. By Lemma 12.1, $L^{\wedge} = \mathrm{Ann}_R(M)$ annihilates every factor of an ascending composition series of $\mathrm{IE}_R(A)$. Obviously,

$$L^{\wedge} = \varprojlim \{L/L^n \mid n \in \mathbb{N}\},$$

and, therefore, $L/\mathrm{Ann}_{DG}(A) \leq L^{\wedge}$, so that P annihilates every factor of an ascending composition series of $\mathrm{IE}_R(A)$. Given $x \in H = C_G(M)$, $x - 1 \in L \leq L^{\wedge}$. In particular, $\mathrm{IE}_R(A)$ is an $F\langle y_1 \rangle$-hypercentral module. Let $p = \mathrm{char}\, F$. Then the additive group of $\mathrm{IE}_R(A)$ is a p-group. Since G has finite section rank, H has a finitely generated subgroup X such that H/X is a periodic p-divisible group. Let

$$X = \langle x_1 \rangle \times \cdots \times \langle x_n \rangle$$

for some $n \in \mathbb{N}$. If $m \in \mathbb{N}$, we put $B_m = \Omega_{P,m}(E)$ so that $E = \bigcup_{m \in \mathbb{N}} B_m$. Look at B_1 as an FG-module, where $F = D/P$. We may think of B_1 as an $F\langle y_1 \rangle$-module, where y_1 is an infinite cyclic group, and y_1 acts on B_1 by $ay_1 = ax_1$, $a \in B_1$. Since $\mathrm{IE}_R(A)$ is an $(x_1 - 1)$-module, $B_1 = \bigcup_{m \in \mathbb{N}} \Omega_{S,m}(B_1)$, where $S = (y_1 - 1)F\langle y_1 \rangle$. Put

$$C_1 = B_1, \text{ and } C_2 = \Omega_{S,1}(B_1) = B_1 \cap C_E(x_1, x_2).$$

In this case we think of C_2 as an $F\langle y_2 \rangle$-module, where $\langle y_2 \rangle$ is an infinite cyclic group, and y_2 acts on C_2 by $ay_2 = ax_2$, $a \in C_2$. Since $\mathrm{IE}_R(A)$ is an $(x_2 - 1)$-module, $C_2 = \bigcup_{m \in \mathbb{N}} \Omega_{U,m}(C_2)$, where $U = (y_2 - 1)F\langle y_2 \rangle$. Proceeding in this way, we put

$$C_3 = B_1 \cap C_E(x_1, x_2, x_3), \ldots, C_{n-1} = B_1 \cap C_E(x_1, \ldots, x_{n-1})$$

and

$$C_n = B_1 \cap C_E(x_1, \ldots, x_n).$$

Since $C_G(C_n) \geq X$, and G/H is a p'-group (see L.A. Kurdachenko, J. Otal and I.Ya. Subbotin [157, Theorem 2.3]), $G/C_G(C_n)$ is a periodic p'-group by Lemma 14.6. By Corollary 7.30, C_n is a semisimple FG-module, and so, by Corollary 4.3, there exists some FG-submodule W such that $C_n = M \oplus W$. Since E is an essential extension of M, $W = \langle 0 \rangle$, and so $C_n = M$, ando so $C_{n-1} = \bigcup_{m \in \mathbb{N}} \Omega_{V,m}(C_{n-1})$, where $V = (y_n - 1)F\langle y_n\rangle$. By Proposition 1.8, C_{n-1} is an artinian FG-module. Iterating the argument, we obtain that B_1 is an artinian FG-module. Since $E = \bigcup_{m\in\mathbb{N}} B_m$, it suffices to apply Proposition 1.8 to obtain that E is an artinian DG-module in this case. □

Lemma 14.8. *Let F be a locally finite field of characteristic p, and let G be an abelian group of finite special rank. Suppose that A is a monolithic artinian FG-module with monolith M such that $C_G(A) = \langle 1\rangle$, and put $H = C_G(M)$. Suppose further that H has an infinite cyclic subgroup $\langle g\rangle$ such that $G/\langle g\rangle$ is a periodic p'-group. If E is the FG-injective envelope of A, then E has an ascending chain of submodules*

$$M = M_0 \le M_1 \le \cdots \le M_k \le \cdots \le \bigcup_{k\in\mathbb{N}} M_k = E$$

such that

$$M_{k+1}/M_k \cong M \quad and \quad M_{k+1}(g-1) = M_k$$

for every $k \in \mathbb{N}$.

Proof. We claim that $E = E(g-1)$. For, suppose the contrary, that is, E is not $(g-1)$-divisible. By Theorem 14.7, E is an artinian FG-module, and so there exists some $m \in \mathbb{N}$ such that $B = E(g-1)^m = E(g-1)^{m+1}$. In other words, B is $(g-1)$-divisible. Assume that $B \neq \langle 0\rangle$. Since E is artinian, $E(g-1)^{m-1}/B$ has a non-zero simple FG-submodule, lets say $(aFG+B)/B$. Since $a(g-1) \in B = B(g-1)$, there exists some $b \in B$ such that $a(g-1) = b(g-1)$. It follows that $(a-b)(g-1) = 0$, that is, $a - b \in C_E(\langle g\rangle)$. Hence $G/C_G((a-b)FG)$ is a periodic p'-group. By Corollary 7.30, $(a-b)FG$ is a semisimple FG-module. Since E is a monolithic FG-module, $M \le (a-b)FG$. By the choice of a we have $a - b \notin B$, and so $a - b \notin M$. By Corollary 4.3, there exists an FG-submodule W such that $(a-b)FG = M \oplus W$, and, since E is an essential extension of M, $W = \langle 0\rangle$, a contradiction. Hence $B = E(g-1)^m = \langle 0\rangle$. In this case $g^t \in C_G(A) = \langle 1\rangle$, where $t = p^{m-1}$, and we come to another contradiction, proving our claim. It follows that the mapping $\gamma : e \mapsto e(g-1)$, $e \in E$, is an epimorphism. Put

$$M_1 = M\gamma^{-1}, \quad M_{n+1} = M_n\gamma^{-1}, \text{ and } C = \bigcup_{n\in\mathbb{N}} M_n, \quad n \in \mathbb{N}.$$

By construction, $C = C(g-1)$. We must have $C = E$, otherwise, proceeding as above, we would find another contradiction. Hence $E = \bigcup_{n\in\mathbb{N}} M_n$, as required. □

Lemma 14.9. *Let F be a locally finite field of characteristic p, and let G be an abelian group of finite special rank. Suppose that A is a monolithic artinian FG-module with monolith M such that $C_G(A) = \langle 1\rangle$, and put $H = C_G(M)$. Suppose further that H has a torsion-free finitely generated subgroup*

$$L = \mathbf{Dr}_{s\in S} \langle g_s\rangle$$

such that G/L is a periodic p'-group. Let E be the injective envelope of A. If S_1 is a non-empty subset of S, then $C_E(\mathbf{Dr}_{s \in S \setminus S_1} \langle g_s \rangle)$ is $(g_t - 1)$-divisible for every element $t \in S_1$.

Proof. We proceed by induction on $s_1 = |S_1|$. Suppose first that $s_1 = 1$, that is, $S_1 = \{t\}$, for some $t \in S$. Put

$$C = C_E(\mathbf{Dr}_{i \in S \setminus S_1} \langle g_i \rangle),$$

and suppose that C is not $(g_t - 1)$-divisible. By Theorem 14.7, E is an artinian FG-module, and so there exists some $m \in \mathbb{N}$ such that $B = C(g_t - 1)^m = C(g_t - 1)^{m+1}$. As in the proof of Lemma 14.8, $B = \langle 0 \rangle$. Let

$$Q = G/(\mathbf{Dr}_{i \in S \setminus S_1} \langle g_i \rangle),$$

and think of C as an FQ-module. Let E_1 be the FQ-injective envelope of C. By Theorem 14.7, E_1 is artinian, and, by Lemma 14.8, $E_1 = E_1(g_t - 1)$. We now think of E_1 as an FG-module such that $\mathbf{Dr}_{i \in S \setminus S_1} \langle g_i \rangle \leq C_G(E_1)$. Suppose we are given a monomorphism $\phi : C \longrightarrow E_1$, and put $U = E \oplus E_1$ and $V = \{a - a\phi \mid a \in C\}$. Obviously, V is an FG-submodule of U, and $V \cap E = \langle 0 \rangle = V \cap E_1$. Therefore,

$$(E + V)/V \cong E, \text{ and } (E_1 + V)/V \cong E_1,$$

i.e. E and E_1 can be embedded in U/V. By the definition of V, $(C + V)/V \leq (E_1 + V)/V$. Put

$$C_1/V = C_{U/V}(\mathbf{Dr}_{i \in S \setminus S_1} \langle g_i \rangle).$$

Since $(E_1 + V)/V \leq C_1/V$,

$$C_1/V = ((E_1 + V)/V) + ((E + V)/V) \cap (C_1/V)).$$

If $a \in E$ and $a + V \in ((E + V)/V) \cap (C_1/V)$, then $a(g_i - 1) \in V$, for each $i \in S$. Since E is an FG-submodule of U, $a(g_r - 1) \in V \cap E = \langle 0 \rangle$, for every $r \in S$. Thus

$$((E + V)/V) \cap (C_1/B) \leq (C + V)/V \leq (E_1 + V)/V$$

and hence

$$C_1/V = (E_1 + V)/V \cong E_1/(E_1 \cap V) \cong E_1.$$

We have already noted that $E_1 = E_1(g_t - 1)$. Let W/V be an FG-submodule of U/V, and pick some $V \neq w + V \in W/V$. Then $w + V = (a_1 + V) + (a_2 + V)$, for suitable $a_1 \in E$ and $a_2 \in E_1$. If $a_2 = 0$, then $w + V \in (E + V)/V$. Suppose that $a_2 \neq 0$. Since E_1 is an essential extension of $C\phi$, $a_2 FG \cap C\phi \neq \langle 0 \rangle$, and so there exists some $x \in FG$ such that $0 \neq a_2 x \in C\phi$. Then there exists some $a_3 \in C$ such that $a_2 x = a_3 \phi$. But $a_3 - a_3 \phi \in V$ whence $a_3 + V = a_3 \phi + V = a_2 x + V$. Thus,

$$(w + V)x = (a_1 x + V) + (a_2 x + V) = a_1 x + a_3 + V \in (E + V)/V.$$

Thus, $(W/V) \cap ((E+V)/V) \neq \langle 0 \rangle$, and hence U/V is an essential extension of $(E+V)/V \cong E$. Since E is FG-injective, every essential extension of E is equal to E. Then $(E+V)/V = U/V$ and $E \cong U/V$. It follows that C is $(g_t - 1)$-divisible, as required. The general result can be deduced by applying induction, and proceeding in the same way. □

We are now in a position to describe the injective envelopes.

Lemma 14.10. *Let F be a locally finite field of characteristic p, and let G be an abelian group of finite special rank. Suppose that A is a monolithic artinian FG-module with monolith M such that $C_G(A) = \langle 1 \rangle$, and put $H = C_G(M)$. Suppose further that H has a torsion-free finitely generated subgroup $L = \langle g_1 \rangle \times \cdots \times \langle g_n \rangle$ such that G/L is a periodic p'-group. If E is the FG-injective envelope of A, then E has a finite series of submodules*

$$E = E_0 \geq E_1 \geq \cdots \geq E_{n-1} \geq E_n = M$$

satisfying the following properties:

(1) $E_1 = C_E(g_1)$, $E_2 = C_E(g_1, g_2)$, ..., $E_{n-1} = C_E(g_1, \ldots, g_{n-1})$.

(2) *Any term E_j has an ascending series of submodules*

$$E_{j+1} = M_{j+1,0} \leq M_{j+1,1} \leq \cdots \leq M_{j+1,k} \leq \cdots \leq \bigcup_{k \in \mathbb{N}} M_{j+1,k} = E_j$$

such that

$$M_{j+1,k+1}/M_{j+1,k} \cong E_{j+1}, \text{ and } M_{j+1,k+1}(g_{j+1} - 1) = M_{j+1,k}$$

for every $k \in \mathbb{N}$.

Proof. By Lemma 14.9, $E_{n-1} = E_{n-1}(x_n - 1)$. Thus the mapping $\phi_1 : a \mapsto a(x_n - 1)$, $a \in E_{n-1}$ is an epimorphism. Put

$$M_{n,0} = M, M_{n,1} = M_{n,0}\phi_1^{-1}, M_{n,k+1} = M_{n,k}\phi_1^{-1}, k \in \mathbb{N}.$$

Clearly,

$$\text{Ker } \phi_1 = E_{n-1} \cap \text{Ann}_E(x_n - 1) = C_E(x_1, \ldots, x_n).$$

Therefore $G/C_G(\text{Ker } \phi_1)$ is a periodic p'-group. By Corollary 7.30, there exists an FG-submodule U such that $\text{Ker } \phi_1 = M \oplus U$. Since E is an essential extension of A, $U = \langle 0 \rangle$, and so $\text{Ker } \phi_1 = M$. Thus $M_{n,1}(x_n - 1) = M_{n,0} = M$, and $M_{n,1}/M_{n,0} \cong M$. Since $E_{n-1}/\text{Ker } \phi_1 \cong E_{n-1}$, we proceed in the same way to get $M_{n,k+1}(x_n - 1) = M_{n,k}$, and $M_{n,k+1}/M_{n,k} \cong M$, for every $k \in \mathbb{N}$. Define $V = \bigcup_{k \in \mathbb{N}} M_{n,k}$ so that $V(x_n - 1) = V$. Suppose that $V \neq E_{n-1}$. Proceeding in the same way as in the proof of Lemma 14.8, we see that there exists some FG-submodule W such that $E_{n-1} = V \oplus W$, and, since $M \leq V$, we conclude that $W = \langle 0 \rangle$, and so $E_{n-1} = \bigcup_{k \in \mathbb{N}} M_{n,k}$. Consider the submodule E_{n-2}. By Lemma

14.9, $E_{n-2} = E_{n-2}(x_{n-1} - 1)$. As above, the mapping $\phi_2 : a \mapsto a(x_{n-1} - 1)$, $a \in E_{n-2}$ is an epimorphism. Note that

$$\text{Ker } \phi_2 = E_{n-2} \cap C_E(x_{n-1}) = C_E(x_1, \ldots, x_{n-2}) \cap C_E(x_{n-1})$$
$$= C_E(x_1, \ldots, x_{n-1}) = E_{n-1}.$$

Put

$$M_{n-1,0} = E_{n-1}, M_{n-1,k+1} = M_{n-1,k}\phi_2^{-1}, k \in \mathbb{N}.$$

Define $U = \bigcup_{k\in\mathbb{N}} M_{n-1,k}$, and suppose that $U \neq E_{n-2}$. By Theorem 14.7, E is an artinian FG-module, and so E_{n-2}/U has a non-zero simple FG-submodule, lets say $(aFG+U)/U$. By Corollary 12.2, $a(x_{n-1} - 1) \in U$, and, by construction, $U = U(x_{n-1}-1)$. Therefore, there exists some $u \in U$ such that $a(x_{n-1}-1) = u(x_{n-1}-1)$. It follows that $(a - u)(x_{n-1} - 1) = 0$, that is, $a - u \in \text{Ker } \phi_2 = E_{n-1} \leq U$. Thus $a \in U$, which contradicts the choice of a. Then $U = E_{n-2}$. By construction, we also have that $M_{n-1,k+1}(x_{n-1} - 1) = M_{n-1,k}$, and $M_{n-1,k+1}/M_{n-1,k} \cong E_{n-1}$, for every $k \in \mathbb{N}$. Applying similar arguments to $E_1, \ldots, E_{n-2}$, we construct the submodules $M_{j,k}$, $1 \leq j \leq n - 3$, and $k \in \mathbb{N}$ in an analogous way. $\square$

Theorem 14.11. *Let D be a Dedekind Z-domain, and let G be an abelian group of finite section rank. Suppose that A is a monolithic artinian DG-module with monolith M, and let $P = \text{Ann}_D(M)$. If E is a DG-injective envelope of A, then:*

(1) *E is an artinian P-module over DG.*

(2) *$\Omega_{P,1}(E)$ is the FG-injective envelope of $\Omega_{P,1}(A)$, where $F = D/P$.*

(3) *E is the D-divisible envelope of $\Omega_{P,1}(E)$.*

Proof. (1) follows from the proof of Lemma 14.8.

(2) We claim that $\Omega_{P,1}(E)$ is FG-injective. To show this, put $L = \text{Ann}_{DG}(A)$, $T = \text{Ann}_{FG}(A)$, $R = DG/L$, and $Q = FG/T \cong DG/S$, where S is the preimage of T in DG under the canonical map. Clearly $S \geq L$. Let B, and C be a pair of Q-modules such that $B \leq C$, and let $\phi : B \longrightarrow \Omega_{P,1}(E)$ be a Q-homomorphism. We may think of B, and C as DG-modules such that $\text{Ann}_{DG}(B) \cap \text{Ann}_{DG}(C) \geq S$. Then ϕ can be extended to a DG-homomorphism $\sigma : C \longrightarrow E$. Since $(\text{Im } \sigma)P = \langle 0 \rangle$, $\text{Im } \sigma \leq \Omega_{P,1}(E)$, and so σ is in fact an FG-homomorphism, which shows our claim. Hence $\Omega_{P,1}(E)$ has an FG-injective envelope U of $\Omega_{P,1}(A)$. Then there exists some FG-submodule V such that $\Omega_{P,1}(E) = U \oplus V$, and it follows that $U = \langle 0 \rangle$ since E is an essential extension of A. Therefore, $\Omega_{P,1}(E)$ is an FG-injective envelope of $\Omega_{P,1}(A)$.

(3) By Proposition 7.5, E is D-divisible so that E is a D-divisible envelope of $\Omega_{P,1}(E)$. $\square$

Corollary 14.12. *Let D be a Dedekind Z-domain, and let G be an abelian group of finite section rank. Suppose that A is a monolithic artinian DG-module with monolith M, and put $P = \text{Ann}_D(M)$. Then A is DG-injective if and only if A is D-divisible, and $\Omega_{P,1}(A)$ is FG-injective, where $F = D/P$.*

Proof. By Theorem 14.11, the conditions are necessary. Conversely, assume that A is D-divisible, and $\Omega_{P,1}(A)$ is FG-injective. Let E be the DG-injective envelope of A. Since $\Omega_{P,1}(A)$ is FG-injective, there exists some FG-submodule B such that $\Omega_{P,1}(E) = \Omega_{P,1}(E) \oplus B$. It follows that $B = \langle 0 \rangle$ since E is an essential extension of A. Hence $\Omega_{P,1}(E) = \Omega_{P,1}(A)$. Since E and A are D-divisible, $E = A$, and we are done. $\qquad\square$

The description of the injective envelope of a monolithic DG-module can be obtained in a similar way to that of the case of a field. We simply state here that result without a proof.

Theorem 14.13. *Let D be a Dedekind Z-domain, and let G be an abelian group of finite special rank. Suppose that A is a monolithic artinian DG-module with monolith M such that $C_G(A) = \langle 1 \rangle$, and put $H = C_G(M)$. Let $P = \operatorname{Ann}_D(A)$, and let p be the characteristic of the field D/P. Suppose further that H has a torsion-free finitely generated subgroup $L = \langle g_1 \rangle \times \cdots \times \langle g_n \rangle$ such that G/L is a periodic p'-group. If E is the DG-injective envelope of A, then E has a finite series of submodules*

$$E = E_0 \geq E_1 \geq \cdots \geq E_{n-1} \geq E_n = M$$

satisfying the following properties:

(1) $E_1 = C_E(g_1)$, $E_2 = C_E(g_1, g_2)$, ..., $E_{n-1} = C_E(g_1, \ldots, g_{n-1})$, *and* E_n *is the DQ-injective envelope of $Soc_{DG}(A)$, where $Q = G/C_G(Soc_{DG}(A))$.*

(2) *Any term E_j has an ascending series of submodules*

$$E_{j+1} = M_{j+1,0} \leq M_{j+1,1} \leq \cdots \leq M_{j+1,k} \leq \cdots \leq \bigcup_{k \in \mathbb{N}} M_{j+1,k} = E_j$$

such that

$$M_{j+1,k+1}/M_{j+1,k} \cong E_{j+1}, \text{ and } M_{j+1,k+1}(g_{j+1} - 1) = M_{j+1,k}$$

for every $k \in \mathbb{N}$.

Let A be an artinian P-module over DG, where $P \in \operatorname{Spec}(D)$, and suppose that $p = \operatorname{char}(D/P)$. As we mentioned above, the problem of the description of A can be *wild* if $|G/G^p| \geq p^2$. If $|G/G^p| \leq p$, the corresponding description is contained in L.A. Kurdachenko [150]. Note that some of the results of this chapter can be translated to artinian modules over a ring with the form FG, where char $F = 0$ (see L.A. Kurdachenko [148]).

Chapter 15

The injective envelopes of simple modules over group rings

As we have already mentioned, one of the most important aspects of the study of artinian modules is to obtain information about their injective envelopes. In particular, the following question is very important: *when is an injective envelope of an artinian module likewise artinian?* The presence of this property would allow us to obtain the desired description in the best possible way. In this chapter, we will obtain some classes of group rings RG having the property that the injective envelope of every artinian RG-module also is artinian. Actually, the injective envelope of an artinian module is equal to the injective envelope of its socle. Moreover, injective envelopes work well with finite direct sums, that is,

$$\mathrm{IE}_R(M_1 \oplus \cdots \oplus M_t) = \mathrm{IE}_R(M_1) \oplus \cdots \oplus \mathrm{IE}_R(M_t).$$

These remarks allow us to reduce our investigation to the consideration of the injective envelope of a simple module. For commutative rings an important result was obtained by E. Matlis [183], who proved that if R is a noetherian commutative ring, the injective envelope of each artinian R-module is likewise artinian. The group rings that appear to be closer to noetherian commutative rings are group rings of the form RG, in which R is noetherian, and G is a finitely generated nilpotent group. In fact, R.L. Snider [269] proved that if G is a finitely generated nilpotent group, then the injective $\mathbb{Z}G$-envelope of a simple module is artinian. In this chapter, we consider a more general situation: the case of a group ring having the form DG, where G is a finitely generated nilpotent group, and D is a Dedekind domain. The reader will also be able to find proofs that are different from Snider's.

Lemma 15.1 (P. Hall [96])**.** *Let G be a finitely generated nilpotent group. If $\zeta(G)$ is finite, then G is finite.*

Proof. Let

$$\langle 1 \rangle = C_0 \leq C_1 \leq \cdots \leq C_n = G$$

be the upper central series of G, and suppose that $|C_1| = m$. We show by induction on n that the subgroups $C_2, \ldots, C_n$ are all finite. Suppose that $G = \langle g_1, \ldots, g_k \rangle$. Given $g \in G$, the mapping $\phi_g : x \mapsto [x, g]$, $x \in C_2$ is a homomorphism such that $\operatorname{Ker} \phi_g = C_G(g) \cap C_2$, and $\operatorname{Im} \phi_g = [G, g] \leq \zeta(G)$. Therefore, $|C_2/\operatorname{Ker} \phi_g| \leq m$. Since

$$(C_G(g_1) \cap C_2) \cap \cdots \cap (C_G(g_k) \cap C_2) = \zeta(G),$$

$|C_2/\zeta(G)| \leq m^k$ by Remak's theorem. Thus C_2 is finite.

Suppose that $n > 2$, and that we have already proved that C_{n-1} is finite. Since $\zeta(G/C_{n-2}) = C_{n-1}/C_{n-2}$, repeating the arguments given in the above paragraph, we obtain that G/C_{n-2} is finite. Hence G is finite, as required. $\qquad\square$

Proposition 15.2 ([96]). *Let A be a simple FG-module, where F is a locally finite field, and G is a finitely generated nilpotent group. Then $\dim_F(A)$ is finite.*

Proof. By L.A. Kurdachenko, J. Otal and I.Ya. Subbotin [157, Corollary 1.21], $\zeta(G/C_G(A))$ is periodic. Since a finitely generated nilpotent group satisfies Max, it follows that $\zeta(G/C_G(A))$ is finite. By Lemma 15.1, $G/C_G(A)$ is finite. Therefore $\dim_F(A)$ is finite. $\qquad\square$

As the following result shows, in the case of finitely generated abelian-by-finite groups the restrictions on the field could be removed.

Proposition 15.3 ([96]). *Let A be a simple FG-module, where F is an arbitrary field, and G is a finitely generated abelian-by-finite group. Then $\dim_F(A)$ is finite.*

Proof. We suppose first that G is abelian. We have $A = aFG$, for each $0 \neq a \in A$, so that $A \cong FG/L$, where $L = \operatorname{Ann}_{FG}(a)$. Since A is simple, L is a maximal ideal of FG. Put $G = \langle g_1, \ldots, g_n \rangle$. We look at FG as a homomorphic image of the polynomial ring $F[X_1, \ldots, X_{2n}]$ in $2n$ variables with coefficients in F, where the homomorphism is the extension of the correspondence

$$X_j \longrightarrow a_j,\ 1 \leq j \leq n,\ \text{and } X_j \longrightarrow a_j^{-1},\ n+1 \leq j \leq 2n.$$

In other words, $A \cong F[X_1, \ldots, X_{2n}]/S$, where S is a maximal ideal of $F[X_1, \ldots, X_{2n}]$. In particular, $S \neq \langle 0 \rangle$, and hence the field $F[X_1, \ldots, X_{2n}]/S$ is a finitely generated algebraic extension of F, that is, finite dimensional over F. Thus $\dim_F(A)$ is finite.

Suppose now that G has a normal abelian subgroup U of finite index. By Theorem 5.5, A contains a simple FU-submodule B, and there exists a finite subset S such that $A = \bigoplus_{g \in S} Bg$. By Lemma 5.4, each Bg is a simple FU-submodule. So, $\dim_F(Bg)$ is finite by the result shown in the above paragraph. Since S is finite, $\dim_F(A)$ is finite. $\qquad\square$

However, we have

Lemma 15.4 ([96]). *Let F be a field, and let G be a group. Suppose that G has a normal subgroup H such that there exists a simple FH-module B with $\dim_F(B)$ infinite. Then there exists a simple FG-module A such that $\dim_F(A)$ is infinite.*

Proof. We have $B \cong_{FH} FH/Q$ for some maximal right ideal Q of FH. Let S be a transversal to H in G. Since H is a normal subgroup of G,

$$P = \sum_{g \in S} Qg = \bigoplus_{g \in S} Qg$$

is a right ideal of FG. Since $P \cap FH = Q$, $P \neq FG$. Hence there is a maximal right ideal P_1 of FG such that $P \leq P_1$. Since $FG = \bigoplus_{g \in S} FHg$, P_1 does not include FH. Now $FH \cap P_1$ contains $FH \cap P = Q$, and $FH \cap P_1$ is a right ideal of FH. Then $FH \cap P_1 = Q$ since Q is a maximal right ideal of FH. Hence,

$$(FH + P_1)/P_1 \cong FH/(FH \cap P_1) = FH/Q,$$

which is infinite dimensional over F. Thus FG/P_1 is infinite dimensional over F. Since P_1 is a maximal right ideal of FG, the FG-module $A = FG/P_1$ is simple, as required. □

Proposition 15.5 ([96])**.** *Suppose that F is a non locally finite field, and let G be a polycyclic group which has no abelian subgroups of finite index. Then there exists a simple FG-module A such that $\dim_F(A)$ is infinite.*

Proof. The group G has a finite series

$$\langle 1 \rangle = G_0 \trianglelefteq G_1 \trianglelefteq \cdots \trianglelefteq G_n = G$$

in which every factor G_j/G_{j-1} is cyclic. We are arguing by induction on n. Put $H = G_{n-1}$ so that $G = \langle H, y \rangle$ for some $y \in G$. We claim that we may assume that H is abelian-by-finite. For, otherwise, by the induction hypothesis, there exists a simple FH-module B such that $\dim_F(B)$ is infinite, and it suffices to apply Lemma 15.4. Suppose then that H has a normal abelian subgroup U of finite index. In this case G/H has to be infinite. Put $m = |H/U|$. Then H^m is a characteristic subgroup of H, and so H^m is G-invariant. Since H/H^m is a finitely generated periodic soluble group, it is finite. On the other hand, $H^m \leq U$, so that H^m is abelian. Since H^m is finitely generated, the periodic part of H^m is finite, and so there exists some $t \in \mathbb{N}$ such that H^{mt} is torsion-free, and once more, H/H^{mt} is finite. In other words, without loss of generality, we may suppose that U is G-invariant and torsion-free. Since H/U is finite, $G/C_G(H/U)$ is finite. Put $k = |C_G(H/U)|$, and define $z = y^k$ and $L = \langle U, z \rangle$. Then L is a normal subgroup of G, and L/U is infinite. We claim that $U = C_L(U)$. For, otherwise $C_L(U)$ is a normal abelian subgroup of L having finite index; so, since G/L is finite, G is abelian-by-finite, a contradiction, which shows our claim: $C_L(U) = U$.

Suppose that

$$U = \langle u_1 \rangle \times \cdots \times \langle u_r \rangle,$$

and for each $1 \leq s \leq r$, put

$$U_s = \langle u_1 \rangle \times \cdots \times \langle u_{s-1} \rangle \times \langle u_{s+1} \rangle \times \cdots \times \langle u_r \rangle.$$

Suppose, if possible, that $N_L(U_s) \neq U$ for each $1 \leq s \leq r$. Then there exists some $q \in \mathbb{N}$ such that

$$z^q \in N_L(U_1) \cap \cdots \cap N_L(U_r).$$

Take an index $1 \leq j \leq r$. Then

$$z^{-q} u_j z^q \in U_1 \cap \cdots \cap U_{j-1} \cap U_{j+1} \cap \cdots \cap U_r = \langle u_j \rangle.$$

Since u_j has infinite order, $[u_j, z^{2q}] = 1$. Hence, $[U, z^{2q}] = \langle 1 \rangle$, a contradiction. Thus, there exists some $1 \leq s \leq r$ such that $N_L(U_s) = U$. For this value s, we write $u = u_s$ and $V = U_s$, so that $U = \langle V, u \rangle$ and U/V is infinite cyclic.

Since F is not locally finite, the multiplicative group $U(F) = F \setminus \langle 0 \rangle$ contains an element λ having infinite order. If $j \in \mathbb{Z}$, we define the mapping $\rho_j : U \longrightarrow F$ in this way:

for any $w \in U$, we put $w\rho_j = \lambda^{a(j,w)}$, where $a(j,w)$ is an integer such that
$$z^j w z^{-j} \in u^{a(j,w)} V.$$

This definition gives immediately that $a(j, ww_1) = a(j,w) + a(j,w_1)$, and hence, $(ww_1)\rho_j = (w\rho_j)(w_1\rho_j)$ for all $w, w_1 \in U$. In other words, ρ_j is a homomorphism of U in $U(F)$. Suppose that $\rho_j = \rho_i$ for some $j < i$. Since λ has infinite order in $U(F)$, $a(j,w) = a(i,w)$ for all $w \in U$. Thus $[w, z^{i-j}] \in V$ for all $w \in U$, and hence, $z^{i-j} \in N_L(V)$, which is impossible by the construction of V. Thus, we have just shown that $\rho_j \neq \rho_i$ if $j \neq i$.

Now let A be a vector space over F with a basis $\{a_j \mid j \in \mathbb{Z}\}$, and define the action of L on A by

$$a_j z = a_{j+1}, \text{ and } a_j w = a_j(w\rho_j) \text{ for any } w \in U, j \in \mathbb{Z}.$$

It is easy to see that this action makes A into an FL-module. Since $\rho_j \neq \rho_i$ for all $j \neq i$, the FU-submodules of A are only the direct sum of the one-dimensional subspaces $a_j F$. Hence, if B is a non-zero FL-submodule of A, B must contain some a_j. Then B contains $a_i = a_j z^{i-j}$ for all i, that is $B = A$. Hence A is a simple FL-module such that $\dim_F(A)$ is infinite, and it suffices to apply Lemma 15.4. $\square$

In connection with Propositions 15.2, 15.3 and 15.5, the following question naturally arises: *Let F be a field. Determine groups G for which every simple FG-module has finite dimension over F.* The following interesting papers should be mentioned in connection with this problem: B. Hartley [108], R.L. Snider [270], B.A.F. Wehrfritz [281, 282, 283].

Lemma 15.6. *Let D be a Dedekind domain, and let G be a polycyclic-by-finite group. Suppose that A is a finitely generated monolithic DG-module with monolith M. If $\operatorname{Ann}_D(M) = P \neq \langle 0 \rangle$, then $\operatorname{Ann}_D(A) = P^n$, for some $n \in \mathbb{N}$.*

Proof. Since M is a simple DG-submodule of A, $P \in \operatorname{Spec}(D)$. Let T be the D-periodic part of A. Since A is monolithic, T is equal to the P-component of A. But DG is a noetherian ring (see D.S. Passman [217, Theorem 10.2.7]), A is a noetherian DG-module by Lemma 1.1. Then there is some $n \in \mathbb{N}$ such that $\Omega_{P,n}(A) = \Omega_{P,n+m}(A)$ for each $m \in \mathbb{N}$. This means that $TP^n = \langle 0 \rangle$, and it follows that there exists a D-submodule C such that $A = T \oplus C$ (see I. Kaplansky [123]). Thus $AP^n \leq C$, and $AP^n \cap T = \langle 0 \rangle$. Since $M \leq T$, this means that $A = T$. □

Lemma 15.7. *Let F be a locally finite field, and let G be a finitely generated nilpotent group. If A is a finitely generated monolithic FG-module, then $G/C_G(A)$ is finite, and hence, $\dim_F(A)$ is finite.*

Proof. Let M be the FG-monolith of A. There is no loss if we assume that $C_G(A) = \langle 1 \rangle$. According to L.A. Kurdachenko, J. Otal and I.Ya. Subbotin [157, Corollary 1.21], $\zeta(G/C_G(M))$ is periodic. It follows that $\zeta(G/C_G(M))$ is finite, and, by Lemma 15.1, so is $G/C_G(M)$. We claim that $\zeta(G)$ is periodic. Otherwise, there exists some $z \in \zeta(G) \cap C_G(M)$ of infinite order since $G/C_G(M)$ is finite. Put $J = F \langle z \rangle$ so that J is a principal ideal domain, and we can think of A as a JG-module. By the choice of z, $\mathrm{Ann}_J(M) = (z-1)J$. By Lemma 15.6, there exists some $n \in \mathbb{N}$ such that $A(z-1)^n = \langle 0 \rangle$. For every $k \in \mathbb{N}$ we have

$$(z-1)^k = z^k - C_1^k z^{k-1} + C_2^k z^{k-2} - \cdots + (-1)^k,$$

where C_1^k, C_2^k, ... are the binomial coefficients. Let $p = \mathrm{char}\, F$, and choose $s \in \mathbb{N}$ such that $t = p^s \geq n$. If $a \in A$, then $0 = a(z-1)^t = az^t + (-1)^t a$, and so $az^t = (-1)^{t+1}a$. If $p = 2$, then $a = -a$, and hence, $az^t = a$. If $p > 2$, then again $az^t = a$. In any case, $z^t \in C_G(A) = \langle 1 \rangle$, a contradiction. Hence, $\zeta(G)$ is periodic. Then it follows that $\zeta(G)$ is finite, and it suffices to apply Lemma 15.1 to conclude that G is finite, as required. □

Proposition 15.8. *Let D be a Dedekind Z-domain, G a locally (polycyclic-by-finite) group of finite 0-rank. Suppose that A is a finitely generated DG-module such that $C_G(A) = \langle 1 \rangle$. If the set of all maximal DG-submodules of A is finite, then A is D-periodic, and $\zeta(G)$ is periodic.*

Proof. Suppose that

$$A = a_1 DG + \cdots + a_n DG.$$

According to L.A. Kurdachenko, J. Otal and I.Ya. Subbotin [157, Theorem 1.15], A is D-periodic, and, then $\mathrm{Ann}_D(a_j) \neq \langle 0 \rangle$ for every index $1 \leq j \leq n$. Since D is an integral domain,

$$L = \mathrm{Ann}_D(A) = \mathrm{Ann}_D(a_1) \cap \cdots \cap \mathrm{Ann}_D(a_n) \neq \langle 0 \rangle.$$

By Theorem 6.3, $L = P_1^{k_1} \cdots P_t^{k_t}$, where $P_1, \ldots, P_t \in \mathrm{Spec}(D)$ are pairwise different prime ideals, and $k_1, \ldots, k_t \in \mathbb{N}$. We proceed by induction on $k(L) = k_1 + \cdots + k_t$. If $k(L) = 1$, then L is a maximal ideal of D, so D/L is a locally finite field. By L.A. Kurdachenko, J. Otal and I.Ya. Subbotin [157, Corollary 1.21], $\zeta(G)$ is periodic.

Suppose that $k(L) > 1$, and we have already proved our assertion for modules B such that $k(\mathrm{Ann}_D(B)) < k(L)$. Put $Z = \zeta(G)$, and $B = AP_1$. Then

$\mathrm{Ann}_D(B) = P_1^{k_1-1} \cdots P_t^{k_t}$, so that $k(\mathrm{Ann}_D(B)) < k(L)$. Moreover, $\mathrm{Ann}_D(A/B) = P_1$, so that $k(\mathrm{Ann}_D(A/B)) < k(L)$. Applying induction, $\zeta(G/C_G(B))$, and $\zeta(G/C_G(A/B))$ are periodic, and so are $Z/(Z \cap C_G(B))$, and $Z/(Z \cap C_G(A/B))$. If $U = C_G(B) \cap C_G(A/B) \cap Z$, then, by Remak's theorem, Z/U is periodic. Put

$$A_0 = A, \ A_1 = AP_1, \ A_2 = AP_1^2, \ldots, A_{k_1} = AP_1^{k_1}, \ A_{k_1+1} = AP_1^{k_1}P_2, \ldots,$$

$$A_{k_1+k_2} = AP_1^{k_1}P_2^{k_2}, \ldots, A_{k_1+\cdots+k_t} = AP_1^{k_1} \cdots P_t^{k_t} = \langle 0 \rangle.$$

We have just constructed a descending chain of submodules

$$A = A_0 \geq A_1 \geq \cdots \geq A_{k_1+\cdots+k_t} = \langle 0 \rangle.$$

Given any factor A_j/A_{j+1} of this series, we put $P_m = \mathrm{Ann}_D(A_j/A_{j+1})$ $(m = m(j))$. Then we may consider A_j/A_{j+1} as an F_mG-module, where $F_m = D/P_m$. Since F_m is a locally finite field, the additive group of A_j/A_{j+1} is an elementary abelian p_m-group, where $p_m = \mathrm{char}\ F_m$. Hence every factor of the above series $\{A_j \mid 0 \leq j \leq k(L)\}$ is elementary abelian, and it readily follows that the additive group of A is periodic. Actually, this group is bounded, that is $sA = \langle 0 \rangle$, where $s = p_1^{k_1} \cdots p_t^{k_t}$. Pick $z \in U$, and $a \in A$. Then there exists some $b \in B$ such that $az = a + b$. If $n \in \mathbb{N}$, then $az^n = a + nb$, and so, in particular, $az^s = a + sb = a$. This means that U is periodic. Then Z is periodic, because Z/U is periodic. $\square$

Corollary 15.9. *Let D be a Dedekind Z-domain, and let G be a finitely generated nilpotent group. Suppose that A is a finitely generated monolithic DG-module with the DG-monolith M. Then $\mathrm{Ann}_D(A) = P^n$ for some $n \in \mathbb{N}$, where $P = \mathrm{Ann}_D(\mu_{DG}(A)) \in \mathrm{Spec}(D)$, and $G/C_G(A)$ is finite. In particular, A has finite D-composition series.*

Proof. There is no loss if we suppose that $C_G(A) = \langle 1 \rangle$. According to L.A. Kurdachenko, J. Otal and I.Ya. Subbotin [157, Corollary 1.16], $P = \mathrm{Ann}_D(M) \in \mathrm{Spec}(D)$. By Lemma 15.6, $\mathrm{Ann}_D(A) = P^n$, for some $n \in \mathbb{N}$. Put $A_j = \Omega_{P,j}(A)$, for $j \in \mathbb{N}$. Pick $y \in P \setminus P^2$. The mapping $\phi : a \mapsto ay$, $a \in A_2$ is a DG-endomorphism of A_2 and so $\mathrm{Ker}\ \phi$ and $\mathrm{Im}\ \phi$ are DG-submodules. Clearly, $\mathrm{Im}\ \phi = A_2y \leq A_1$ and $\mathrm{Ker}\ \phi = \{a \in A_2 \mid ay = 0\} \geq A_1$. By Proposition 6.13, $P = yD + P^2$, and so $\mathrm{Ker}\ \phi = \{a \in A_2 \mid aP = \langle 0 \rangle\} = A_1$. Hence, we may deduce $A_1 \cong A_2/A_1$. By Lemma 15.7, $\dim_F(A_1)$ is finite so that $\dim_F(A_2/A_1)$ is likewise finite. It follows that A_2 has finite D-composition series. Since $A_n = A$, we may apply induction to obtain that A has finite D-composition series. Then A has finite DG-composition series. By Proposition 15.8, $\zeta(G)$ is periodic, and hence finite. It suffices to apply Lemma 15.1 to obtain that G is finite, as required. $\square$

Corollary 15.10. *Let D be a Dedekind Z-domain, and let G be a finitely generated nilpotent group. Suppose that A is a monolithic DG-module with the DG-monolith M. Then A satisfies the following conditions:*

(1) *A is a P-module, where $P = \mathrm{Ann}_D(M) \in \mathrm{Spec}(D)$.*

(2) *A is the last term of its upper socular series.*

(3) *The socular height of A is at most ω, the first infinite ordinal.*

Proof. According to L.A. Kurdachenko, J. Otal and I.Ya. Subbotin [157, Corollary 1.16], $P = \operatorname{Ann}_D(M) \in \operatorname{Spec}(D)$. Let T be the D-periodic part of A. Since A is monolithic, T is equal to the P-component of A. Suppose that B is a finitely generated DG-submodule of A. By Corollary 15.9, B is D-periodic. It follows that $T = A$, and thus A is a P-module. Since B has finite composition DG-series, it suffices to apply Lemma 4.9. □

Lemma 15.11. *Let D be a Dedekind Z-domain, and let G be a finitely generated nilpotent group. Suppose that A is a monolithic DG-module with DG-monolith M, and put $H = C_G(M)$. Let*

$$\langle 0 \rangle = A_0 \leq A_1 = M \leq A_2 \leq \cdots \leq A_n \leq \cdots \leq \bigcup_{n \in \mathbb{N}} A_n = A$$

be an ascending DG-composition series of A. Then $H \leq C_G(A_n/A_{n-1})$ for every $n \in \mathbb{N}$.

Proof. Suppose that there exists some $m \in \mathbb{N}$ such that $C_G(A_m/A_{m-1})$ does not include H. Let $b \in A_n \setminus A_{n-1}$, and put $B = bDG$, and $C = B \cap A_{n-1}$. Then $G/C_G(B/C)$ does not include H. By Corollary 15.9, B has finite DG-composition series. By Theorem 10.21, B has the $Z - DK$-decomposition, that is, $B = U \oplus Z$, where $Z = \zeta_{DH}^\infty(B)$ is the upper hypercenter of B, and $U = \zeta_{DH}^*(B)$ is the unique maximal hypereccentric DG-submodule of B. Since H is a normal subgroup of G, U, and Z are DG-submodules of B. By the choice of H, we have $U \neq \langle 0 \rangle$. On the other hand, $H = C_G(M)$, so that $M \leq Z$, and hence, $U = \langle 0 \rangle$ since A is monolithic, a contradiction. □

Lemma 15.12. *Let F be a locally finite field, and let G be a finite nilpotent group. If A is a monolithic FG-module, then A is an artinian FG-module.*

Proof. Let M be the FG-monolith of A, and put $H = C_G(M)$. Then G/H is a p'-group, where $p = \operatorname{char} F$ (see, for example, L.A. Kurdachenko, J. Otal and I.Ya. Subbotin [157, Theorem 3.1]). By Corollary 15.10, A has an ascending series of FG-submodules

$$\langle 0 \rangle = A_0 \leq A_1 = M \leq A_2 \leq \cdots \leq A_n \leq \cdots \leq \bigcup_{n \in \mathbb{N}} A_n = A$$

with simple FG-factors. By Lemma 15.11, $H \leq C_G(A_j/A_{j-1})$, for every $j \in \mathbb{N}$. We proceed by induction on $|H|$. If $H = \langle 1 \rangle$, by Corollary 5.15, $A = M$, and so A is a simple FG-module. Suppose now that $H \neq \langle 1 \rangle$, and pick $1 \neq z \in H \cap \zeta(G)$. Since A is FH-hypercentral, A is likewise $F\langle z \rangle$-hypercentral; furthermore, the length of an upper $F\langle z \rangle$-central series of A is at most ω. Let

$$\langle 0 \rangle = C_0 \leq C_1 \leq C_2 \leq \cdots \leq C_n \leq \cdots \leq \bigcup_{n \in \mathbb{N}} C_n = C$$

be the upper $F \langle z \rangle$-central series of A. Since $z \in \zeta(G)$, every term C_n is an FG-submodule of A. The FG-submodule $C_1 = \mathrm{Ann}_A(z - 1)$ is monolithic, and $z \in H \cap C_G(C_1)$. In particular, $|H \cap C_G(C_1)| < |H|$. By induction, C_1 is an artinian FG-module. Hence, by Proposition 1.8, the FG-module A is artinian. □

Lemma 15.13. *Let F be a locally finite field, and let G be a finitely generated nilpotent group. If A is a monolithic FG-module, then A is an artinian FG-module.*

Proof. Let M be the FG-monolith of A, and put $H = C_G(M)$. Then G/H is a p'-group, where $p =$ char F (see L.A. Kurdachenko, J. Otal and I.Ya. Subbotin [157, Theorem 3.1]). By Corollary 15.10, A has an ascending series of FG-submodules

$$\langle 0 \rangle = A_0 \leq A_1 = M \leq A_2 \leq \cdots \leq A_n \leq \cdots \leq \bigcup_{n \in \mathbb{N}} A_n = A$$

with simple FG-factors. By Lemma 15.11, $H \leq C_G(A_j/A_{j-1})$, for every $j \in \mathbb{N}$. We proceed by induction on $r_0(H)$. If $r_0(H) = 0$, then H is finite, and our assertion follows from Lemma 15.12. Suppose that $r_0(H) > 0$. By Lemma 15.1, there exists some $z \in H \cap \zeta(G)$ having infinite order. Since A is FH-hypercentral, A is likewise $F \langle z \rangle$-hypercentral; furthermore, the length of an upper $F \langle z \rangle$-central series of A is at most ω. Let

$$\langle 0 \rangle = C_0 \leq C_1 \leq C_2 \leq \cdots \leq C_n \leq \cdots \leq \bigcup_{n \in \mathbb{N}} C_n = C$$

be the upper $F \langle z \rangle$-central series of A. Since $z \in \zeta(G)$, every term C_n is an FG-submodule of A. The FG-submodule $C_1 = \mathrm{Ann}_A(z - 1)$ is monolithic, and $z \in H \cap C_G(C_1)$. In particular, $r_0(H \cap C_G(C_1)) < r_0(H)$. By induction, C_1 is an artinian FG-module, and so is A by Proposition 1.8. □

Theorem 15.14. *Let D be a Dedekind Z-domain, and let G be a finitely generated nilpotent group. If A is a monolithic DG-module, then A is an artinian DG-module.*

Proof. If M is the DG-monolith of A, according to L.A. Kurdachenko, J. Otal and I.Ya. Subbotin [157, Corollary 1.16], $P = \mathrm{Ann}_D(M) \in \mathrm{Spec}(D)$. By Corollary 15.10, A is a P-module. Put $B = \Omega_{P,1}(A)$, and look at B as an FG-module, where $F = D/P$ is a locally finite field. By Lemma 15.13, B is an artinian FG-module, and so A is an artinian DG-module by Proposition 1.8. □

Corollary 15.15. *Let D be a Dedekind Z-domain, and let G be a finitely generated nilpotent-by-finite group. If A is a monolithic DG-module, then A is an artinian DG-module.*

Proof. Let M be the DG-monolith of A, and let H be a nilpotent normal subgroup of G having finite index. Then H is finitely generated (see, for example, [234, Theorem 1.41]). By Theorem 5.5, there exist a simple DH-submodule B of M, and a finite subset S of G such that $M = \bigoplus_{g \in S} Bg$. Let U be a DH-submodule of A which is maximal under $B \cap U = \langle 0 \rangle$. Then A/U is a monolithic DH-module with DH-monolith $(B+U)/U$. Since H is a finitely generated nilpotent subgroup, by Theorem 15.14, A/U is an artinian DH-module. Let T be a transversal to H in G. Put $V = \bigcap_{x \in T} Ux$ so that V is a DG-submodule. If $V \neq \langle 0 \rangle$, then $M \leq V$, and so $B \leq V$. On the other hand, $V \leq U$, and $U \cap B = \langle 0 \rangle$. This contradiction shows that $V = \langle 0 \rangle$. By Remak's theorem,

$$A \hookrightarrow A_1 = \bigoplus_{x \in T} A/Ux.$$

Further, for every $x \in T$, $A/Ux = Ax/Ux \cong A/U$ is an artinian DH-module. By Lemma 1.2, A_1 and A are artinian DH-modules. Thus, A is an artinian DG-module, as required. $\qquad\square$

Corollary 15.16. *Let D be a Dedekind Z-domain, and let G be a finitely generated nilpotent-by-finite group. If M is a simple DG-module, then $IE_{DG}(M)$ is an artinian DG-module.*

Corollary 15.17. *Let D be a Dedekind Z-domain, and let G be a finitely generated nilpotent-by-finite group. If A is an artinian DG-module, then $IE_{DG}(A)$ is an artinian DG-module.*

As we mentioned at the beginning of this chapter, in the case $D = \mathbb{Z}$, these two last results were shown by R.L. Snider [269] and extended to polycyclic-by-finite groups by I.M. Musson [195], who proved the following facts.

Theorem 15.18. *The following results hold:*

(1) *Let G be a polycyclic-by-finite group, and suppose that A is a finitely generated $\mathbb{Z}G$-module such that $|G/C_G(A)|$ is finite, and $pA = \langle 0 \rangle$ for some prime p (in particular, this happens if A is simple). Then $IE_{\mathbb{Z}G}(A)$ is an artinian $\mathbb{Z}G$-module.*

(2) *Let F be a field of characteristic $p > 0$, and let G be a polycyclic-by-finite group. If A is a finite dimensional FG-module (that is, $\dim_F(A)$ is finite), then $IE_{FG}(A)$ is an artinian FG-module.*

(3) *Let F be a non-locally finite field, and let G be a polycyclic-by-finite group. Then the injective envelope of every simple FG-module is artinian if, and only if G is abelian-by-finite.*

In connection with some parts of the above result, S. Donkin [58] proved:

Theorem 15.19. *The following results hold:*

(1) *Let F be a field, and let G be a polycyclic-by-finite group. Suppose that A is a locally finite dimensional FG-module (that is, $\dim_F(aFG)$ is finite for every $a \in A$). Then every essential extension of A is locally finite dimensional.*

(2) *Let F be a field of characteristic 0, and let G be a polycyclic-by-finite group. If A is a finite dimensional FG-module, then every essential extension of A is artinian, and $\mathrm{End}_{FG}(IE_{FG}(A))$ is a noetherian ring.*

The proofs of these results depend on applications from both the theory of polycyclic group rings and the theory of Hopf algebras, and we omit them here.

To finish, it is worth mentioning the following generalization of a result obtained by I.M. Musson [199].

Theorem 15.20. *Let F be a field of characteristic 0, and let G be a soluble-by-finite torsion-free group of finite special rank. If A is a finite dimensional FG-module, then every essential extension of A is artinian, and $\mathrm{End}_{FG}(IE_{FG}(A))$ is a noetherian ring.*

Chapter 16

Quasifinite modules

As we have seen in previous chapters, there are not too many cases in which artinian modules can be satisfactorily described, although many problems require the investigation of some specific artinian modules. We have not dealt with this question in this book in full. In this chapter, however, we consider one of the most important types of artinian modules, namely, the quasifinite modules. These modules appear in the following way. Suppose that A is an artinian DG-module, and let $\mathcal{U}$ be the family of all infinite submodules of A. Choose a minimal element M of $\mathcal{U}$. Then either M is a minimal (infinite), and hence, a simple submodule of A or M is infinite and not simple, but every proper submodule of M is finite. D.I. Zaitsev introduced this type of modules in connection with the study of the complementability of normal subgroups [296]. These modules also appeared in other group-theoretical researches, for example, in the study of groups with the weak maximal or minimal conditions for normal subgroups (L.A. Kurdachenko [145, 147], D.I. Zaitsev, L.A. Kurdachenko and A.V. Tushev [312]). In [299], D.I. Zaitsev initiated the investigation of modules over integral group rings in which all proper submodules are finite. Besides these modules, there are many types of Dedekind domains for which the unique finite module is the zero module. Therefore the condition of being a finite submodule is a reasonable change to the condition of being a finitely generated submodule. In other words, we come to a module A over a group ring DG with the property:

> *every proper DG-submodule of A is finitely generated as a D-submodule.*

In this setting, the two following situations appear:

(1) A has a proper DG-submodule B such that A/B is a simple DG-module;

(2) every proper DG-submodule of A is finitely generated as a D-submodule, and A is the union of its proper DG-submodules.

The first case can be reduced to the case of finitely generated D-modules and simple DG-modules. Thus, the second case remains more interesting. To study it, we consider the following concept. Let R be a ring, and let G be a group. An

RG-module A is said to be *a quasifinite RG-module* if A satisfies the following conditions:

(1) A is not finitely generated as a D-module.

(2) If B is a proper RG-submodule of A, then B is finitely generated as a D-submodule.

(3) A is the union of its proper DG-submodules.

The case in which $R = F$ is a finite field was considered by D.I. Zaitsev [299] while the case in which $R = F$ is an arbitrary field was considered by L.A. Kurdachenko and I.Ya. Subbotin in [172]; the case in which $R = D$ is a Dedekind domain was considered in the paper L.A. Kurdachenko [152]. In this chapter, we collect the basic results of these papers.

If $R = \mathbb{Z}$ and $G = \langle 1 \rangle$, then a Prüfer p-group gives us an example of a quasifinite module. Let F be a field, $G = \langle x \rangle$ be an infinite cyclic group $D = F \langle x \rangle$. Then a Prüfer P-module is an example of a quasifinite FG-module for every maximal ideal P of D. As we will see later, the Prüfer P-modules play a very important role in the structure of quasifinite modules.

Lemma 16.1. *Let A be a quasifinite RG-group, where R is a ring, and G is a group. Then A cannot be decomposed as a direct sum of two proper RG-submodules. In particular, $Soc_R(A)$ is a proper submodule of A.*

Proof. This lemma is obvious. $\qquad\qquad\qquad\qquad\qquad\qquad\qquad\qquad\qquad\square$

Corollary 16.2. *Let D be a Dedekind domain, G a group, and A a quasifinite DG-module. Then $Soc_{DG}(A)$ is a proper submodule of A.*

Lemma 16.3 (L.A. Kurdachenko [152]). *Let A be a quasifinite RG-module, where R is an integral domain, and G is a group. Then either A is R-divisible or there exists some $0 \neq x \in R$ such that $Ax = \langle 0 \rangle$.*

Proof. If $Ax = A$ for every element $0 \neq x \in R$, then A is R-divisible. Therefore suppose that there exists some $0 \neq x \in R$ such that $Ax \neq A$. The mapping $\phi : a \mapsto ax$, $a \in A$ is an RG-endomorphism of A such that $\text{Im } \phi = Ax$ and $\text{Ker } \phi = \text{Ann}_A(x)$. We claim that $\text{Ker } \phi = A$. For, otherwise $\text{Ker } \phi$ is proper, and so it is finitely generated as an R-module. Since $\text{Im } \phi = Ax \neq A$, $\text{Im } \phi$ is likewise a finitely generated R-module. Since

$$A/\text{Ker } \phi \cong \text{Im } \phi,$$

A is a finitely generated R-module. This contradiction shows that $\text{Ker } \phi = A$ so that $Ax = \langle 0 \rangle$, as required. $\qquad\qquad\qquad\qquad\qquad\qquad\qquad\square$

Further, we need to establish a result on the structure of finitely generated modules over Dedekind domains. Before doing so, we need to establish the uniqueness of the direct decompositions there involved.

Theorem 16.4. *Let P be a maximal ideal of a Dedekind domain D, and suppose that A is a P-module over D. Given two direct decompositions of A as direct sum of cyclic submodules,*

$$A = \bigoplus_{\lambda \in \Lambda} a_\lambda D = \bigoplus_{\gamma \in \Gamma} b_\gamma D,$$

then there are an automorphism ϕ of A, and a bijection $\theta : \Gamma \longrightarrow \Lambda$ such that $b_\gamma \phi = a_{\gamma\theta}$, for every $\gamma \in \Gamma$.

Proof. For every $n \in \mathbb{N}$, we put

$$\Lambda_n = \{\lambda \in \Lambda \mid P^n = \operatorname{Ann}_D(a_\lambda)\}, \text{ and}$$
$$\Gamma_n = \{\gamma \in \Gamma \mid P^n = \operatorname{Ann}_D(b_\gamma)\}.$$

It suffices to show that $|\Lambda_n| = |\Gamma_n|$, for every $n \in \mathbb{N}$. Define $C_n = \Omega_{P,1}(A) \cap AP^{n-1}$. We have

$$AP^{n-1} = (\bigoplus_{\lambda \in \Lambda_n} a_\lambda P^{n-1}) \oplus (\bigoplus_{\lambda \in \Lambda \setminus \Delta} a_\lambda P^{n-1})$$

and

$$AP^{n-1} = (\bigoplus_{\gamma \in \Gamma_n} b_\gamma P^{n-1}) \oplus (\bigoplus_{\gamma \in \Gamma \setminus \Upsilon} b_\gamma P^{n-1}),$$

where $\Delta = \Lambda_1 \cup \cdots \cup \Lambda_n$, and $\Upsilon = \Gamma_1 \cup \cdots \cup \Gamma_n$. Therefore,

$$C_n = (\bigoplus_{\lambda \in \Lambda_n} a_\lambda P^{n-1}) \bigoplus C_{n+1} = (\bigoplus_{\gamma \in \Gamma_n} b_\gamma P^{n-1}) \bigoplus C_{n+1}.$$

Hence, $|\Lambda_n| = |\Gamma_n| = \dim_{D/P}(C_{n+1}/C_n)$. $\qquad\square$

Lemma 16.5. *Let R be an integral domain, and suppose that A is a finitely generated R-module. If A is R-periodic, then $\operatorname{Ann}_R(A) \neq \langle 0 \rangle$.*

Proof. Let M be a finite subset of A such that $A = MR$. Obviously,

$$\prod_{a \in M} \operatorname{Ann}_R(a) \leq \bigcap_{a \in M} \operatorname{Ann}_R(a) \leq \operatorname{Ann}_R(A).$$

Since R has no zero-divisors, $\bigcap_{a \in M} \operatorname{Ann}_R(a) \neq \langle 0 \rangle$. Hence, $\operatorname{Ann}_R(a) \neq \langle 0 \rangle$, as required. $\qquad\square$

Proposition 16.6. *Let D be a Dedekind domain, and suppose that A is a finitely generated periodic D-module. Then:*

(1) *A is a direct sum of finitely many cyclic submodules.*

(2) *If $A = \bigoplus_{\lambda \in \Lambda} a_\lambda D = \bigoplus_{\gamma \in \Gamma} b_\gamma D$, where the sets Λ and Γ are finite, and for every $\lambda \in \Lambda$ and $\gamma \in \Gamma$ we have $\operatorname{Ann}_D(a_\lambda D) = P_\lambda^{k_\lambda}$, and $\operatorname{Ann}_D(b_\gamma D) = P_\gamma^{s_\gamma}$ for certain maximal ideals P_λ and P_γ of D, then there exist an automorphism ϕ of A, and a bijection $\theta : \Gamma \longrightarrow \Lambda$ such that $b_\gamma \phi = a_{\gamma\theta}$, for every $\gamma \in \Gamma$.*

Proof. By Lemma 16.5, $\mathrm{Ann}_D(A) \neq \langle 0 \rangle$. Then A is a direct sum of a finitely many cyclic submodules (D.W. Sharpe and P. Vamos [265, Theorem 6.14]), which gives (1). The assertion (2) follows from Corollary 6.25 and Theorem 16.4. □

Suppose that A is a finitely generated module over a Dedekind domain D, and put $T = t_D(A)$. By Lemma 1.1, A is a noetherian D-module, and, in particular, T is finitely generated. As the finitely generated factor-module A/T is D-torsion-free, our next step is the consideration of finitely generated torsion-free modules.

Lemma 16.7. *If L is an ideal of an integral domain R, there is a $\mathbb{Z}$-isomorphism $\theta : L^{-1} \longrightarrow \mathrm{Hom}_R(L, R)$ such that $a(x\theta) = xa$, for every $x \in L^{-1}$ and $a \in L$.*

Proof. Pick $x \in L^{-1}$, and define a mapping $\phi_x : L \longrightarrow R$ by $a\phi_x = xa$, for every $a \in L$. Obviously,

$$(a + b)\phi_x = x(a + b) = xa + xb = a\phi_x + b\phi_x,$$

and

$$(ca)\phi_x = x(ca) = c(xa) = c(a\phi_x)$$

so that ϕ_x is an R-homomorphism. This defines a map θ by the rule $x\theta = \phi_x$, for every $x \in L^{-1}$. Since

$$a\phi_{x+y} = (x + y)a = xa + ya = a\phi_x + a\phi_y = a(\phi_x + \phi_y),$$

we have $(x + y)\theta = x\theta + y\theta$, and, therefore, θ is a $\mathbb{Z}$-homomorphism. If $x \in \mathrm{Ker}\ \theta$, $a\phi_x = 0$ for each $a \in A$. Since R has no zero-divisors, $Lx = \langle 0 \rangle$ implies $x = 0$. Thus, θ is a monomorphism.

Finally, given $\chi \in \mathrm{Hom}_R(L, R)$, let $0 \neq a, b \in L$. Then

$$(a\chi)b = (ab)\chi = (ba)\chi = (b\chi)a,$$

so that $(a\chi)a^{-1} = (b\chi)b^{-1}$. It follows that $(a\chi)a^{-1}$ is constant when a runs through $L \setminus \langle 0 \rangle$. Put $(a\chi)a^{-1} = x$. Thus, $a\chi = x \cdot a$ provided $0 \neq a \in L$. Since $0\chi = 0 = x \cdot 0$, $\chi = \phi_x$. Furthermore, $a\chi \in R$, i.e. $ax \in R$, for every $a \in A$. Therefore, $x \in L^{-1}$, which shows that $\chi = (\phi_x)\theta$. Consequently, θ is a $\mathbb{Z}$-isomorphism. □

Recall that a module A over a ring R is said to be *hereditary* if A, and all of its submodules are projective. It is obvious that submodules of hereditary modules are hereditary. A ring R is said to be *hereditary* if all of its right ideals are projective modules.

Proposition 16.8. *Let L be a non-zero ideal of an integral domain R. Then the R-module L is projective if and only if L is an invertible ideal. In particular, R is a heredity ring if and only if every non-zero ideal of R is invertible.*

Proof. Suppose that L is invertible. Then there are some $a_1, \ldots, a_m \in L$, and $x_1, \ldots, x_m \in L^{-1}$ such that

$$1 = a_1 x_1 + \cdots + a_m x_m.$$

Let C be a free R-module with basis $\{c_1, \ldots, c_m\}$. There is an R-homomorphism $\pi : C \longrightarrow L$ such that $c_j \pi = a_j$, for every $1 \leq j \leq m$ (see D.S. Passman [219, Thoerem 2.4]). Define $\alpha : L \longrightarrow C$ by

$$a\alpha = c_1(a x_1) + \cdots + c_m(a x_m)$$

for each $a \in L$. Since R is commutative, α is an R-homomorphism. Further, if $a \in L$ we have

$$\begin{aligned}
a(\alpha\pi) = (a\alpha)\pi &= (c_1 a x_1 + \cdots + c_m a x_m)\pi \\
&= (c_1\pi)a x_1 + \cdots + (c_m\pi)a x_m \\
&= a_1 a x_1 + \cdots + a_m a x_m = a(a_1 x_1 + \cdots + a_m x_m) = a \cdot 1 = a,
\end{aligned}$$

and so $a\pi = \varepsilon_L$. By Lemma 7.15, $C = \operatorname{Im}\alpha \oplus \operatorname{Ker}\pi$, and $L \cong \operatorname{Im}\alpha$. It follows that L is a projective R-module (D.S. Passman [219, Thoerem 2.8]).

Conversely, suppose that L is a projective R-module. Then there exists a free R-module C_1 such that $C_1 = L \oplus A$, for some R-submodule A (see D. S. Passman [219, Theorem 2.8]). Let $\{c_\lambda | \lambda \in \Lambda\}$ be a free basis of C_1. If $a \in L$, then

$$a = \sum_{\lambda \in \Lambda} c_\lambda \tau_\lambda(a),$$

for some $\tau_\lambda(a) \in R$, $\lambda \in \Lambda$. If b is another element of L, then

$$a + b = \sum_{\lambda \in \Lambda} c_\lambda \tau_\lambda(a) + \sum_{\lambda \in \Lambda} c_\lambda \tau_\lambda(b) = \sum_{\lambda \in \Lambda} c_\lambda(\tau_\lambda(a) + \tau_\lambda(b)).$$

Since

$$a + b = \sum_{\lambda \in \Lambda} c_\lambda \tau_\lambda(a + b)$$

and $\{c_\lambda \mid \lambda \in \Lambda\}$ is a free basis of C_1,

$$\tau_\lambda(a + b) = \tau_\lambda(a) + \tau_\lambda(b),$$

for every $\lambda \in \Lambda$. Similarly, if $d \in R$,

$$\tau_\lambda(da) = d\tau_\lambda(a),$$

for every $\lambda \in \Lambda$. Hence, the mapping $a \mapsto \tau_\lambda(a)$ is an R-homomorphism for every $\lambda \in \Lambda$. By Lemma 16.7, for each $a \in L$, there are elements $y_\lambda \in L^{-1}$ such that $\tau_\lambda(a) = y_\lambda a$, for every $\lambda \in \Lambda$. In other words

$$a = \sum_{\lambda \in \Lambda} c_\lambda(y_\lambda a).$$

Let $0 \neq b \in L$, and put $\Lambda_b = \mathrm{Supp}((\tau_\lambda(b))_{\lambda \in \Lambda})$. Then Λ_b is finite, say $\Lambda_b = \{\lambda_1, \ldots, \lambda_k\}$. In particular, $\tau_\lambda(b) = 0$, for every $\lambda \in \Lambda \setminus \Lambda_b$. This means that $y_\lambda b = 0$, for every $\lambda \in \Lambda \setminus \Lambda_b$, and therefore $y_\lambda = 0$, for every $\lambda \in \Lambda \setminus \Lambda_b$. Hence,

$$b = \sum_{\lambda \in \Lambda} c_\lambda(y_\lambda b).$$

Let ρ be the projection mapping of C_1 on L. Since ρ is an R-homomorphism,

$$b = b\rho = (\sum_{\lambda \in \Lambda} c_\lambda(y_\lambda b))\rho = \sum_{\lambda \in \Lambda} b_\lambda(y_\lambda b).$$

Multiplying both sides of this equation by y^{-1} (R is a subring of its field of fractions), we obtain, then

$$1 = \sum_{\lambda \in \Lambda} b_\lambda y_\lambda.$$

Thus, $R = LL^{-1}$. $\square$

Lemma 16.9. *Let A be a finitely generated D-module, where D is a Dedekind domain. Suppose that there exists some $a \in A$ such that $\mathrm{Ann}_D(a) = \langle 0 \rangle$. Then A has a D-submodule B such that $a \notin B$ and that satisfies the following conditions:*

(1) *There exists some D-submodule C such that $A = B \oplus C$.*

(2) *There exists an ideal L of D such that $C \cong_D L$.*

Proof. We define a mapping $\theta : D \longrightarrow A$ by $x\theta = ax$, for every $x \in D$. Obviously, θ is a D-homomorphism, and, since $\mathrm{Ker}\, \theta = \mathrm{Ann}_D(a) = \langle 0 \rangle$, θ is in fact a monomorphism. Let F be the field of fractions of D. Since F is D-divisible, F is an injective D-module by Theorem 7.9. Therefore, if $\iota : D \longrightarrow F$ is the canonical embedding, then there exists a D-homomorphism $\phi : A \longrightarrow F$ such that $\theta\phi = \iota$. Then

$$1 = 1\iota = 1(\theta\phi) = (1\theta)\phi = a\phi$$

and so ϕ is a non-zero homomorphism.

Moreover, $A\phi$ is a fractional ideal of D. It follows that there is an ideal L of D such that $A\phi \cong_D L$. Since D is a Dedekind domain, the ideal L is invertible, and so L is a projective D-module by Theorem 7.9. It follows that $A = B \oplus C$, where $B = \mathrm{Ker}\, \phi$ and $C \cong_D L$. Since $a\phi \neq 0$, we have $a \notin B$, and we are done. $\square$

Corollary 16.10. *Let A be a finitely generated D-module, where D is a Dedekind domain. If A is D-torsion-free, then*

$$A = C_1 \oplus \cdots \oplus C_m,$$

where $C_1, \ldots, C_m$ are D-isomorphic to ideals of D. In particular, every finitely generated D-torsion-free D-module is projective.

Proof. Let M be a maximal D-free subset of A, and put

$$E = \bigoplus_{a \in M} aD.$$

By Lemma 1.1, E is finitely generated, and, in particular, M is finite, let us say $M = \{a_1, \ldots, a_m\}$. By Lemma 16.9, $A = C_1 \oplus B_1$, where $a_1 \notin B_1$, and C_1 is D-isomorphic to some ideal of D. Since A/B_1 is D-torsion-free, $a_1 D \cap B_1 = \langle 0 \rangle$. If M_1 is a maximal D-free subset of B_1, then $M_1 \cup \{a_1\}$ is a maximal D-free subset of A. Since D is an integral domain, $|M_1| = m - 1$, so that $M_1 = \{b_1, \ldots, b_{m-1}\}$. Applying Lemma 16.9 to B_1, we obtain a decomposition $B_1 = C_2 \oplus B_2$, where C_2 is D-isomorphic to some ideal of D, and $b_1 \notin B_2$. Since A/B_2 is D-torsion-free, $(a_1 D \oplus b_1 D) \cap B_2 = \langle 0 \rangle$. Proceeding in the same way, we finally get a decomposition

$$A = C_1 \oplus \cdots \oplus C_m \oplus E_1,$$

where $C_1, \ldots, C_m$ are isomorphic to ideals of D. Since $r_0(C_1 \oplus \cdots \oplus C_m) = m = r_0(A)$, we have that $E_1 = \langle 0 \rangle$; hence,

$$A = C_1 \oplus \cdots \oplus C_m,$$

as required. □

Corollary 16.11. *Let A be a finitely generated D-torsion-free D-module, where D is a Dedekind domain. Then there exists a non-zero ideal L of D such that, if $m = r_0(A)$,*

$$A \cong_D \underbrace{D \oplus \cdots \oplus D}_{m-1} \oplus L.$$

Moreover, L is determined up to multiplication by a non-zero constant in the field of fractions of D.

Proof. This follows at once from Corollary 16.10 and D.S. Passman [219, Theorem 7.7]. □

Corollary 16.12. *Let A be a finitely generated D-module, where D is a Dedekind domain. Then $A = T \oplus C$, where $T = t_D(A)$, and C is a finitely generated D-torsion-free submodule.*

Proof. This follows at once from Corollary 16.10 and the standard properties of projective modules. □

We are now in a position to prove the basis structure theorem of finitely generated modules over Dedekind domains.

Theorem 16.13. *Let A be a finitely generated D-module, where D is a Dedekind domain. Then the following assertions hold:*

(1) $A = T \oplus C$, *where $T = t_D(A)$ is the D-periodic part of A and C is a finitely generated D-torsion-free submodule.*

(2) $T = A_1 \oplus \cdots \oplus A_k$, where $\{A_1, \ldots, A_k\}$ are primary cyclic submodules.

(3) If $T = B_1 \oplus \cdots \oplus B_t$ where every B_j is a primary cyclic submodule, then $k = t$, and there are a permutation $\sigma \in S_k$, and an automorphism ϕ of T such that $B_j\phi = A_{j\sigma}$, for every $1 \le j \le k$.

(4) If $m = r_0(C)$, then there exists a non-zero ideal L of D such that

$$C \cong_D \underbrace{D \oplus \cdots \oplus D}_{m-1} \oplus L.$$

Moreover, L is determined up to multiplication by a non-zero constant in the field of fractions of D.

Proof. Apply Corollary 16.12, Proposition 16.6, and Corollary 16.11. $\square$

Now, we may continue the study of quasifinite modules.

Lemma 16.14 ([152]). *Let A be a quasifinite DG-module, where D is a Dedekind domain and G is a group. If A is D-periodic, then there exists some $P \in \mathrm{Spec}(D)$ such that A is a P-module, and either $AP = \langle 0 \rangle$ or*

$$A = C_1 \oplus \cdots \oplus C_n,$$

where $C_1, \ldots, C_n$ are Prüfer P-modules.

Proof. Since A is D-periodic,

$$A = \bigoplus_{P \in \pi} A_P,$$

where $\pi = \mathrm{Ass}_D(A)$. Since every A_P is a DG-submodule, by Lemma 16.1, there exists some $P \in \mathrm{Spec}(D)$ such that $A = A_P$. Consider $A_1 = \Omega_{P,1}(A)$. If $A = A_1$, then $AP = \langle 0 \rangle$. Otherwise A_1 is finitely generated over D. In particular, $\dim_{D/P} A_1$ is finite and L.A. Kurdachenko, J. Otal and I.Ya. Subbotin [157, Lemma 5.6] yields that A is D-artinian. By Lemma 16.3, A is D-divisible and L.A. Kurdachenko, J. Otal and I.Ya. Subbotin [157, Theorem 5.7] shows that A is a direct sum of finitely many Prüfer P-modules. $\square$

Lemma 16.15 ([152]). *Suppose that D is a Dedekind domain that is not a field. Let G be a group, and let A be a quasifinite DG-module. Then we have:*

(1) *If D is not a local ring, then A is D-periodic.*

(2) *If D is a local ring, and A is not D-periodic, then A is D-torsion-free and divisible. Moreover, if F is the field of fractions for D, then A as an FG-module is simple, and $\dim_F A$ is finite.*

Proof. Put $T = t_D(A)$. Suppose first that $T \ne \langle 0 \rangle$. By Lemma 16.3, either $Ax = \langle 0 \rangle$ for some $0 \ne x \in D$ or A is D-divisible. In the first case $A = T$. In the second one, A is D-divisible so that T is also D-divisible by Lemma 7.18. Also, by Theorem 16.13, T is not finitely generated over D, and it follows that $T = A$. Suppose now that $T = \langle 0 \rangle$. By (QF 3), A has a proper DG-submodule B, which *a fortiori* is finitely generated over D. In fact, B can be chosen such that B has minimal D-rank. Let C be a D-pure envelope of B so that C is a D-divisible DG-submodule of A. By Theorem 7.25, C can not be finitely generated as D-submodule so that $C = A$. Since A/B is D-periodic and D-divisible, by Theorem 7.25

$$A/B = \bigoplus_{P \in \mathrm{Spec}(D)} A_P/B,$$

where A_P/B is a a direct sum of Prüfer P-modules. If D is not a local ring, A/B cannot be a P-module for some $P \in \mathrm{Spec}(D)$, and we obtain a contradiction with Lemma 16.14.

Let now D be a local ring. Since A is D-divisible and D-torsion-free, we can consider A as an FG-module. Since A is a pure envelope of a finitely generated D-submodule, $\dim_F A$ is finite.

Finally, let U be a non-zero FG-submodule of A. Then $B \cap U$ is a non-zero DG-submodule of B. By the choice of B we have $r_0(B) = r_0(B \cap U)$. This means that a pure envelope of $B \cap U$ coincides with the pure envelope of B, that is, with A. In other words, A is a simple FG-module. $\square$

The structure of simple FG-modules A of finite dimension can be studied in a satisfactory way. Remark that the most interesting cases appear when D is a field, and when D is a not a field but A is D-periodic. Let D be a Dedekind domain. If $P \in \mathrm{Spec}(D)$, and $n \in \mathbb{N}$, we denote by $\phi_n : D/P^n \longrightarrow D/P^{n+1}$ the canonical epimorphism, and we put

$$D(P^\infty) = \varprojlim \{D/P^n, \ \phi_n \mid n \in \mathbb{N}\}.$$

Proposition 16.16 (L.A. Kurdachenko and H. Smith [168]). *Let D be a Dedekind domain. If $P \in \mathrm{Spec}(D)$, and A is a Prüfer P-module, then the ring of endomorphisms of A is isomorphic to $D(P^\infty)$. Moreover, $D(P^\infty)$ is a principal ideal domain, and the set of all non-zero ideals of $D(P^\infty)$ is $\{D(P^\infty)P^n \mid n \in \mathbb{N}\}$.*

Proof. Let $y \in P \setminus P^2$, and consider the mapping $a \mapsto ay$, $a \in A$. For $n \in \mathbb{N}$, put $A_n = \Omega_{P,n}(A)$ so that $A_{n+1}y = A_n$. Since A_1 is simple, there exists some $0 \neq a_1$ such that $A_1 = a_1 D$ and $\mathrm{Ann}_D(a_1) = P$. Since $A_2 y = A_1$, there exists some a_2 such that $a_1 = a_2 y$. Thus, $a_2 \notin A_1$, and therefore $a_2 D = A_2$ and $\mathrm{Ann}_D(a_2) = P^2$. Similarly, we choose elements $\{a_n \mid n \in \mathbb{N}\}$ such that

$$A = \bigcup_{n \in \mathbb{N}} a_n D, \ A_n = a_n D \leq a_{n+1} D = A_{n+1},$$

$$a_1 y = 0, \ a_{n+1} y = a_n, \text{ and } \mathrm{Ann}_D(a_n) = P^n.$$

In particular, $a_n A \cong D/P^n$, for every $n \in \mathbb{N}$.

Pick $\phi \in \text{Ann}_D(A)$. Then $(a_n\phi)P^n = \langle 0 \rangle$, and so $a_n\phi \in a_nD$. It follows that there exists some $\alpha_n \in D$ such that $a_n\phi = a_n\alpha_n$. If $\beta_n \in D$ and $a_n\alpha_n = a_n\beta_n$, then $a_n(\alpha_n - \beta_n) = 0$, that is $\alpha_n - \beta_n \in \text{Ann}_D(a_n) = P^n$. In particular, $\alpha_n + P^n = \beta_n + P^n$, for every $n \in \mathbb{N}$. Define a mapping

$$\Xi : \text{End}_D(A) \longrightarrow \prod_{n \in \mathbb{N}} D/DP^n$$

by $\phi\Xi = (a_n + P^n)_{n \in \mathbb{N}}$. We have

$$a_n\alpha_n = a_n\phi = (a_{n+1}y)\phi = (a_{n+1}\phi)y = (a_{n+1}\alpha_{n+1})y = a_n\alpha_{n+1}.$$

It follows that $a_n(\alpha_{n+1} - \alpha_n) = 0$, that is $\alpha_{n+1} - \alpha_n \in \text{Ann}_D(a_n) = P^n$, which gives $\alpha_n + P^n = \alpha_{n+1} + P^n$. It follows that $\phi\Xi \in D(P^\infty)$. Let $\psi \in \text{End}_D(A)$, and suppose that $\psi\Xi = (\beta_n + P^n)_{n \in \mathbb{N}}$. One easily obtains that

$$a_n(\phi + \psi) = a_n(\alpha_n + \beta_n), \text{ and } a_n(\phi\psi) = a_n(\alpha_n\beta_n).$$

It follows that

$$(\phi + \psi)\Xi = \phi\Xi + \psi\Xi, \text{ and } (\phi\psi)\Xi = \phi\Xi \cdot \psi\Xi$$

and so Ξ is a homomorphism. If $(\gamma_n + P^n)_{n \in \mathbb{N}} \in D(P^\infty)$, then we define a mapping χ by $a_n\chi = a_n\gamma_n$, for every $n \in \mathbb{N}$. It is easy to see that χ is an endomorphism of A, and $\chi\Xi = (\gamma_n + P^n)_{n \in \mathbb{N}}$. Hence, $\text{Im } \Xi = D(P^\infty)$. Obviously, $\text{Ker } \Xi = \langle 0 \rangle$, and so it follows that

$$\text{End}_D(A) \cong \text{Im } \Xi = D(P^\infty).$$

Suppose that $(\lambda_n + P^n)_{n \in \mathbb{N}} \in U(D(P^\infty))$. Then there is $(\mu_n + P^n)_{n \in \mathbb{N}} \in D(P^\infty)$ such that

$$(\lambda_n + P^n)_{n \in \mathbb{N}}(\mu_n + P^n)_{n \in \mathbb{N}} = (1 + P^n)_{n \in \mathbb{N}}.$$

Thus, $\lambda_n\mu_n + P^n = 1 + P^n$, for every $n \in \mathbb{N}$, and, in particular $\lambda_1 \notin P$.

Conversely, suppose we are given some $(\nu_n + P^n)_{n \in \mathbb{N}} \in D(P^\infty)$ such that $\nu_1 \notin P$. Since P is a maximal ideal of D, $P + \nu_1D = D$. We have $\nu_n + P^m = \nu_m + P^m$, for every pair $m \leq n$, and then $\nu_n \notin P$ for every $n \in \mathbb{N}$. Again $P + \nu_nD = D$. By Lemma 6.7, $\nu_nD + P^n = D$. Hence, there exist some $\tau_n \in D$, and $\xi_n \in P^n$ such that $\nu_n\tau_n + \xi_n = 1$. Therefore,

$$1 + P^n = \nu_n\tau_n + \xi_n + P^n$$
$$= \nu_n\tau_n + P^n = (\nu_n + P^n)(\tau_n + P^n).$$

Since the inverse element is unique, $\tau_n + P^m = \tau_m + P^m$, for every pair $m \leq n$, and so $(\tau_n + P^n)_{n \in \mathbb{N}} \in D(P^\infty)$. Hence,

$$U(D(P^\infty)) = \{(\nu_n + P^n)_{n \in \mathbb{N}} \in D(P^\infty) \mid \nu_1 \notin P\}.$$

Suppose now we are given a non-zero element $(\beta_n + P^n)_{n\in\mathbb{N}} \in D(P^\infty)$, and some $t \in \mathbb{N}$ such that $\beta_1, \ldots, \beta_t \in P$ but $\beta_{t+1} \notin P$. Note that $D(P^\infty)P^t = D(P^\infty)y^t$. Indeed, let $(\sigma_n + P^n)_{n\in\mathbb{N}} \in D(P^\infty)$, $x \in P^t$. By Proposition 6.13, $P^t = Dy^t + P^{t+m}$, for every $m \in \mathbb{N}$, and then there exist some $z_{t,m} \in D$, and $w_{t,m} \in P^{t+m}$ that $x = y^t z_{t,m} + w_{t,m}$. Then

$$((\sigma_n + P^n)_{n\in\mathbb{N}})x = ((\sigma_n + P^n)x)_{n\in\mathbb{N}}$$
$$= (\sigma_n x + P^n)_{n\in\mathbb{N}} = (\sigma_n(y^t z_{t,m} + w_{t,m}) + P^n)_{n\in\mathbb{N}}$$
$$= (0,\ldots,0,\sigma_{t+1}(y^t z_{t,1} + w_{t,1}) + P^{t+1},\ldots,\sigma_{t+m}(y^t z_{t,m} + w_{t,m}) + P^{t+m},\ldots)$$
$$= (0,\ldots,0,\sigma_{t+1}y^t z_{t,1} + P^{t+1},\ldots,\sigma_{t+m}y^t z_{t,m} + P^{t+m},\ldots)$$

because $w_{t,m} \in P^{t+m}$, for every $m \in \mathbb{N}$. Since $D(P^\infty)P^t = D(P^\infty)y^t$,

$$(\beta_n + P^n)_{n\in\mathbb{N}} = ((\gamma_n + P^n)_{n\in\mathbb{N}})y^t,$$

where $(\gamma_n + P^n)_{n\in\mathbb{N}} \in U(D(P^\infty))$. Since the D-module $D(P^\infty)$ is torsion-free (see D.G. Northcott [207, 9.10, Proposition 14]), the expression

$$(\beta_n + P^n)_{n\in\mathbb{N}} = ((\gamma_n + P^n)_{n\in\mathbb{N}})y^t,$$

where $(\gamma_n + P^n)_{n\in\mathbb{N}} \in U(D(P^\infty))$, is unique. If $(\delta_n + P^n)_{n\in\mathbb{N}}$ is another non-zero element of $D(P^\infty)$, again

$$(\delta_n + P^n)_{n\in\mathbb{N}} = ((\eta_n + P^n)_{n\in\mathbb{N}})y^m,$$

where $(\eta_n + P^n)_{n\in\mathbb{N}} \in U(D(P^\infty))$. Thus,

$$(\beta_n + P^n)_{n\in\mathbb{N}}(\delta_n + P^n)_{n\in\mathbb{N}} = ((\gamma_n + P^n)_{n\in\mathbb{N}})((\eta_n + P^n)_{n\in\mathbb{N}})y^{t+m} \neq 0.$$

Consequently, $D(P^\infty)$ is an integral domain.

Finally, let I be a non-zero ideal of $D(P^\infty)$, and pick $0 \neq (\beta_n + P^n)_{n\in\mathbb{N}} \in I$. Then

$$(\beta_n + P^n)_{n\in\mathbb{N}} = ((\gamma_n + P^n)_{n\in\mathbb{N}})y^t,$$

where $(\gamma_n + P^n)_{n\in\mathbb{N}} \in U(D(P^\infty))$. It follows that $D(P^\infty)y^t \leq I$. Note that

$$D(P^\infty)y^t = D(P^\infty)P^t = \varprojlim\{D/P^{t+m} \mid m \in \mathbb{N}\}$$

and $D(P^\infty)/D(P^\infty)P^t \cong D/P^t$. By Proposition 6.13, the set of all ideals of D/P^t is

$$\{D/P^t, P/P^t, \ldots, P^{t-1}/P^t, \langle 0 \rangle\},$$

hence, $I/D(P^\infty)P^t$ is isomorphic to P^{t-k}/P^t for some $k \in \mathbb{N}$. Therefore $I = D(P^\infty)P^{t-k} = D(P^\infty)y^{t-k}$. This means that $D(P^\infty)$ is a principal ideal domain, and the set of all non-zero ideals of $D(P^\infty)$ is equal to

$$\{D(P^\infty)P^n \mid n \in \mathbb{N}\},$$

and all has been proved. $\qquad\qquad\qquad\qquad\qquad\qquad\qquad\qquad\qquad\qquad\qquad\square$

Corollary 16.17 ([152]). *Let D be a Dedekind domain that is not a field, G a group, and A a quasifinite DG-module. If A is D-periodic, then $G/C_G(A)$ can be embedded in $GL_n(D(P^\infty))$ where $n = \dim_{D/P} \Omega_{P,1}(A)$, and $\{P\} = \mathrm{Ass}_D(A)$.*

Proof. By Lemma 16.14,
$$A = C_1 \oplus \cdots \oplus C_n,$$
where $C_1, \ldots, C_n$ are Prüfer P-modules. It suffices to apply Proposition 16.16. $\square$

Proposition 16.18 ([152]). *Let D be a Dedekind domain, G a group, H a normal subgroup of G of finite index, $\{g_1, \ldots, g_t\}$ a transversal to H in G, and A a quasifinite DG-module such that $C_G(A) = \langle 1 \rangle$. If A is D-periodic, then A has a DH-submodule B satisfying the following properties:*

(i) *Bg_j is a quasifinite DH-submodule for each $1 \le j \le t$.*

(ii) *$A = Bg_1 + \cdots + Bg_t$.*

(iii) *$H \hookrightarrow H/(g_1^{-1}C_H(B)g_1) \times \cdots \times H/(g_t^{-1}C_H(B)g_t)$.*

Proof. Let $\mathcal{M}$ be the set of DH-submodules E of A that are not finitely generated as D-submodules. If U is a proper DG-submodule, then U is finitely generated over D. Since U is D-periodic, by Theorem 16.13, U has a finite D-composition series so that U certainly has a finite DG-composition series. It follows that A is an artinian DG-module. By Theorem 5.2, A is an artinian DH-module as well. Hence, $\mathcal{M}$ has a minimal element B. By the choice of B, every proper DH-submodule of B is finitely generated over D. We claim that B is a quasifinite DH-submodule. Otherwise, B has a proper DH-submodule C such that B/C is a simple DH-module. Put
$$C_1 = CDG = Cg_1 + \cdots + Cg_t.$$
Thus, C_1 is finitely generated as a D-submodule. Since A is a quasifinite DG-module, A has an ascending series of proper DG-submodules
$$C_1 \le C_2 \le \cdots \le C_n \le \cdots$$
such that
$$A = \bigcup_{n \in \mathbb{N}} C_n.$$
It follows that
$$B = \bigcup_{n \in \mathbb{N}} (B \cap C_n).$$
Since B/C is a simple DH-module, either $(B/C) \cap (C_n/C) = B/C$ or this intersection is zero. Since B is not finitely generated over D, the same is true for B/C. Since each C_n is finitely generated as a D-submodule, $(B/C) \cap (C_n/C) = \langle 0 \rangle$, that is $B \cap C_n = C$, which gives $B = C$, a contradiction. This contradiction shows our claim. For every $1 \le i \le t$, each mapping $a \mapsto ag_i$, $a \in A$ is a D-isomorphism that applies isomorphically the DH-submodules among themselves.

Therefore $Bg_1, \ldots, Bg_t$ are quasifinite DH-submodules. Since $Bg_1 + \cdots + Bg_t$ is a DG-submodule, which is not finitely generated over D,

$$A = Bg_1 + \cdots + Bg_t.$$

Since every $C_H(Bg_i) = g_i^{-1} C_H(B) g_i$,

$$g_1^{-1} C_H(B) g_1 \cap \cdots \cap g_t^{-1} C_H(B) g_t = C_H(A) = \langle 1 \rangle,$$

and, therefore, the embedding (3) is guaranteed by Remak's theorem. $\square$

Lemma 16.19 ([152]). *Let D be a Dedekind domain that is not a field, G a locally soluble group, and A a quasifinite DG-module such that $C_G(A) = \langle 1 \rangle$. If A is D-periodic, then G is abelian-by-finite.*

Proof. By Corollary 16.17, $\mathrm{Ass}_D(A) = \{P\}$, where $P \in \mathrm{Spec}(D)$, and

$$A = C_1 \oplus \cdots \oplus C_n,$$

where $C_1, \ldots, C_n$ are Prüfer P-module, $1 \le j \le n$. Let $R = D(P^\infty), K$ be a field of fraction for R, and let R_0 denote the R-module K/R. Put $A^* = \mathrm{Hom}_R(A, R_0)$. Then A^* is a free R-module, and $r_0(A^*) = n$ (B. Hartley [106, Lemma 1.2]). By [106, Lemma 2.1] $B = (A^*) \bigotimes_R K$ is a simple KG-module. By Zassenhaus's theorem (see, for example, B.A.F. Wehrfritz [280, Theorem 3.7]) G is soluble, and Maltsev's theorem (see, for example, B.A.F. Wehrfritz [280, Lemma 3.5]), G is abelian-by-finite. $\square$

Lemma 16.20 ([152]). *Let D be a Dedekind domain that is not a field, suppose that G is a group having an abelian normal subgroup U of finite index, and let A be a quasifinite DG-module with $C_G(A) = \langle 1 \rangle$. If A is D-periodic, then the periodic part T of U has finite special rank.*

Proof. By Corollary 16.17, $\mathrm{Ass}_D(A) = \{P\}$, where $P \in \mathrm{Spec}(D)$, and

$$A = C_1 \oplus \cdots \oplus C_n,$$

where $C_1, \ldots, C_n$ are Prüfer P-modules, $1 \le j \le n$. By Proposition 16.18, A includes a quasifinite DU-submodule B satisfying the conditions (i)-(iii). Let $R = D(P^\infty)$, K be the field of fractions for R, and let R_0 denote the R-module K/R. Put $B^* = \mathrm{Hom}_R(B, R_0)$. According to B. Hartley [106, Lemma 2.1], $C = (B^*) \bigotimes_R K$ is a simple KU-module, $\dim_K C = r_0(B) = n$. Then the periodic part of $U/C_U(B)$ is locally cyclic (see, for example, L.A. Kurdachenko, J. Otal and I.Ya. Subbotin [157, Theorem 2.3]). By Proposition 16.18,

$$U \hookrightarrow U/(g_1^{-1} C_U(B) g_1) \times \cdots \times U/(g_t^{-1} C_U(B) g_t),$$

where $\{g_1, \ldots, g_t\}$ is a transversal to U in G, so that, T is a subgroup of finite special rank. $\square$

Lemma 16.21 ([152]). *Let R be a ring, G a group, and A a quasifinite RG-module such that $C_G(A) = \langle 1 \rangle$. If $1 \neq x \in \zeta(G)$, then $A = A(x - 1)$.*

Proof. Since the mapping $\phi : a \mapsto a(x - 1)$, $a \in A$ is an RG-endomorphism of A, Im $\phi = A(x - 1)$, and Ker $\phi = C_A(x)$ are RG-submodules. Since $x \notin C_G(A)$, $C_A(x) \neq A$. It follows that $C_A(x)$ is finitely generated over R. Since

$$A(x - 1) \cong A/C_A(x),$$

$A(x - 1)$ cannot be finitely generated over R. This means that $A = A(x - 1)$, as required.   □

Corollary 16.22 ([152]). *Let D be a Dedekind domain of characteristic $p > 0$, G a group, and A a quasifinite DG-module such that $C_G(A) = \langle 1 \rangle$. Then G has no non-identity normal finite p-subgroups.*

Proof. Suppose the contrary, and let P a non-identity normal finite p-subgroup of G. Let S be a minimal G-invariant subgroup of P. Then S is abelian. If $H = C_G(S)$, then $|G : H|$ is finite, and $S \leq \zeta(H)$. By Proposition 16.18, A has a quasifinite DH-submodule B satisfying the conditions (1)–(3) of that statement. We claim that $C_H(B)$ does not include S. For, otherwise $S \leq C_H(B)$, so $S = g^{-1}Sg \leq g^{-1}C_H(B)g = C_H(Bg)$ for every $g \in G$, and, by above (2), $S \leq C_G(A) = \langle 1 \rangle$, a contradiction that shows our claim. Pick $x \in S \backslash C_H(B)$ of prime order. Since char $D = p$, the additive group of B is an elementary abelian p-subgroup, and it follows that the natural semidirect product $B \rtimes \langle x \rangle$ is a nilpotent p-group. In particular, $[B, x] = B(x - 1) \neq B$. By Lemma 16.21, $B(x - 1) = B$, and, then we have just found a final contradiction that proves the result.   □

Theorem 16.23 ([152]). *Let D be a Dedekind domain that is not a field, G a locally soluble group, and A a quasifinite DG-module such that $C_G(A) = \langle 1 \rangle$. Suppose that A is D-periodic, then the following assertions hold:*

(1) $\mathrm{Ass}_D(A) = \{P\}$, *where $P \in \mathrm{Spec}(D)$.*

(2) $A = C_1 \oplus \cdots \oplus C_n$, *where $C_1, \ldots, C_n$ are Prüfer P-modules.*

(3) G *has an abelian normal subgroup U of finite index.*

(4) *The periodic part of U has finite special rank.*

(5) *If char $D = p > 0$, then $O_p(G) = \langle 1 \rangle$.*

Proof. (1) and (2) follow from Lemma 16.14, (3) from Lemma 16.19 and (4) from Lemma 16.20. Suppose that $O_p(G) \neq \langle 1 \rangle$. By (3) and (4), $O_p(G)$ has finite special rank, hence, $O_p(G)$ has a non-identity finite G-invariant subgroup. But this is a contradiction to Corollary 16.22.   □

Suppose that A is a quasifinite DG-module. As we mentioned above, the study of the structure of A falls into two cases:

(1) either A is D-periodic and D-divisible, or

(2) there exists some $P \in \mathrm{Spec}(D)$ such that $\mathrm{Ann}_D(A) = P$.

Note that we already dealt with case (1) in Theorem 16.23. Therefore, we focus on the case (2), that is we consider a quasifinite DG-module A such that $\mathrm{Ann}_D(A) = P$ is a maximal ideal. If $F = D/P$ is the corresponding residue field, A can be naturally made into an FG-module; so we need to study quasifinite FG-modules, where F is a field.

Lemma 16.24 (L.A. Kurdachenko and I.Ya. Subbotin [172]). *Let A be a quasifinite FG-module, where F is a field, and G is a group. Then $\mathrm{Soc}_{FG}(A)$ is a non-zero proper submodule of A.*

Proof. Since A is not simple, A has a non-zero proper FG-submodule B. Then $\dim_F B$ is finite, and therefore B has a non-zero simple FG-submodule S. It follows that $\mathrm{Soc}_{FG}(A) \neq \langle 0 \rangle$. It suffices to apply Corollary 16.2. □

Lemma 16.25 ([172]). *Let F be a field, G a group, and A a quasifinite DG-module such that $C_G(A) = \langle 1 \rangle$. If H is a non-identity finite normal subgroup of G, then $\mathrm{Soc}_{FH}(A) = A$.*

Proof. Suppose the contrary, that is, $B = \mathrm{Soc}_{FH}(A) \neq A$. Since H is normal, B is an FG-submodule. It follows that $\dim_F B$ is finite. Then there exists a vector F-subspace C such that

$$A = B \oplus C.$$

Define

$$C_0 = \bigcap_{h \in H} Ch.$$

Since

$$A/Ch = Ah/Ch \cong A/C,$$

$\dim_F(A/Ch) = \dim_F(A/C) = \dim_F B$ is finite. It follows that C_0 has finite codimension, and, in particular, $C_0 \neq \langle 0 \rangle$. Pick $0 \neq c_1 \in C_0$, and put $C_1 = c_1 FH$. Since H is finite, $\dim_F C_1$ is finite. Therefore C_1 has a non-zero simple FH-submodule C_2. We have $C_2 \leq \mathrm{Soc}_{FH}(A) = B$, and, on other hand, $C_2 \leq C_0$ and $C_0 \cap B = \langle 0 \rangle$, a contradiction. □

Corollary 16.26 ([172]). *Let F be a field, G a group, and A a quasifinite DG-module such that $C_G(A) = \langle 1 \rangle$. If H is a non-identity finite normal subgroup of G, then $C_A(H) = \langle 0 \rangle$, and $A(\omega FH) = A$.*

Proof. By Lemma 16.25,

$$A = \bigoplus_{n \in \mathbb{N}} C_n,$$

where every C_n is a simple FH-submodule. Since $C_n(\omega FH)$ is an FH-submodule of C_n, either $C_n(\omega FH) = C_n$ or $C_n(\omega FH) = \langle 0 \rangle$. Therefore

$$A = C_A(H) \bigoplus A(\omega FH).$$

By Lemma 16.1, $C_A(H) = \langle 0 \rangle$, and so $A(\omega FH) = A$.   □

Corollary 16.27 ([172]). *Let F be a field, G a group, and A a quasifinite FG-module such that $C_G(A) = \langle 1 \rangle$. If H is a non-identity finite normal subgroup of G, and B is a non-zero proper FG-submodule of A, then $C_H(B) = \langle 1 \rangle$.*

Corollary 16.28 ([172]). *Let F be a field, G a group, and A a quasifinite FG-module such that $C_G(A) = \langle 1 \rangle$. Suppose that H is a non-identity normal subgroup of G that has an ascending series of G-invariant subgroups*

$$\langle 1 \rangle = H_0 \leq H_1 \leq \cdots \leq H_\alpha \leq H_{\alpha+1} \leq H_\gamma = H$$

with finite factors. If B is a non-zero proper FG-submodule of A, then $C_H(B) = \langle 1 \rangle$.

Proof. We proceed by induction on α. If $\alpha = 1$, then it suffices to apply Corollary 16.27. Let $\alpha > 1$, and suppose we have $H_\beta \cap C_H(B) = \langle 1 \rangle$, for every ordinal $\beta < \alpha$. Put $C_\alpha = H_\alpha \cap C_H(B)$. If α is a limit ordinal, then

$$H_\alpha = \bigcup_{\beta < \alpha} H_\beta,$$

so that

$$C_\alpha = \bigcup_{\beta < \alpha} (C_\alpha \cap H_\beta) = \bigcup_{\beta < \alpha} C_\beta = \langle 1 \rangle.$$

Suppose now that α is not a limit. Put $L = H_{\alpha-1}$, and assume that $C_\alpha \neq \langle 1 \rangle$. Then $L \cap C_\alpha = \langle 1 \rangle$, and so

$$C_\alpha \cong C_\alpha/(L \cap C_\alpha) \cong (L + C_\alpha)/L \leq H_\alpha/L.$$

Thus, C_α is finite, which contradicts Corollary 16.26 since $B \leq C_A(C_\alpha)$. Hence, $C_\alpha = \langle 1 \rangle$. By induction, matching $\alpha = \gamma$, we obtain $C_H(B) = \langle 1 \rangle$, as required.   □

Let G be a group. A normal subgroup H of G is said to be *the hyperfinite radical* of G if it satisfies the following conditions:

(1) H possesses an ascending series of G-invariant subgroups

$$\langle 1 \rangle = H_0 \leq H_1 \leq \cdots \leq H_\alpha \leq H_{\alpha+1} \leq \cdots \leq H_\gamma = H,$$

every factor of which is finite.

(2) G/H has no non-identity normal finite subgroups.

We denote the hyperfinite radical by $HF(G)$.

Corollary 16.29 ([172]). *Let F be a field, G a group, and A a quasifinite DG-module such that $C_G(A) = \langle 1 \rangle$. Then $HF(G)$ is an abelian-by-finite subgroup of finite special rank. Moreover, if $\operatorname{char} F = p > 0$, then $O_p(HF(G)) = \langle 1 \rangle$.*

Proof. Put $H = HF(G)$. Let B be a proper non-zero FG-submodule of A. By Corollary 16.28, $C_H(B) = \langle 1 \rangle$ so that we can think of H as a subgroup of $GL(n, F)$, where $n = \dim_F B$. By a result due to Kargapolov (see B.A.F. Wehrfritz [280, Corollary 9.31]), H is soluble-by-finite. By Maltsev's theorem (see B.A.F. Wehrfritz [280, Theorem 3.6]) H has a normal subgroup S of finite index such that $g^{-1}Sg \leq T(n, F_1)$, where F_1 is a finite field extension of F. Put

$$U = (g^{-1}Sg) \cap UT(n, F_1), \quad V = gUg^{-1},$$

so that V is normal in S. If $\operatorname{char} F = p > 0$, then $UT(n, F_1)$ is a bounded nilpotent p-subgroup. Suppose that $U \neq \langle 1 \rangle$. Then $O_p(H) \neq \langle 1 \rangle$. By the definition of H, this means that H has a finite non-identity G-invariant p-subgroup in contradiction with Corollary 16.22. If $\operatorname{char} F = 0$, then $UT(n, F_1)$ is a torsion-free nilpotent subgroup, and hence, $U = \langle 1 \rangle$ in this case. Further,

$$T(n, F_1)/UT(n, F_1) \cong \underbrace{U(F_1) \times \cdots \times U(F_1)}_{n}$$

Since the periodic subgroups of $U(F_1)$ are locally cyclic (see G. Karpilovsky [131, Proposition 4.4.1]), then S is an abelian subgroup of finite special rank. $\square$

Corollary 16.30 ([172]). *Let F be a field, G a hypercentral group, and A a quasifinite FG-module such that $C_G(A) = \langle 1 \rangle$. Then the periodic part T of G is an abelian-by-finite p'-subgroup of finite special rank, where $p = \operatorname{char} F$.*

Proof. Indeed, in a hypercentral group, the periodic part is known to be equal to the hyperfinite radical. $\square$

Put $S = \operatorname{Soc}_{FG}(A)$. By Lemma 16.24, S is a proper non-zero FG-submodule of A. In particular, $\dim_F S = s$ is finite. If $C_G(S) = \langle 1 \rangle$, then we think of G as a subgroup of $GL_s(F)$ (in fact, it is a completely reducible subgroup of $GL_s(F)$).

Lemma 16.31 ([152]). *Let F be a field, G a locally soluble group, and A a quasifinite FG-module such that $C_G(A) = \langle 1 \rangle$. If $C_G(\operatorname{Soc}_{FH}(A)) = \langle 1 \rangle$, then G is abelian-by-finite.*

Proof. As we mentioned above, G is a completely reducible subgroup of $GL_s(F)$, where $s = \dim_F \operatorname{Soc}_{FG}(A))$. Again, by Zassenhaus's theorem (see B.A.F. Wehrfritz [280, Theorem 3.7]), G is soluble, and, by Maltsev's theorem (see B.A.F. Wehrfritz [280, Lemma 3.5]), G is abelian-by-finite. $\square$

Lemma 16.32 ([152]). *Let F be a field, G an abelian group, and A a quasifinite FG-module such that $C_G(A) = \langle 1 \rangle$. Given $1 \neq x \in G$, we consider A as a $D(x)$-module, where $D(x) = F \langle t_x \rangle$ is the group algebra of an infinite cyclic group $\langle t_x \rangle$ over F whose action is induced by $at_x = ax$, for every $a \in A$. Then there exists some $x \in G$ such that $\mathrm{Ann}_{D(x)}(A) = \langle 0 \rangle$.*

Proof. We recall that $D(x)$ is a principal ideal domain. By Lemma 16.15, A is a $D(x)$-periodic module, and, by Lemma 16.14, either $\mathrm{Ann}_{D(x)}(A) \neq \langle 0 \rangle$ or A is $D(x)$-divisible. Suppose that $\mathrm{Ann}_{D(x)}(A) \neq \langle 0 \rangle$ for every $x \in G$. Put

$$S_1 = \mathrm{Soc}_{FG}(A), \text{ and } S_2/S_1 = \mathrm{Soc}_{FG}(A/S_1).$$

If $0 \neq a \in S_2$, then $aFG \cap S_1 \neq \langle 0 \rangle$. Hence, there exists some $u \in FG$ such that $0 \neq au \in S_1$. We write

$$u = a_1 x_1 + \cdots + a_m x_m,$$

where $a_1, \ldots, a_m \in F$, and $x_1, \ldots, x_m \in G$. By our assumption, there exists $P_1 \in \mathrm{Spec}(D(x_1))$ such that $AP_1 = \langle 0 \rangle$. Put $F_1 = D(x_1))/P_1$, and think of A as an $F_1 G$-module. Since A has a non-zero annihilator in $D(x_1)$, A has a non-zero annihilator P_2 in $F_1 \langle t_2 \rangle$, where $t_2 = t_{x_2}$. By Lemma 16.14, $P_2 \in \mathrm{Spec}(F_1 \langle t_2 \rangle)$, which means that the $F \langle x_1, x_2 \rangle$-module S_2 is semisimple, and homogeneous. Using the same arguments, after finitely many steps we obtain that the $F \langle x_1, \ldots, x_m \rangle$-module S_2 is semisimple and homogeneous. But in this case $aF \langle x_1, \ldots, x_m \rangle$ is a simple submodule so that $aF \langle x_1, \ldots, x_m \rangle \cap S_1 = \langle 0 \rangle$, which contradicts the choice of a. This contradiction proves the result. $\square$

Theorem 16.33 ([152]). *Let F be a field, G a locally soluble group, and A a quasifinite FG-module such that $C_G(A) = \langle 1 \rangle$. Put $S = \mathrm{Soc}_{FG}(A)$. If $C_G(S) = \langle 1 \rangle$, then the following assertions hold:*

(1) *G has an abelian normal subgroup U of finite index.*

(2) *The periodic part of U has finite special rank*

(3) *If $\mathrm{char}\, F = p > 0$, then $O_p(G) = \langle 1 \rangle$.*

(4) *U contains an element x of infinite order, and A has a quasifinite FU-submodule B such that $\mathrm{Ass}_{F\langle x \rangle}(B) = \{P\}$, for some $P \in \mathrm{Spec}(F \langle x \rangle)$, and*

$$B = C_1 \oplus \cdots \oplus C_n,$$

where $C_1, \ldots, C_n$ are Prüfer P-modules.

(5) *$A = B \oplus Bg_1 \oplus \cdots \oplus Bg_t$, where $\{1, g_1, \ldots, g_t\}$ is a transversal to U in G.*

Proof. The assertion (1) follows from Lemma 16.31, while (2) and (3) follow from Corollary 16.29. Let B be a quasifinite FU-submodule satisfying the conditions (1)-(3) of Proposition 16.18. By Lemma 16.32, $U \setminus C_U(B)$ contains an element

x such that $\mathrm{Ann}_{F\langle x\rangle}(B) \neq \langle 0\rangle$. Clearly $|x|$ is infinite. By Lemma 16.14, B is a P-module, for some $P \in \mathrm{Spec}(F\langle x\rangle)$, and

$$B = C_1 \oplus \cdots \oplus C_n,$$

where $C_1, \ldots, C_n$ are Prüfer P-modules. Finally, (5) follows from Proposition 16.18. $\qquad\square$

Lemma 16.34 ([152]). *Let F be a field, G a hypercentral group, and A a quasifinite FG-module such that $C_G(A) = \langle 1\rangle$. If $C_G(Soc_{FG}(A)) \neq \langle 1\rangle$, then $C_G(Soc_{FG}(A)) \cap \zeta(G)$ contains an element x of infinite order such that A is $F\langle x\rangle$-periodic, and $\mathrm{Ass}_{F\langle x\rangle}(A) = \{(x-1)F\langle x\rangle\}$.*

Proof. Put $S_1 = Soc_{FG}(A)$, and $Z = C_G(S_1) \cap \zeta(G)$. Then $Z \neq \langle 1\rangle$, and, by Corollary 16.27, Z is torsion-free. Pick $1 \neq x \in Z$. Then the mapping $\phi : a \mapsto a(x-1)$, $a \in A$ is an FG-endomorphism of A, and so $\mathrm{Ker}\,\phi = C_A(x)$, and $\mathrm{Im}\,\phi = A(x-1)$ are FG-submodules. Put $S_2/S_1 = Soc_{FG}(A/S_1)$. By Lemma 16.3, S_2/S_1 is a proper FG-submodule, and it follows that

$$S_2/S_1 = E_1/S_1 \oplus \cdots \oplus E_k/S_1,$$

where $E_1/S_1, \ldots, E_k/S_1$ are simple FG-submodules. For every $1 \leq j \leq k$, since

$$E_j\phi \cong E_j/\mathrm{Ker}\,\phi = E_j/S_1,$$

$E_j\phi$ is simple so that $E_j\phi \leq S_1$. Thus, $S_2(x-1) \leq S_1$. Put

$$S_3/S_2 = Soc_{FG}(A/S_2), \ldots, S_{n+1}/S_n = Soc_{FG}(A/S_n), \ldots.$$

Similarly, $S_{n+1}(x-1) \leq S_n$, for every $n \in \mathbb{N}$. Since $\dim_F(\bigcup_{n\in\mathbb{N}} S_n)$ is infinite, we deduce

$$\bigcup_{n\in\mathbb{N}} S_n = A.$$

It follows that A is $F\langle x\rangle$-periodic. Moreover, $\mathrm{Ass}_{F\langle x\rangle}(B) = \{(x-1)F\langle x\rangle\}$. $\qquad\square$

Theorem 16.35 ([152]). *Let F be a field, G a hypercentral group, and A a quasifinite FG-module such that $C_G(A) = \langle 1\rangle$. Put $S = Soc_{FG}(A)$. If $C_G(S) \neq \langle 1\rangle$, then the following assertions hold:*

(1) *G is abelian-by-finite.*

(2) *The periodic part T of a group G is a p'-group of finite special rank where $p = \mathrm{char}\,F$.*

(3) *$T \cap \zeta(G)$ is locally cyclic.*

(4) *$C_G(S) \cap \zeta(G)$ contains an element x of infinite order such that A is $F\langle x\rangle$-periodic, and $\mathrm{Ass}_{F\langle x\rangle}(A) = \{P\}$ where $P = (x-1)F\langle x\rangle$.*

(5) $A = C_1 \oplus \cdots \oplus C_n$, where $C_1, \ldots, C_n$ are Prüfer P-modules.

Proof. By Lemma 16.34, $C_G(S) \cap \zeta(G)$ contains an element x of infinite order such that A is $F\langle x \rangle$-periodic, and $\mathrm{Ass}_{F\langle x \rangle}(B) = \{P\}$, where $P = (x-1)F\langle x \rangle$. If $D = F\langle x \rangle$, then D is a principal ideal domain, and it suffices to apply Theorem 16.23. $\qquad\square$

We note that this result can be extended to FC-hypercentral groups.

To finish, we are dealing with results of D.I. Zaitsev that concern quasifinite FG-modules, that are carried out when F is a finite field. In this case, it is possible to obtain more information concerning the periodic subgroups of the group G considered.

Lemma 16.36 (D.I. Zaitsev [299])**.** *Let F be a finite field, G a group, and A a quasifinite FG-module such that $C_G(A) = \langle 1 \rangle$. If $\zeta(G)$ is infinite, then G has a finitely generated subgroup H such that A is a quasifinite FH-module.*

Proof. Let T be the periodic part of $Z = \zeta(G)$, and suppose that B is a proper FG-submodule of A. Since F is finite, B is likewise finite. By Corollary 16.28, $C_T(B) = \langle 1 \rangle$. If follows that T is finite, and hence, $\zeta(G)$ contains an element x of infinite order. Think of A as a DG-module, where $D = F\langle x \rangle$. By Lemma 16.15 and Lemma 16.14,

$$A = A_1 \oplus \cdots \oplus A_n,$$

where $A_1, \ldots, A_n$ are Prüfer P-modules for some $P \in \mathrm{Spec}(F\langle x \rangle)$, $1 \leq j \leq n$. Put $C_m = \Omega_{P,m}(A)$, $m \in \mathbb{N}$. Then $A = \bigcup_{m \in \mathbb{N}} C_m$. Besides, C_m is finite for any m.

Let $\mathfrak{G}$ be the family of all finitely generated subgroups, containing the element x. Since A is an artinian $F\langle x \rangle$-module, then A is an artinian FK-module for every subgroup $K \in \mathfrak{G}$. Therefore A includes a quasifinite FK-submodule $A(K)$. Clearly, $C_1 \cap A(K) \neq \langle 0 \rangle$. If B is an F-subspace of A, then put

$$\mathfrak{G}(B) = \{K \in \mathfrak{G} \mid C_1 \cap A(K) = B\}.$$

Since C_1 is finite, there is a finite F-subspace B_1 such that $\mathfrak{G}(B_1)$ is a local system for G. Since A is a P-module, $C_2 \cap A(K) \neq B_1$, $K \in \mathfrak{G}(B_1)$. There is a finite F-subspace B_2 such that $\mathfrak{G}(B_2)$ is a local system for G. Proceeding in this way, we construct a strictly ascending chain of F-subspaces

$$B_1 < B_2 < \cdots < B_j < \cdots,$$

and a descending chain of local systems for G

$$\mathfrak{G}_1 \supseteq \mathfrak{G}_2 \supseteq \cdots \supseteq \mathfrak{G}_j \supseteq \cdots$$

such that $\mathfrak{G}_j = \mathfrak{G}(B_j)$ (that is, $B_j = C_j \cap A(K)$ for every $K \in \mathfrak{G}_j, j \in \mathbb{N}$). Let $B = \bigcup_{j \in N} B_j$, and $b \in B, g \in G$. Then $b \in B_j$ for some $i \in \mathbb{N}$. Since $\mathfrak{G}_j$ is a local system for G, $g \in K$ for some subgroup $K \in \mathfrak{G}_j$. Since $A(K)$ is an FK-submodule,

$B_j = C_j \cap A(K)$ is an FK-submodule too. It follows that $bg \in B_j \leq B$, i.e. B is an FG - submodule. Since B is infinite, then $B = A$. Thus, $C_1 \leq B_j$ for some $j \in \mathbb{N}$. If $K \in \mathfrak{G}_j$, then

$$C_1 \cap A(K) = (C_1 \cap C_j) \cap A(K)$$
$$= C_1 \cap (C_j \cap A(K)) = C_1 \cap B_j = C_1.$$

On the other hand, $K \in \mathfrak{G}_j \subseteq \mathfrak{G}_1$. Thus, $C_1 \cap A(K) = B_1$; that is, $C_1 \cap B_1$. Hence

$$\Omega_{P,1}(A) = C_1 = B_1 \leq A(K)$$

for every $K \in \mathfrak{G}_1$. By Lemmas 16.15 and 16.14 $A(K)$ is a D-divisible P-module, so that the equation $\Omega_{P,1}(A) = \Omega_{P,1}(A(K))$ implies $A(K) = A$. Thus, we may choose every subgroup from the family $\mathfrak{G}_1$ as a candidate for H . $\qquad\square$

Theorem 16.37 ([299]). *Let F be a finite field, G a group, and A a quasifinite FG-module such that $C_G(A) = \langle 1 \rangle$. If $\zeta(G)$ is infinite, then the following assertions hold:*

(1) *If S is a periodic normal subgroup of G, then S is finite.*

(2) *If S is a periodic subgroup of G, then S has a bounded nilpotent p-subgroup of finite index, where $p = \operatorname{char} F$.*

Proof. As in the previous result, we deduce that there exists some $x \in \zeta(G)$ of infinite order such that

$$A = C_1 \oplus \cdots \oplus C_n,$$

where $C_1, \ldots, C_n$ are Prüfer P-modules for some $P \in \operatorname{Spec}(F\langle x \rangle)$. Together with Proposition 16.16 this implies that G is isomorphic with some subgroups of $GL_n(D(P^\infty))$. Since $D(P^\infty)$ is a principal ideal domain, there exists the field of fractions K for $D(P^\infty)$, i.e. G can be embedded in $GL(n, K)$. By Schur's theorem (see, for example, B.A.F. Wehrfritz [280, Theorem 9.1]) every periodic subgroup of G is locally finite.

Let S be a periodic normal subgroup of G, and define $H = S \times \langle x \rangle$. Since A is an artinian $F\langle x \rangle$-module, A is an artinian FH-module, and so A has a quasifinite FH-submodule B. Suppose that $x^t \in C_H(B)$ for some $t \neq 0$. Then $B \leq C_A(x^t)$, and, in particular, $C_A(x^t)$ is infinite. Note that $C_A(x^t)$ is an FG-submodule, because $x^t \in \zeta(G)$. This means that $A = C_A(x^t)$, and then $x^t \in C_G(A) = \langle 1 \rangle$. Since $|x|$ is infinite, this is a contradiction that shows $\langle x \rangle \cap C_H(B) = \langle 1 \rangle$. Thus, $C_H(B) \leq S$. Since $C_H(B)$ is normal in H, $H^* = H/C_H(B)$ is infinite. By Lemma 16.36, there exists a finitely generated subgroup Q such that $x \in Q$, and B is a quasifinite $F(Q^*)$-module, where $Q^* = QC_H(B)/C_H(B)$. Thus, $Q/\langle x \rangle$ is a finitely generated locally finite group, and so it is finite. Thus, $Q = \langle x \rangle \times U$, where $U = Q \cap S$ is finite. Obviously,

$$B_1 = C_B(x) \neq \langle 0 \rangle .$$

Since $U^* = UC_H(B)/C_H(B)$ is a finite normal subgroup of Q^*, we may apply Corollary 16.27 to the quasifinite $F(Q^*)$-module B, and its submodule B_1 to obtain

$$U^* \cap C_{H^*}(B_1) = U^* \cap C_{Q^*}(B_1) = \langle 1 \rangle .$$

In other words,

$$C_H(B) = UC_H(B) \cap C_H(B_1) = C_H(B)(U \cap C_H(B_1)) = C_H(B)C_U(B_1),$$

so that $C_U(B_1) \le C_H(B)$. Since this remains true for every finite subgroup $V \ge U$, $C_U(B_1) \le C_H(B)$. Since B_1 is a finite FH-submodule, $S/C_S(B_1)$ is finite, and the inclusion $C_H(B) \le S$, gives further that $S/C_H(B)$ is finite as well. By the finiteness of $\Omega_{P,1}(A)$, and $\Omega_{P,1}(B)$, there are elements $g_1, \ldots, g_t$ such that

$$\Omega_{P,1}(A) = \Omega_{P,1}(B)g_1 + \cdots + \Omega_{P,1}(B)g_t.$$

Thus,

$$E = Bg_1 + \cdots + Bg_t$$

is an $F\langle x \rangle$-divisible P-module since so is P, and we have $\Omega_{P,1}(E) = \Omega_{P,1}(A)$. Since A is an $F\langle x \rangle$-divisible P-module, $E = A$. Since S is a normal subgroup of G, $|S : C_H(B)| = |S : C_H(Bg_j)|$ for every $1 \le j \le n$. It follows that

$$|S : (C_H(Bg_1) \cap \cdots \cap C_H(Bg_t))|$$

is finite. Since

$$C_H(Bg_1) \cap \cdots \cap CH(Bg_t) \le C_G(A) = \langle 1 \rangle ,$$

S is finite.

Now let S be a periodic subgroup of G. Again we put $H = S \times \langle x \rangle$. Let B_1 be a quasifinite FH-submodule of A, B_2/B_1 a quasifinite FH-submodule of A/B_1, and so on. Proceeding as above, we show that the series

$$\langle 0 \rangle = B_0 \le B_1 \le B_2 \le \cdots \le B_n \le \cdots .$$

must be finite, that is there exists some $n \in \mathbb{N}$ such that $B_n = A$. For every $1 \le j \le n$, we define $H_j = C_H(B_j/B_{j-1})$. As above, we find that $H_j \le S$. By (1), every periodic normal subgroup of H/H_j is finite. In particular, S/H_j is finite. Define

$$S_0 = H_1 \cap \cdots \cap H_n$$

so that S/S_0 is finite, and moreover S_0 acts trivially on all factors of the series

$$\langle 0 \rangle = B_0 \le B_1 \le \cdots \le B_n = A.$$

It suffices to apply, for example, O.H. Kegel and B.A.F. Wehrfritz [134, Theorem 1.C.1], to get that S_0 is a bounded nilpotent p-subgroup, as required.   □

Chapter 17

Some applications: splitting over the locally nilpotent residual

We recall that if G is a group and N is a normal subgroup of G, it is said that G *splits over* N if there exists a subgroup H of G such that $G = N \rtimes H$ (that is, $G = NH$, and $N \cap H = \langle 1 \rangle$). Such an H is said to be *a complement* to N in G. If all complements to N are conjugate, then it is said that G *conjugately splits over* N; the results involving this are known in general as splitting theorems.

In the theory of finite soluble groups, splitting theorems play a very significant role. We mention here the result of W. Gaschütz [83] and E. Schenkman [253], asserting that if the nilpotent residual L of the finite group G is abelian, then G conjugately splits over L, is one of the most important achievements in this direction. Here it is decisive that the finiteness of G implies the existence of the Z-decomposition in L, that is, $L = Z \oplus E$, where $Z = \zeta^\infty_{\mathbb{Z}G}(L)$, and $E = \mathfrak{J}E^\infty_{\mathbb{Z}G}(L)$. By the choice of L, we have $L = \mathfrak{J}E^\infty_{\mathbb{Z}G}(L)$, and, in particular, $L = [L, G]$, and $L \cap \zeta(G) = \langle 1 \rangle$. The latter two conditions are the most common restrictions involved in almost all splitting theorems.

Since the condition Min-G appears as an extension of the finiteness, we are naturally drawn to the following situation: *the locally nilpotent residual L of the group G is abelian and is an artinian $\mathbb{Z}G$-module.* The theorems about splitting of a group over its abelian generalized nilpotent radical are very useful in many different investigations. Among these results we choose a theorem proven by D.J.S. Robinson [246]. In [246] the author's proof is based on homological methods. In this chapter we decided to give a group theoretical proof of this theorem. The following proof is based on the ideas and results of D.I. Zaitsev [304, 307, 308] and M.J. Tomkinson [275]. This fact aptly illustrates the effectiveness of the artinian condition.

Let G be a group. Suppose that $H_1, \ldots, H_n$ are normal subgroups of G and put

$$H = H_1 \cap \cdots \cap H_n.$$

By Remak's theorem, the mapping $g \mapsto (gH_1, \ldots, gH_n)$, $g \in G$, defines a homomorphism

$$\rho : G \hookrightarrow G/H_1 \times \cdots \times G/H_n.$$

In particular, G/H is isomorphic to some subgroup of $G/H_1 \times \cdots \times G/H_n$. If $\mathrm{Im}\,\rho = G/H_1 \times \cdots \times G/H_n$, we will write

$$G/H = G/H_1 \times \cdots \times G/H_n.$$

Lemma 17.1 (M.J. Tomkinson [275]). *Let G be a group, H a normal subgroup of G having finite index and $T = \{g_1, \ldots, g_n\}$ a transversal to H in G. Suppose that A is a $\mathbb{Z}G$-module and B is a proper $\mathbb{Z}H$-submodule of A. For each set $\mathcal{J}$ of subsets of T, form*

$$B(\mathcal{J}) = \sum_{S \in \mathcal{J}} \left(\bigcap_{g \in S} Bg \right)$$

and choose $B(\mathcal{J}_0)$ to be a maximal member of the set

$$\{ B(\mathcal{J}) \mid B \leq B(\mathcal{J}) \text{ and } B(\mathcal{J}) \neq A \}.$$

Then

$$B_0 = \bigcap_{g \in G} B(\mathcal{J}_0)g$$

is a $\mathbb{Z}G$-submodule of A and

$$A/B_0 = \bigoplus_{g \in S} A/B(\mathcal{J}_0)g,$$

for some subset S of T.

Proof. Clearly, we may write

$$B_0 = B(\mathcal{J}_0)g_1 \cap \cdots \cap B(\mathcal{J}_0)g_n.$$

We show, by induction on m, that

$$A/(B(\mathcal{J}_0)g_1 \cap \cdots \cap B(\mathcal{J}_0)g_m)$$

is a direct sum of certain $A/B(\mathcal{J}_0)g$. So we assume that

$$A/(B(\mathcal{J}_0)g_1 \cap B(\mathcal{J}_0)g_{m-1}) = \bigoplus_{g \in M} A/B(\mathcal{J}_0)g$$

for some subset M of T. If $B(\mathcal{J}_0)g_m \geq B(\mathcal{J}_0)g_1 \cap \cdots \cap B(\mathcal{J}_0)g_{m-1}$, then

$$A/(B(\mathcal{J}_0)g_1 \cap \cdots \cap B(\mathcal{J}_0)g_{m-1}) = \bigoplus_{g \in M} A/B(\mathcal{J}_0)g.$$

Otherwise, by maximality of $B(\mathcal{J}_0)$ and, consequently, of $B(\mathcal{J}_0)g_m$, we have

$$B(\mathcal{J}_0)g_m + (B(\mathcal{J}_0)g_1 \cap \cdots \cap B(\mathcal{J}_0)g_{m-1}) = A,$$

and so

$$A/B_0 = \bigoplus_{g \in S} A/B(\mathcal{J}_0)g,$$

where $S = M \cup \{g_m\}$. □

Lemma 17.2 ([275]). *Let G be a group, H a non-identity normal subgroup of G and A a non-zero $\mathbb{Z}G$-module. Assume further that the following conditions hold:*

(i) $HC_G(A)/C_G(A) \le FC_\infty(G/C_G(A))$.

(ii) *A is an artinian $\mathbb{Z}G$-module.*

(iii) $A = A(\omega\mathbb{Z}H)$.

Then G has a normal subgroup K and A has a proper $\mathbb{Z}G$-submodule Q satisfying the following conditions:

(1) $K/C_K(A/Q)$ *is finitely generated.*

(2) $K/C_K(A/Q) \le FC(G/C_K(A/Q)) \cap HC_K(A/Q)/C_K(A/Q)$.

(3) $A/Q = (A/Q)(\omega\mathbb{Z}K)$.

Moreover, if H is locally soluble, then $K/C_K(A/Q)$ is the normal closure of a single element of $G/C_K(A/Q)$, and is either a finite elementary abelian p-group for some prime p or a free abelian group of finite 0-rank. Further, if H is locally nilpotent, we can choose K so that $K/C_K(A/Q) \le \zeta(HC_K(A/Q)/C_K(A/Q))$.

Proof. Since A is an artinian $\mathbb{Z}G$-module, it suffices to deduce the general case from the case when the result holds for every proper $\mathbb{Z}G$-submodule of A; that is, we may assume that the latter is true. There is no loss if we assume that $C_G(A) = \langle 1 \rangle$. By Lemma 3.15, $H \cap FC(G) \ne \langle 1 \rangle$. Pick $1 \ne x \in H \cap FC(G)$, and put $F = \langle x \rangle^G$. Then F is a finitely generated subgroup; that is, $F = \langle x_1, \ldots, x_n \rangle$. If $A = A(\omega\mathbb{Z}F)$, then it suffices to choose $K = F$ and $B = \langle 0 \rangle$.

Therefore, we may suppose that $A_1 = A(\omega\mathbb{Z}F) \ne A$. Since $C_G(A) = \langle 1 \rangle$, $A_1 \ne \langle 0 \rangle$. Put $U = C_G(F)$ so that G/F is finite. By Theorem 5.2, A is an artinian $\mathbb{Z}U$-module. Let $V = H \cap U = C_H(F)$.

Suppose first that $A_1(\omega\mathbb{Z}V) \ne A_1$. Let $\eta : A \longrightarrow A/A_1(\omega\mathbb{Z}V)$ be the canonical epimorphism. Since $V \le C_G(A_1\eta)$, $A\eta(y - 1) \le A\eta(\omega\mathbb{Z}F) \le A_1\eta$ for every $y \in F$. Thus, $V \le C_G(A\eta(y - 1))$. From the isomorphism $A\eta(y - 1) \cong_{\mathbb{Z}V} A\eta/C_{A\eta}(y)$ we obtain the inclusion $V \le C_G(A\eta/C_{A\eta}(y))$ for every $y\eta F$, and therefore, $V \le C_G(A\eta/C_{A\eta}(F))$.

If we suppose that $C_{A\eta}(F) = A\eta$, then $A(\omega\mathbb{Z}F) \le A_1(\omega\mathbb{Z}V)$, and so $A_1 = A(\omega\mathbb{Z}F) = A_1(\omega\mathbb{Z}V)$ contrary to our assumption. Hence, the factor-module $A^* = A\eta/C_{A\eta}(F))$ is non-zero. The equation $A = A(\omega\mathbb{Z}H)$ implies that $A\eta = A\eta(\omega\mathbb{Z}V)$, and $A^* = A^*(\omega \mathbb{Z} V)$. Also $V \le C_G(A^*)$, and so $H/C_H(A^*)$ is finite. Therefore, in this case, we may choose $K = H$.

Suppose now that $A_1(\omega \mathbb{Z} V) = A_1$. Since A_1 is a proper $\mathbb{Z} G$-submodule, by our initial assumption, G has a normal subgroup L, and A_1 has a proper $\mathbb{Z} G$-submodule C such that $A_1/C = (A_1/C)(\omega \mathbb{Z} L)$, and $L/C_L(A_1/C)$ is a finitely generated subgroup of $FC(G/C_L(A_1/C)) \cap (VC_L(A_1/C)/C_L(A_1/C))$. Let

$$D_1 = A(x_1 - 1), \dots, D_n = A(x_n - 1).$$

Then every D_j is a $\mathbb{Z} U$-submodule, and

$$A_1 = A(\omega \mathbb{Z} F) = D_1 + \cdots + D_n.$$

Let $\nu : A \longrightarrow A/C$ be the canonical epimorphism. Then

$$A_1 \nu = D_1 \nu + \cdots + D_n \nu.$$

Let k be the largest integer such that

$$A_1 \nu = D_k \nu + \cdots + D_n \nu.$$

Then the $\mathbb{Z} U$-module $A_1 \nu/(D_{k+1}\nu + \cdots + D_n\nu)$ is non-zero, and

$$A_1 \nu/(D_{k+1}\nu + \cdots + D_n\nu) \cong_{\mathbb{Z} U} D_k\nu/(D_k\nu \cap (D_{k+1}\nu + \cdots + D_n\nu)).$$

Since $A_1 = A_1 \nu(\omega \mathbb{Z} L)$,

$$(D_k\nu/(D_k\nu \cap (D_{k+1}\nu + \cdots + D_n\nu)))(\omega \mathbb{Z} L)$$
$$= D_k\nu/(D_k\nu \cap (D_{k+1}\nu + \cdots + D_n\nu)).$$

This means that D_k has a non-zero $\mathbb{Z} U$-image D_k^{**} such that $D_k^{**}(\omega \mathbb{Z} L) = D_k^{**}$. Using the isomorphism

$$D_k = A(x_k - 1) \cong_{\mathbb{Z} U} A/C_A(x_k)$$

we see that A has a proper $\mathbb{Z} U$-submodule B such that $D_k^{**} \cong_{\mathbb{Z} U} A/B$, and so $(A/B)(\omega \mathbb{Z} L) = A/B$. Now we form the $\mathbb{Z} G$-submodule B_0 as we did in Lemma 17.1. Then

$$B_0 = \bigcup_{g \in G} B(\mathcal{J}_0)g, \ B(\mathcal{J}_0) \ge B, \text{ and}$$

$$A/B_0 = \bigoplus_{g \in S} A/B(\mathcal{J}_0)g,$$

where S is a subset of a transversal T to U in G. Since $(A/B)(\omega \mathbb{Z} L) = A/B$, $A(\omega \mathbb{Z} L) + B = A$, and so $A(\omega \mathbb{Z} L) + Bg = A$ for every $g \in G$. From $B(\mathcal{J}_0)g \ge Bg$, $(A/B(\mathcal{J}_0)g)\omega \mathbb{Z} L) = A/B(\mathcal{J}_0)g$, and, since A/B_0 is the direct sum of certain modules $A/B(\mathcal{J}_0)g$, we have $A/B_0 = (A/B_0)(w\mathbb{Z}L)$. Put $Q = B_0$ so that $A/Q = (A/Q)(w\mathbb{Z}L)$.

Let $\xi : A \longrightarrow A/Q$ be the canonical epimorphism. Since $C_K(A_1\xi)$ acts trivially on $A_1\xi$, so does it on $D_k\xi$. From the isomorphism $D_k^{**} \cong_{\mathbb{Z}U} A/B$ we see that $C_K(A_1\xi)$ acts trivially on A/B. It follows that $C_K(A_1\xi)$ acts trivially on every $A/B(\mathcal{J}_0)g$, and so on $A\xi = A/B_0$.

Since $C_L(A\xi) = C_L(A/B_0) \geq C_L(A_1\xi)$, and $L/C_L(A_1\chi)$ is a finitely generated subgroup of $FC(G/C_L(A_1\xi)) \cap (VC_L(A_1\xi)/C_L(A_1\xi))$, we see that $L/C_L(A\xi)$ is a finitely generated subgroup of $FC(G/C_L(A\xi)) \cap (VC_L(A\xi)/C_L(A\xi))$. The subgroup L has a finite series of G-invariant subgroups

$$C_L(A) = L_0 \leq L_1 \leq \cdots \leq L_s = L$$

such that L_{j+1}/L_j is the normal closure of a single element, and is either a finite abelian group or a free abelian group of finite 0-rank, $1 \leq j \leq s-1$. If H is locally nilpotent, we will choose a series like this in such a way that every factor L_{j+1}/L_j is H-central. Choose m minimal such that $A/Q = (A/Q)(w\mathbb{Z}L_m)$. Let

$$\psi : A/Q \longrightarrow (A/Q)/(A/Q)(w\mathbb{Z}L_m)$$

be the canonical epimorphism, and let χ be the composition of ξ and ψ. Put $R = L_m$. If $R/C_R(A\chi)$ is infinite, then there exists an integer t such that $K/C_R(A\chi) = (R/C_R(A\chi))^t$ is free abelian. If $A\chi = (A\chi)(w\mathbb{Z}K)$, then we are done. If $A\chi \neq (A\chi)(w\mathbb{Z}K)$, then we consider $(A\chi)/(A\chi)(w\mathbb{Z}K)$. We have

$$(A\chi)/(A\chi)(w\mathbb{Z}K) = ((A\chi)/(A\chi)(w\mathbb{Z}K))(w\mathbb{Z}R), \text{ and}$$
$$C_R((A\chi)/(A\chi)(w\mathbb{Z}K)) \geq K,$$

so that $R/C_R((A\chi)/(A\chi)(w\mathbb{Z}K))$ is finite. In other words, it suffices to consider the following case: G has a normal subgroup L, and A has a proper $\mathbb{Z}G$-submodule Q satisfying that $L/C_L(A/Q)$ is a finite abelian group, and $A/Q = (A/Q)(w\mathbb{Z}L)$. In this case, L has a finite series of G-invariant subgroups

$$C_L(A) = L_0 \leq L_1 \leq \cdots \leq L_r = L$$

such that L_{j+1}/L_j is the normal closure of a single element, and is an elementary abelian group, $1 \leq j \leq r-1$. Choosing an ℓ minimal such that $A/Q = (A/Q)(w\mathbb{Z}L_\ell)$ we can define $(A/Q)/(A/Q)(w\mathbb{Z}L_\ell)$ to be the required $\mathbb{Z}G$-image of A, and put $K = L_\ell$. $\square$

Lemma 17.3. *Let G be a locally nilpotent group, and let A be a $\mathbb{Z}G$-module. Suppose that for every finite subset M of A, and every finitely generated subgroup K of G we have that every K-chief factor of $M\mathbb{Z}K$ is K-central. Then $C_A(K) \neq \langle 0 \rangle$.*

Proof. Let $\mathcal{R}$ be a $\mathbb{Z}G$-chief series of A, and consider an arbitrary factor C of this series. Then C is a simple $\mathbb{Z}G$-module, and so it follows that either C is an elementary abelian p-group for some prime p or C is $\mathbb{Z}$-divisible. We consider the natural semidirect product $L = C \rtimes K$. If $B = M\mathbb{Z}K$, then $D = B \rtimes K$ is finitely generated soluble. Let H be an arbitrary normal subgroup of D of finite index. Since every D-chief factor of BH/H is central in D, and D/BH is nilpotent, D/H is nilpotent. By a result due to D.J.S. Robinson (see [235, Theorem 10.51]), D is likewise nilpotent. Elementary properties of nilpotent groups give that $B \cap \zeta(D) \neq \langle 1 \rangle$. It follows that $C_A(K) \neq \langle 0 \rangle$. $\qquad\square$

Lemma 17.4. *Let G be a group, H a normal subgroup of G, and A a $\mathbb{Z}G$-module. Assume that the following conditions hold:*

(i) *The natural semidirect product $A \rtimes H$ is locally nilpotent.*

(ii) *$HC_G(A)/C_G(A) \leq FC_\infty(G/C_G(A))$.*

(iii) *H is locally nilpotent.*

Then every chief $\mathbb{Z}G$-factor of A is H-central.

Proof. We suppose the contrary; that is, there exists a $\mathbb{Z}G$-chief factor W of A such that W is H-eccentric. Let

$$\nu : A \leftrightarrow A/C_G(W)$$

be the canonical epimorphism. By Lemma 3.15, $\zeta(H\nu) \cap FC(G\nu) \neq \langle 1 \rangle$. Pick $1 \neq x\nu \in \zeta(H\nu) \cap FC(G\nu)$, and put $Q\nu = \langle x\nu \rangle^{G\nu}$. Then $Q\nu$ is a finitely generated abelian subgroup. By the choice of $Q\nu$, the index $|G\nu : C_{G\nu}(Q\nu)|$ is finite. Put $X\nu = C_{G\nu}(Q\nu)$. Then $Q\nu \leq \zeta(G\nu)$. By Theorem 5.5,

$$W = W_1 \oplus \cdots \oplus W_m$$

for some simple $\mathbb{Z}(X\nu)$-submodules $W_1, \ldots, W_m$. Since $Q\nu$ is finitely generated, by (i), it is not hard to see that $C_{W_1}(Q\nu), \ldots, C_{W_m}(Q\nu) \neq \langle 1 \rangle$. Since every $C_{W_j}(Q\nu)$ is a $\mathbb{Z}(X\nu)$-submodule of W_j, $C_{W_j}(Q\nu) = W_j$. This holds for every $1 \leq j \leq m$, so that we conclude that either $C_W(Q\nu) = W$ or $Q \leq C_G(W)$, a contradiction. This contradiction shows that $H \leq C_G(W)$, as required. $\qquad\square$

Lemma 17.5. *Let G be a group, H and L normal subgroups of G, and A a $\mathbb{Z}G$-module. Suppose that $H \leq L$, and A has a $\mathbb{Z}L$-submodule B such that the natural semidirect product $B \rtimes H$ is a locally nilpotent group. Then for every finite subset X of G the natural semidirect product $(\sum_{x \in X} Bx) \rtimes H$ is locally nilpotent.*

Proof. Since L is normal in G, Bx is a $\mathbb{Z}G$-submodule of G. Let F be an arbitrary finitely generated subgroup of H. Put $K = \langle F^x \mid x, x^{-1} \in X \rangle$, so that K is likewise finitely generated. Let M be an arbitrary finite subset of B, and put $D = M\mathbb{Z}K$. Since $B \rtimes H$ is locally nilpotent, D has a finite upper $\mathbb{Z}K$-central series

$$\langle 0 \rangle = D_0 \leq D_1 \leq \cdots \leq D_k = D.$$

By the construction of K, the series

$$\langle 0 \rangle = D_0 x \leq D_1 x \leq \cdots \leq D_k x = D_x$$

consists of $\mathbb{Z}K$-submodules. Let $g \in K$. Then $xgx^{-1} = y \in K$; that is, $xg = yx$. If $1 \leq j \leq k$, we have

$$D_j x(g - 1) = D_j(xg - x) = D_j(yx - x) = D_j(y - 1)x \leq D_{j-1}x.$$

This shows that $\{D_j x \mid 0 \leq j \leq k\}$ is an upper $\mathbb{Z}K$-central series of D_x. Moreover,

$$\langle 0 \rangle = (D_0 + D_0 x) \leq (D_1 + D_1 x) \leq \cdots \leq (D_k + D_k x) = D + D_x$$

is an upper $\mathbb{Z}K$-central series of $D + D_x$. Proceeding by induction on $|X|$, we obtain that $(\sum_{x \in X} D_x)$ is $\mathbb{Z}K$-nilpotent. Then $(\sum_{x \in X} D_x)$ is $\mathbb{Z}F$-nilpotent, and it follows that $(\sum_{x \in X} B_x) \rtimes H$ is locally nilpotent. $\square$

Lemma 17.6. *Let G be a group, H a normal subgroup of G, and A a $\mathbb{Z}G$-module. Assume that the following conditions hold:*

(i) *A has an artinian $\mathbb{Z}G$-submodule B.*

(ii) *Every chief $\mathbb{Z}G$-factor of B is H-eccentric.*

(iii) *The natural semidirect product $(A/B) \rtimes H$ is a locally nilpotent group.*

(iv) *$H \leq FC_\infty(G)$.*

(v) *$C_G(A) = \langle 1 \rangle$.*

Then A includes a $\mathbb{Z}G$-submodule C such that $A = B \oplus C$.

Proof. Suppose the contrary; that is, every $\mathbb{Z}G$-submodule X such that $A = B + X$ satisfies that $B \cap X \neq \langle 0 \rangle$. Choose in A a $\mathbb{Z}G$-submodule M maximal under $B \cap M = \langle 0 \rangle$. Replacing M by A/M, we may assume that $M = \langle 0 \rangle$. This means that every non-zero $\mathbb{Z}G$-submodule of A has a non-zero intersection with B. Put

$$\mathfrak{M} = \{U \mid U \text{ is a } \mathbb{Z}G\text{-submodule such that } B \text{ does not include } U\},$$
$$\mathcal{S} = \{U \cap B \mid U \in \mathfrak{M}\}.$$

Clearly, $\mathcal{S} \neq \emptyset$. Since B is an artinian $\mathbb{Z}G$-module, $\mathcal{S}$ has a minimal element L. Let D be a $\mathbb{Z}G$-submodule of A such that $D \cap B = L$. For every non-zero $\mathbb{Z}G$-submodule E of D such that $E \in \mathfrak{M}$ we have $B \cap E \neq \langle 0 \rangle$. By the choice of D, it follows that $E \cap B = D \cap B$. If we suppose that $E = (E \cap B) + V$ for some proper non-zero $\mathbb{Z}G$-submodule V of E with the property $V \in \mathfrak{M}$, then, from the inclusion $V \cap B \leq D \cap B$, we obtain $V \cap B = D \cap B$; that is, $V = E$.

Clearly, there is no loss if we assume that $C_G(D) = \langle 1 \rangle$. Suppose that $C_H(L) \neq \langle 1 \rangle$. By Lemma 3.15, $C_H(L) \cap \zeta(H) \cap FC(G) \neq \langle 1 \rangle$. Pick $1 \neq x \in C_H(L) \cap \zeta(H) \cap FC(G)$, and put $Q = \langle x \rangle^G$. Then Q is a finitely generated subgroup. By the choice of Q, the index $|G : C_G(Q)|$ is finite. Put $X = C_G(Q)$. Then the mapping $\theta : a \mapsto a(x - 1)$, $a \in D$, is a $\mathbb{Z}X$-endomorphism of A. By the choice of x, we have $L \leq \mathrm{Ker}\ \theta$, so that $D(x - 1) \cong_{\mathbb{Z}X} D/\mathrm{Ker}\ \theta$. The inclusion $H \leq X$ implies that the natural semidirect product $(D/\mathrm{Ker}\ \theta) \rtimes H$ is a locally nilpotent group. Then the natural semidirect product $D(x - 1) \rtimes H$ is likewise a locally nilpotent group. Let T be a transversal to X in G. Since T is finite, by Lemma 17.5, $(\sum_{t \in T} D(x - 1)t) \rtimes H$ is locally nilpotent. Note that $\sum_{t \in T} D(x - 1)t$ is a $\mathbb{Z}G$-submodule of D. By Lemma 17.4, every $\mathbb{Z}G$-chief factor of $\sum_{t \in T} D(x - 1)t$ is H-central. By (ii),

$$\left(\sum_{t \in T} A(x - 1)t \right) \cap B = \langle 0 \rangle .$$

By our assumption this means that $\langle 0 \rangle = \sum_{t \in T} D(x-1)t$, in particular, $D(x-1) = \langle 0 \rangle$, and $x \in C_G(D) = \langle 1 \rangle$. Thus, $C_H(L) = \langle 1 \rangle$.

By Lemma 3.15, $\zeta(H) \cap FC(G) \neq \langle 1 \rangle$. Pick $1 \neq y \in \zeta(H) \cap FC(G)$, and put $Y = \langle y \rangle^G$. Then Y is a finitely generated abelian subgroup; that is, $Y = \langle g_1, \ldots, g_2 \rangle$. By the choice of Q, the index $|G : C_G(Y)|$ is finite. Put $R = C_G(Y)$. Since the natural semidirect product $(D/L) \rtimes H$ is a locally nilpotent group, every $\mathbb{Z}Y$ chief factor of an arbitrary finitely generated $\mathbb{Z}Y$-submodule of (D/L) is Y-central. By Lemma 17.3, $P/L = C_{D/L}(Y) \neq \langle 0 \rangle$. We note that P is a $\mathbb{Z}G$-submodule, because Y is a normal subgroup of G. The mapping

$$\gamma : a \mapsto a(g_1 - 1), \ a \in P,$$

is a $\mathbb{Z}R$-endomorphism of A. If $P(g_1 - 1)/L(g_1 - 1)$ is non-zero, since it is a $\mathbb{Z}R$-image of D/L, the natural semidirect product $P(g_1 - 1)/L(g_1 - 1) \rtimes H$ is a locally nilpotent group (note also that $H \leq R$). By Lemma 17.4, every $\mathbb{Z}R$-chief factor of $P(g_1 - 1)/L(g_1 - 1)$ is H-central. On the other hand, $P(g_1 - 1) \leq L \leq B$ by the choice of g_1. Since the index $|G : R|$ is finite, it is not hard to see that every $\mathbb{Z}R$-chief factor of B is H-eccentric. In particular, every $\mathbb{Z}R$-chief factor of $P(g_1 - 1)$, hence, of $P(g_1 - 1)/L(g_1 - 1)$, is H-eccentric. This contradiction shows the equality $P(g_1 - 1) = L(g_1 - 1)$. In other words,

$$P = C_P(g_1) + L.$$

Applying similar arguments to the $\mathbb{Z}R$-submodule $S = C_P(g_1)$, we obtain the equality $S = C_S(g_2) + (S \cap L)$, which implies that

$$P = C_P(g_1, g_2) + L.$$

Proceeding in this way, after finitely many steps, we obtain

$$P = C_P(Y) + L.$$

But in a beginning of the proof we have already demonstrated that in this case, $P = C_P(Y)$. Since $L \leq P$, $C_H(L) \neq \langle 1 \rangle$, a contradiction. This final contradiction proves the required result. $\square$

Lemma 17.7. *Let G be a group, and let H, B and A be normal subgroups of G. Assume that the following conditions hold:*

(i) *A is abelian, and $A \le B \le C_H(A)$.*

(ii) *$H/B \le FC_\infty(G/B)$.*

(iii) *H/A is locally nilpotent.*

(iv) *A is an artinian $\mathbb{Z}G$-module.*

(v) *Every chief $\mathbb{Z}G$-factor of A is H-eccentric.*

Then G has a G-invariant subgroup L such that $B = A \times L$. Moreover, if S is a subgroup of G such that $G = AS$, then $L \le S$.

Proof. Since $A \le \zeta(B)$, B is locally nilpotent. Choose in H an arbitrary finitely generated subgroup K, and put $D = A \cap K$, and $E = B \cap K$. Then

$$K/D = K/(A \cap K) \cong KA/A$$

is finitely generated nilpotent. Let U/V be a $\mathbb{Z}K$-chief factor of $D \cap [E, E]$. The factor-group K/V is finitely generated abelian-by-nilpotent. By a result due to P. Hall (see [235, Theorem 9.51]), K/V is residually finite. It follows that K has a normal subgroup F of finite index such that $F \cap U = V$. Consider now the finite group K/F. Since $(K/F)/(DF/F)$ is locally nilpotent, DF/F contains the nilpotent residual R/F of K/F. In particular, R/F is abelian. We have above noticed that in this case, $K/F = R/F \rtimes S/F$, for some nilpotent subgroup S/F (W. Gaschütz [83], E. Schenkman [253]). Since $R/F \le DF/F$, and $DF/F \le \zeta(EF/F)$,

$$EF/F = R/F \times (S/F \cap EF/F).$$

Clearly, $(S/F \cap EF/F)$ is a normal subgroup of K/F. The inclusion $U \le [E, E]$ implies that $UF/F \le (S/F \cap EF/F)$. In particular, $[R/F, UF/F] = \langle 1 \rangle$. Since UF/F is a chief factor of K/F, the equation $K/F = (R/F)(S/F)$ implies that UF/F is a chief factor of S/F. From the nilpotency of S/F, we obtain that UF/F is central in S/F; hence, in K/F. Therefore,

$$[K, U] \le F \cap U = V.$$

It follows that every $\mathbb{Z}K$-chief factor of $D \cap [E, E]$ is K-central.

Let W be a $\mathbb{Z}G$-chief factor of $A \cap [B, B]$, and suppose that W is not H-central. Let

$$\nu : A \longrightarrow A/C_G(W)$$

be the canonical epimorphism. By Lemma 3.15, $H\nu \cap FC(G\nu) \ne \langle 1 \rangle$. Pick $1 \ne x\nu \in H\nu \cap FC(G\nu)$, and put $Q\nu = \langle x\nu \rangle^{G\nu}$. Then $Q\nu$ is a finitely generated subgroup; that is, $Q\nu = \langle x_1\nu, \ldots, x_n\nu \rangle$. By the choice of $Q\nu$, the index $|G\nu : C_{G\nu}(Q\nu)|$ is finite. Put $X\nu = C_{G\nu}(Q\nu)$. By Theorem 5.5,

$$W = W_1 \oplus \cdots \oplus W_m$$

for some simple $\mathbb{Z}(X\nu)$-submodules $W_1, \ldots, W_m$. Since $Q\nu$ is finitely generated, by the above paragraph, every $\mathbb{Z}(Q\nu)$-chief factor of W_j is $Q\nu$-central, $1 \leq j \leq m$. By Lemma 17.3, $C_{W_j}(Q\nu) \neq \langle 1 \rangle$. But $C_{W_j}(Q\nu)$ is a $\mathbb{Z}(X\nu)$-submodule of W_j, and therefore, $C_{W_j}(Q\nu) = W_j$. This holds for $1 \leq j \leq m$, and so $C_W(Q\nu) = W$; that is, $Q\nu \leq C_G(W)$, a contradiction. This contradiction shows that $H \leq C_G(W)$; hence, it establishes that every $\mathbb{Z}G$-chief factor of $A \cap [B, B]$ is H-central. If we suppose now that $A \cap [B, B] \neq \langle 1 \rangle$, then we obtain a contradiction with condition (v).

Consider now the factor-group $G/[B, B]$. Its normal subgroup $B/[B, B]$ is abelian. Since $H/A[B, B]$ is locally nilpotent, by Lemma 17.6, B has a G-invariant subgroup $L \geq [B, B]$ such that

$$B/[B, B] = (A[B, B]/[B, B]) \times L/[B, B].$$

This and $A \cap [B, B] = \langle 1 \rangle$ give that $B = A \times L$.

Let S be a subgroup of G such that $G = AS$. Since $A \leq B$, $B = A(S \cap B)$. The subgroup B is normal in G, so that $S \cap B$ is normal in S. The inclusion $B \leq C_H(A)$, and the equation $G = AS$ imply that $S \cap B$ is normal in G. Since

$$B/(B \cap S) = A(B \cap S)/(B \cap S) \cong A/(A \cap S)$$

every G-chief factor of $B/(B \cap S)$ is H-eccentric. On the other hand, we have seen above that every G-chief factor of L is H-central. Since $L \leq B$, it follows that $L(B \cap S)/(B \cap S) = \langle 1 \rangle$; that is, $L \leq B \cap S$. $\qquad\square$

Lemma 17.8 ([275]). *Let G be a group, A a $\mathbb{Z}G$-module. Suppose that $\zeta(G)$ contains an element g such that A is not $\mathbb{Z}\langle g \rangle$-nilpotent. Then A has a $\mathbb{Z}G$-submodule B such that $(A/B)(g - 1) = A/B$, and $C_A(g) \leq B$.*

Proof. There is a descending chain of $\mathbb{Z}G$-submodules

$$A \geq A(g - 1) \geq A(g - 1)^2 \geq \cdots \geq A(g - 1)^n \geq \cdots .$$

Since A is artinian, there exists some m such that

$$A(g - 1)^{m-1} \neq A(g - 1)^m = A(g - 1)^{m+1} = \cdots .$$

If $j \in \mathbb{N}$, put $A_j = A(g - 1)^j$. Then

$$A_{m-1}(g - 1) = A_m = A_m(g - 1).$$

It follows that $A_{m-1} \leq C_A(g) + A_m$, and therefore,

$$(A_{m-1} + C_A(g))/C_A(g) = (A_m + C_A(g))/C_A(g).$$

Since $g \in \zeta(G)$, $C_A(g)$ is a $\mathbb{Z}G$-submodule of A. Proceeding by ordinary induction, after finitely many steps we find a $\mathbb{Z}G$-submodule B such that $C_A(g) \leq B$, and $A = B + A_m$. Thus,

$$(A/B)(g - 1) = ((B + A_m)/B)(g - 1) = ((B + A_m(g - 1))/B$$
$$= ((B + A_m)/B) = A/B,$$

as required. $\qquad\square$

Lemma 17.9. *Let G be a group, and let A be a $\mathbb{Z}G$-module. Suppose that $\zeta(G)$ contains an element g such that $A = A(g-1)$. Let E be any extension of A by G. If B is a $\mathbb{Z}G$-submodule of A, and $C_A(g) \leq B$, then E/B splits over A/B.*

Proof. Let $y \in E$ such that $g = yA$. In this proof, we will denote A multiplicatively; in particular, instead of $A(y-1)$ we will write $[A, y]$. Let x be an arbitrary element of E. We have $y^x = ay$ for some $a \in A$. Since $A = [A, y]$, $A = [A, y^{-1}]$, and so there exists some $b \in A$ such that $a = [b, y^{-1}]$. We have

$$y^x = ay = [b, y^{-1}]y = b^{-1}yby^{-1}y = y^b.$$

Therefore, $xb^{-1} \in C_E(y)$, and $x \in AC_E(y)$. Since $C_E(y) \cap A = C_A(y) = C_A(g)$, we see that $C_E(y)/C_A(y)$ is a complement to $A/C_A(y)$ in $E/C_A(y)$. But $B \leq C_A(y)$; so it is clear, that $BC_E(y)/B$ is a complement to A/B in E/B. $\square$

Lemma 17.10 ([275])**.** *Let G be a group, and let A be a normal subgroup of G. Suppose that there exist G-invariant subgroups $A_1, \ldots, A_n$ of A such that*

$$A_1 \cap \cdots \cap A_n = \langle 1 \rangle, \quad \text{and } A = A/A_1 \times \cdots \times A/A_n.$$

Then we have:

(1) *If every factor A/A_j has a complement S_j/A_j in G/A_j, then A has a complement S in G.*

(2) *If S and R are complements to A in G such that SA_j is conjugate to RA_j for every $1 \leq j \leq n$, then S and R are conjugate in G.*

Proof. Put $A_1 \cap \cdots \cap A_{t-1} = B$, and assume, inductively, that A/B has a complement C/B in G/B. Consider $C \cap S_t$. Certainly $C \cap S_t \cap A = B \cap A_t$. Also,

$$(C \cap S_t)A = (C \cap S_t)BA_t = (C \cap S_t B)A_t$$
$$= (C \cap S_t A)A_t = CA_t = CBA_t = CA = G.$$

Therefore, $(C \cap S_t)/(A_t \cap B)$ is a complement of $A/(A_1 \cap \cdots \cap A_t)$ in $G/(A_1 \cap \cdots \cap A_t)$. Assume now inductively that SB and RB are conjugate. Replacing R by an appropriate conjugate we may assume that $SB = RB$. There is an element $a \in A$ such that $S^a A_t = RA_t$. Since $A = BA_t$ we may assume that $a \in B$, and so $S^a B = RB$. Now

$$R(B \cap A_t) = R((R \cap A)A_t \cap B) = R(RA_t \cap B) = RA_t \cap RB.$$

So

$$(RA_t \cap B) = S^a A_t \cap SB = S^a(A_t \cap B);$$

hence $R/(A_1 \cap \cdots \cap A_t)$ is conjugate to $S/(A_1 \cap \cdots \cap A_t)$. $\square$

Lemma 17.11 ([275])**.** *Let G be a group, and let A be a $\mathbb{Z}G$-module. Suppose that $\zeta(G)$ contains an element g such that $A = A(g-1)$. Let E be a split extension of A by G. If K is another complement to A in E, and B is a $\mathbb{Z}G$-submodule of A such that $C_A(g) \leq B$, then GB and KB are conjugate in E.*

Proof. It is clear that $A \langle g \rangle \cap G = \langle g \rangle$, and $A \langle g \rangle \cap K = \langle ag \rangle$ for some element $a \in A$. Thus, $K \le C_E(ag)$, and $C_E(ag) = KC_A(g)$. Also $C_E(g) = GC_A(g)$.

Since $A = [A, g]$, $A = [A, g^{-1}]$, and so $a = [b, g^{-1}]$ for some element $b \in A$, and we have $ag = [b, g^{-1}]g = g^b$. It follows that

$$KC_A(g) = C_E(ag) = C_E(g^b) = (GB)^b,$$

as required. □

Theorem 17.12 (D.J.S. Robinson [246]). *Let G be a group, H a subgroup of G, and A a $\mathbb{Z}G$-module. Suppose that the following conditions hold:*

(i) *A is an artinian $\mathbb{Z}G$-module.*

(ii) *H is a normal subgroup of G.*

(iii) *H is locally nilpotent.*

(iv) *$HC_G(A)/C_G(A) \le FC_\infty(G/C_G(A))$.*

(v) *$C_A(H) = \langle 1 \rangle$.*

Then every extension E of A by G splits over A, and all complements of A in such an extension are conjugate.

Proof. We note that the condition (iv) implies that $HC_G(A)/C_G(A)$ is hyper-central. Repeating word by word the proof of Theorem 10.21, we obtain that A has the Z-$\mathbb{Z}H$-decomposition; that is,

$$A = \zeta_{\mathbb{Z}H}^\infty(A) \oplus \zeta_{\mathbb{Z}H}^*(A).$$

Condition (v) implies that $\zeta_{\mathbb{Z}H}^\infty(A) = \langle 0 \rangle$. Hence, every non-zero $\mathbb{Z}G$-factor of A is H-eccentric. In particular $A = A(\omega\mathbb{Z}H)$. Since we consider A not only as a module but as a normal subgroup of a group E, we will can write A multiplicatively. In particular, instead of $A(\omega\mathbb{Z}H)$ we will write $[A, H]$.

Existence of complements

Suppose the contrary; that is, for every subgroup X such that $G = AX$, $A \cap X \ne \langle 0 \rangle$. Note that $A \cap X$ is a G-invariant subgroup of A. Let $\mathfrak{M}$ be the set of subgroups X such that $G = AX$, and put $\mathcal{S} = \{X \cap A \mid X \in \mathfrak{M}\}$. Clearly, $\mathcal{S} \ne \emptyset$. Since A is an artinian $\mathbb{Z}G$-module, $\mathcal{S}$ has a minimal element M. Let Y be a subgroup of A such that $A \cap Y = M$. If we suppose that $Y = (A \cap Y)Y_1$ for some subgroup Y_1 of Y, then $G = AY_1$, and therefore, $\langle 1 \rangle \ne A \cap Y_1 \le A \cap Y$, so, by the choice of Y, $A \cap Y_1 = A \cap Y$. This means that $Y = Y_1$. In other words, for every proper subgroup Y_1 of Y we have $G \ne AY_1$. Without loss of generality, we obtain that $E = Y$, and $A = M$. By Lemma 17.2, G has a normal subgroup K, and A has a proper G-invariant subgroup Q such that the following conditions hold:

(i) The factor-group $(K/Q)/C_{K/Q}(A/Q)$ is the normal closure of a single element of $(G/Q)/C_{K/Q}(A/Q)$.

(ii) $(K/Q)/C_{K/Q}(A/Q)$ is either a finite elementary abelian p-group for some prime p or a free abelian group of finite 0-rank.

(iii) $(K/Q)/C_{K/Q}(A/Q)$ lies in the intersection of $FC((G/Q)/C_{K/Q}(A/Q))$, and $\zeta((H/Q)C_{K/Q}(A/Q)/C_{K/Q}(A/Q))$.

(iv) $A/Q = [A/Q, K/Q]$.

Replacing E by E/Q, we may assume that $Q = \langle 1 \rangle$. By Lemma 17.7, $C_K(A) = A \times B$ for some E-invariant subgroup B. We may replace E by E/B to assume $B = \langle 1 \rangle$. Thus, $C_K(A) = A$. Put

$$K = \langle x \rangle^E = \langle x_1, \ldots, x_n \rangle.$$

It is not hard to see that if A is $\mathbb{Z}\langle x_j \rangle$-nilpotent for every $1 \leq j \leq n$, then A is $\mathbb{Z}K$-nilpotent. But in this case, $A \neq [A, K]$. Therefore, there exists some index j such that A is not $\mathbb{Z}\langle x_j \rangle$-nilpotent. Let $U/A = C_{E/A}(K/A)$. By Lemma 17.8, A has a U-invariant subgroup B such that

$$C_A(g) \leq B, \text{ and } [A/B, gB] = A/B.$$

Now we form the G-invariant subgroup B_0 just as in Lemma 17.1. Then

$$B_0 = \bigcap_{g \in G} B(\mathcal{J})^g, \ B(\mathcal{J}_0) \geq B \geq C_A(g),$$

$$\text{and } A/B_0 = \mathrm{Dr}_{g \in S} A/B(\mathcal{J}_0)^g$$

for some subset S of a transversal T to U in G. Note also that

$$B(\mathcal{J}_0)g \geq B^g \geq C_A(g^g) = C_A(g).$$

By Lemma 17.9, $U/B(\mathcal{J}_0)^g$ splits over $A/B(\mathcal{J}_0)^g$. By Lemma 17.10, there is a complement L/B_0 to A/B_0 in U/B_0.

Since $B(\mathcal{J}_0) \geq B$, we have $[A/B(\mathcal{J}_0), x] = A/B$. Denote by $C/B(\mathcal{J}_0)$ the centralizer of x in $A/B(\mathcal{J}_0)$. Then $C/B(\mathcal{J}_0) \neq A/B(\mathcal{J}_0)$. Note that the centralizer of x^g in $A/B(\mathcal{J}_0)^g$ is $C^g/B(\mathcal{J}_0)^g$, so that $C^g/B_0 \geq C_{A/B_0}(x^g)$. Let

$$C_0 = \bigcap_{g \in G} C^g.$$

Then C_0 is a $\mathbb{Z}E$-submodule, and

$$A/C_0 = \mathrm{Dr}_{g \in S} A/C^g.$$

Let L/B_0 and L_1/B_0 be two complements to A/B_0 in U/B_0. By Lemma 17.11, L/C^g and L_1/C^g are conjugate in U. It now follows from Lemma 17.10 that LC_0 and L_1C_0 are conjugate in U. Application of the Frattini argument gives that $E = AN_E(LC_0)$. Since A has no proper supplement in E, we must have $E = N_E(LC_0)$, i.e. LC_0 is normal in E. Hence,

$$[A, LC_0] \leq A \cap LC_0 = (A \cap L)C_0 = C_0 \neq A,$$

contrary to $[A, K] = A$.

This completes the proof that E splits over A.

Conjugacy of complements

Let P_1 and P_2 be two complements to A in E. Let $\mathfrak{C}$ be the set of all E-invariant subgroups W of A such that WP_1 and WP_2 are conjugate. Clearly, $A \in \mathfrak{C}$, so that $\mathfrak{C} \neq \emptyset$. Since A is an artinian $\mathbb{Z}G$-module, $\mathfrak{C}$ has a minimal element M. Suppose that $M \neq \langle 1 \rangle$. Then BP_1 and BP_2 are not conjugate for every proper G-invariant subgroup B of M. Without loss of generality we may assume that $MP_1 = MP_2$, $A = M$, and $E = MP_1$, so that we have $AP_1 = AP_2 = E$, and BP_1 and BP_2 are not conjugate for every proper E-invariant subgroup B of A. By Lemma 17.2, E has a normal subgroup K, and A has a proper E-invariant subgroup Q satisfying the following conditions:

(i) The factor-group $(K/Q)/C_{K/Q}(A/Q)$ is the normal closure of a single element of $(G/Q)/C_{K/Q}(A/Q)$.

(ii) $(K/Q)/C_{K/Q}(A/Q)$ is either a finite elementary abelian p-group for some prime p or a free abelian group of finite 0-rank.

(iii) $(K/Q)/C_{K/Q}(A/Q)$ lies in

$$FC((G/Q)/C_{K/Q}(A/Q)) \cap \zeta((H/Q)C_{K/Q}(A/Q)/C_{K/Q}(A/Q)).$$

(iv) $A/Q = [A/Q, K/Q]$.

We may replace E by E/Q to assume $Q = \langle 1 \rangle$. By Lemma 17.7,

$$C_K(A) = A \times B$$

for some E-invariant subgroup B. Replacing E by the factor-group E/B, we may assume $B = \langle 1 \rangle$. Then $C_K(A) = A$. Put

$$K = \langle x \rangle^E = \langle x_1, \ldots, x_n \rangle.$$

Proceeding as we did in the existence part of this proof, we can choose a U-invariant subgroup B such that

$$C_A(g) \leq B, \text{ and } [A/B, gB] = A/B.$$

Again let $U/A = C_{E/A}(K/A)$. We form the E-invariant subgroup B_0 as in Lemma 17.1. Then

$$B_0 = \bigcap_{g \in G} B(\mathfrak{J}_0)^g, \ B(\mathfrak{J}_0) \geq B \geq C_A(g),$$

$$\text{and } A/B_0 = \mathrm{Dr}_{g \in S} A/B(\mathfrak{J}_0)^g,$$

where S is a subset of a transversal T to U in G. Note also that

$$B(\mathcal{J}_0)^g \geq B^g \geq C_A(g^g) = C_A(g).$$

By Lemma 17.11, $(U \cap P_1)B(\mathcal{J}_0)^g$ is conjugate to $(U \cap P_2)B(\mathcal{J}_0)^g$ in U. By Lemma 17.10, $(U \cap P_1)B_0$ is conjugate to $(U \cap P_2)B_0$ in U. But

$$(U \cap P_1)B_0 = U \cap P_1 B_0, \text{ and } P_j B_0 \leq N_E(U \cap P_j B_0), \ j \in \{1, 2\}.$$

Therefore,

$$N_E(U \cap P_j B_0) = P_j N_A(U \cap P_j)B_0), \ j \in \{1, 2\}.$$

However, $N_A(U \cap P_j)B_0) = C_{A/B_0}(U)$. The inclusion $H \leq U$ implies that A has a non-zero H-central $\mathbb{Z}G$-section. We have already said above that every non-zero $\mathbb{Z}G$-factor of A is H-eccentric. It follows that $N_A(U \cap P_j)B_0) = B_0$, and so $N_E(U \cap P_j B_0) = P_j B_0$. Therefore, $P_1 B_0$ and $P_2 B_0$ are conjugate. This contradiction completes the proof of the theorem. □

Bibliography

[1] J. Alcázar, J. Otal, *Sylow subgroups of groups with Chernikov conjugacy classes*. J. Algebra **110** (1987), 507–513.

[2] F. Anderson, K. Fuller, *Rings and categories of modules*. Springer, 1974.

[3] M.F. Atiyah, I.G. MacDonald, *Introduction to commutative algebra*. Addison-Wesley, 1969.

[4] R. Baer, *Finiteness properties of groups*. Duke Math. J. **15** (1948), 1021–1032.

[5] R. Baer, *Groups with descending chain condition for normal subgroups*. Duke Math. J. **16** (1949), 1–22.

[6] R. Baer, *Auflösbare Gruppen mit Maximalbedingung*. Math. Ann. **129** (1955), 139–173.

[7] R. Baer, *Polyminimaxgruppen*. Math. Ann. **175** (1968), 1–43.

[8] R. Baer, *Durch Formationen bestimmte Zerlegungen von Normalteilern endlicher Gruppen*. J. Algebra **20** (1972), 38–56.

[9] R. Baer, H. Heineken, *Radical groups of finite abelian subgroup rank*. Illinois J. Math. **16** (1972), 533–580.

[10] S.D. Berman, *The group algebras of countable abelian p-groups*. Pub. Math. **14** (1967), 365–405.

[11] S.D. Berman, A.G. Podgorny, *On the representations of a group of type p^∞ over one class of the fields*. Ukrain. Math. J. **29** (1977), 632–637.

[12] S.D. Berman, N.I. Vishnyakova, *On the representations of a group of type p^∞ over a field of characteristic p*. Proc. Moskow Math. Soc. **29** (1973), 87–100.

[13] S.D. Berman, N.I. Vishnyakova, *Locally cyclic modules over a valuation rings*. "Representation theory, group rings and coding theory", Contemporary Math. Amer. Math. Soc. 1989, 85–103.

[14] N. Bourbaki, *Algebre (Polynomes et fractions rationelles; Corps commutatifs)*. Hermann, 1959.

[15] N. Bourbaki, *Algebre commutative (Modules Plats, Localisation)*. Hermann, 1961.

[16] N. Bourbaki, *Algebre commutative (Graduations, filtrations et topologies; Ideaux premier associes et decomposition primaire)*. Hermann, 1961.

[17] N. Bourbaki, *Algebre (Algebre lineaire)*. Hermann, 1962.

[18] N. Bourbaki, *Algebre commutative (Entiers, valuations)*. Hermann, 1964.

[19] N. Bourbaki, *Algebre (Groupes et corps ordonnes; Modules sur les anneaux principaux)*. Hermann, 1964.

[20] N. Bourbaki, *Algebre commutative (Diviseurs)*. Hermann, 1965.

[21] Z.I. Borevich, A.R. Shafarevich, *Number theory*. Academic Press, 1966.

[22] A. A. Bovdi, *The multiplicative group of an integral group rings*. University Of Uzhgorod, 1987.

[23] A. A. Bovdi, *Group rings*. University of Uzhgorod, 1988.

[24] D.C. Brewster, J.C. Lennox, *Subnormality and quasinormality in soluble minimax groups*. J. Algebra **48** (1977), 368–381.

[25] C.J.B. Brookes, *Ideals in group rings of soluble groups of finite rank*. Math. Proc. Cambridge Phil. Soc. **97** (1985), 27–49.

[26] C.J.B. Brookes, *Abelian subgroups and Engel elements of soluble groups*. J. London Math. Soc. **32** (1985), 467–472.

[27] C.J.B. Brookes, *Engel elements of soluble groups*. Bull. London Math. Soc. **16** (1986),, 7–10.

[28] C.J.B. Brookes, *Modules over polycyclic groups*. Proc. London Math. Soc. **57** (1988), 88–108.

[29] C.J.B. Brookes, J.R.J. Groves, *Modules over nilpotent group rings*. J. London Math. Soc. **52** (1995), 467–481.

[30] K.A. Brown, *The structure of modules over polycyclic groups*. Math. Proc. Cambridge Phil. Soc. **89** (1981), 257–283.

[31] K.A. Brown, *The Nullstellensatz for certain group rings*. J. London Math. Soc. **26** (1982), 425–434.

[32] S.U. Chase, *Direct products of modules*. Trans. Amer. Math. Soc. **97** (1960), 437–473

[33] V.S. Charin, *A remark on the minimal condition for subgroups*. Doklady AN SSSR **66** (1949), 575–576.

[34] V.S. Charin, *On the minimal condition for normal subgroups in locally soluble groups*. Mat.Sbornik **33** (1953), 27–36.

[35] V.S. Charin, *On the automorphism groups of certain classes of soluble groups.* Ukrain. Math. J. **5** (1953), 363–369.

[36] V.S. Charin, *On the automorphism groups of nilpotent groups.* Ukrain. Math. J. **6** (1954), 295–304.

[37] V.S. Charin, *On locally soluble groups of finite rank.* Mat. Sb. **41** (1957), 37–48.

[38] C.S. Charin, *On soluble groups of type A4.* Mat. Sb. **52** (1960), 895–914.

[39] V.S. Charin, *On soluble groups of type A3.* Mat. Sb. **54** (1961), 489–499.

[40] V.S. Charin, D.I. Zaitsev, *On groups with finiteness conditions and other restrictions on subgroups.* Ukrain. Math. J. **40** (1988), 277–287.

[41] A.W. Chatters, C.R. Hajarnavis, *Rings with chain condition.* Pitman, 1980.

[42] S.N. Chernikov, *The finiteness conditions in general group theory.* Uspekhi Mat. nauk **14** (1959), 45–96.

[43] S.N. Chernikov, *The groups with prescribed properties of systems of subgroups.* Nauka Moskow, 1980.

[44] G.H. Cliff, S.K. Sehgal, *Group rings whose units form an FC-group.* Math. Z. **161** (1978), 163–168.

[45] A.H. Clifford, *Representations induced in an invariant subgroups.* Annals Math. **38** (1937), 533–550.

[46] J. Cornick, P.H. Kropholler, *Homological finiteness conditions for modules over group algebras.* J. London Math. Soc. **58** (1998), 49–62.

[47] C.W. Curtis, I. Reiner, *Representation theory of finite groups and associative algebras.* John Wiley, 1962.

[48] C.W. Curtis, I. Reiner, *Methods of representation theory, Vol 1.* John Wiley, 1981.

[49] C.W. Curtis, I. Reiner, *Methods of representation theory, Vol 2.* John Wiley, 1981.

[50] G. Cutolo, L.A. Kurdachenko, *Weak chain conditions for non-almost normal subgroups.* London Math. Soc. Lecture Notes Series **211** (1995), 120–130.

[51] R. Dedekind, *Über die Theorie der ganzen algebraischen Zahlen.* Supplement XI to Dirichlet's "Vorlesungen uber Zahlentheorie", 2nd ed., 1871.

[52] M.R. Dixon, *Sylow theory, formations and Fitting classes in locally finite groups.* World Scientific, 1994.

[53] M.R. Dixon, M.J. Evans, H. Smith, *Locally (soluble-by-finite) groups of finite rank.* J. Algebra **182** (1996), 756–769.

[54] M.R. Dixon, M.J. Evans, H. Smith, *Locally (soluble-by-finite) groups with all proper insoluble subgroups of finite rank.* Arch. Math. **68** (1997), 100–109.

[55] M.R. Dixon, L.A. Kurdachenko, N.V. Polyakov, *Locally generalized radical groups satisfying certain rank conditions.* To appear.

[56] K. Doerk, T. Hawkes, *Finite soluble groups.* Walter de Gruyter, 1992.

[57] S. Donkin, *Polycyclic groups, Lie algebras and algebraic groups.* J. reine angew. Math. **326** (1981), 104–123.

[58] S. Donkin, *Locally finite representations of polycyclic-by-finite groups.* Proc. London Math. Soc **44** (1982), 333–348.

[59] Z.Y. Duan, *The structure of noetherian modules over hyperfinite groups.* Math. Proc. Cambridge Phil. Soc. **112** (1992), 21–28.

[60] Z.Y. Duan, *Extensions of abelian-by-hyper(cyclic or finite) groups.* Comm. Algebra **20** (1992), 2305–2321.

[61] Z.Y. Duan, *The decomposition of noetherian modules over hyperfinite groups.* Ricerche Mat. **44** (1995), 65–89.

[62] Z.Y. Duan, *The $\mathfrak{F}$-decomposition of artinian modules over hyperfinite groups.* Proc. Edinburgh Math. Soc. **38** (1995), 117–120.

[63] Z.Y. Duan, *Modules over hyperfinite groups.* "Rings, groups and algebras", Marcel Dekker, 1996, 51–61.

[64] Z.Y. Duan, M.J. Tomkinson, *The decomposition of minimax modules over hyperfinite groups.* Arch. Math. **81** (1993), 340–343.

[65] Z.Y. Duan, L.A. Kurdachenko, I.Ya. Subbotin, *Artinian modules over hyperfinite groups.* South. Asian Bull. Math. **22** (1998), 273–286

[66] C. Faith, *Algebra: Rings, Modules and Categories I.* Springer, 1973.

[67] C. Faith, *Algebra: Rings, Modules and Categories II.* Springer, 1976.

[68] D.R. Farkas, *Noetherian group rings: an exercise in creating folklore and induction.* "Noetherian rings and their applications", Amer.Math. Soc., Math. Surveys and monographs **24** (1987), 89–118.

[69] D.R. Farkas, R.L. Snider, *Group algebras whose simple modules are injective.* Trans. Amer. Math. Soc. **194** (1974), 241–248.

[70] I. Fleischner, *Modules of finite rank over Prüfer rings.* Annals of Math. **65** (1957), 250–254.

[71] E. Formanek, R.L. Snider, *Primitive group rings.* Proc. Amer. Math. Soc. **36** (1972), 357–360.

[72] S. Franciosi, F. de Giovanni, *On maximal subgroups of minimax groups.* Atti. Accad. Naz. Lincei. Mat. Applic. **9** (1995), 23–27.

[73] S. Franciosi, F. de Giovanni, L.A. Kurdachenko, *The Schur property and groups with uniform conjugace classes.* J. Algebra **174** (1995), 823–847.

[74] S. Franciosi, F. de Giovanni, L.A. Kurdachenko, *On groups with many almost normal subgroups.* Annali Mat. **169** (1995), 35–65.

[75] S. Franciosi, F. de Giovanni, L.A. Kurdachenko, *Groups whose proper quotients are FC-groups.* J. Algebra **186** (1996), 544–577.

[76] S. Franciosi, F. de Giovanni, M.J. Tomkinson, *Groups with polycyclic-by-finite conjugacy classes.* Boll. Unione Mat. Italiana **4B** (1990), 35–55.

[77] S. Franciosi, F. de Giovanni, M.J. Tomkinson, *Groups with Chernikov conjugacy classes.* J. Austral. Math. Soc.(series A) **50** (1991), 1–14.

[78] L. Fuchs, *Infinite abelian groups, Vol. 1.* Academic Press, 1970.

[79] L. Fuchs, *Infinite abelian groups, Vol. 2.* Academic Press, 1973.

[80] L. Fuchs, S.B. Lee, *Primary decompositions over domains.* Glasgow Math. J. **38** (1996), 321–326

[81] L. Fuchs, L. Salce, *Modules over valuation domains.* Marcel Dekker, 1985.

[82] A.D. Gardiner, B. Hartley, M.J. Tomkinson, *Saturated formations and Sylow structure in locally finite groups.* J. Algebra **17** (1971), 177–211.

[83] W. Gaschütz, *Zur Erweiterungstheorie der endlichen Gruppen.* J. Reine Angew. Math. **190** (1952), 93–107.

[84] R. Gilmer, *Multiplicative ideal theory.* Marcel Dekker, 1972.

[85] J.S. Golan, T. Head, *Modules and structure of rings.* Marcel Dekker, 1991.

[86] M. González, J. Otal, *P. Hall's covering group and embedding of countable CC-groups.* Comm. Algebra **18** (1990), 3405–3412.

[87] M. González, J. Otal, *Embedding theorems for residually Chernikov CC-groups.* Proc. Amer. Math. Soc. **123** (1995), 2383–2332.

[88] M. González, J. Otal, *The extension of results due to Gorchakov and Tomkinson from FC-groups to CC-groups.* J. Algebra **185** (1996), 314–328.

[89] M. González, J. Otal, J.M. Peña, *CC-groups with periodic central factors.* Manuscripta Math. **69** (1990), 93–103.

[90] Yu.M. Gorchakov, *The groups with finite classes of conjugacy elements.* Nauka Moskow, 1978.

[91] D. Gorenstein, *Finite groups.* Harper Row, 1968.

[92] R.I. Grigorchuk, *Burnside's problem on periodic groups.* Funktsional Analis i prilozh. **14** (1980), 53–54.

[93] R.I. Grigorchuk, *Solved and unsolved problems around one group.* Geometric, Combinatorial and Dynamical Aspects . Progress in Mathematics Series Vol. 248, Birhäuser, pp. 117–218.

[94] K.W. Gruenberg, *Ring theoretic methods and finiteness conditions in infinite soluble group theory.* Lecture Notes Math. **319** (1974), 75–84.

[95] P. Hall, *Finiteness conditions for soluble groups.* Proc. London Math. Soc. **4** (1954), 419–436.

[96] P. Hall, *On the finiteness of certain soluble groups.* Proc. London Math. Soc. **9** (1959), 595–632.

[97] P. Hall, *Nilpotent groups.* Queen Mary College Mathematics Notes, 1969.

[98] P. Hall, *The Frattini subgroup of finitely generated groups.* Proc. London Math. Soc. **11** (1961), 327–352.

[99] P. Hall, B. Hartley, *The stability group of a series of subgroups.* Proc. London Math. Soc. **16** (1966), 1–39.

[100] B. Hartley, *A note on the normalizer conditions.* Proc. Cambridge Phil. Soc. **74** (1973), 11–15.

[101] B. Hartley, *A class of modules over locally finite groups I.* J. Austral. Math. Soc.(series A) **16** (1973), 431–442.

[102] B. Hartley, *A class of modules over locally finite groups II* . J. Austral. Math. Soc.(series A) **19** (1975), 437–469.

[103] B. Hartley, *A class of modules over locally finite groups III.* Bull. Austral. Math. Soc. **14** (1976), 95–110.

[104] B. Hartley, *Splitting over the locally nilpotent residual of locally finite groups.* Quart. Journal Math. **27** (1976), 395–400.

[105] B. Hartley, *Uncountable artinian modules and uncountable soluble groups satisfying Min-n.* Proc. London Math. Soc. **35** (1977), 55–75.

[106] B. Hartley, *A dual approach to Chernikov modules.* Math. Proc. Cambridge Phil. Soc. **82** (1977), 215–239.

[107] B. Hartley, *Injective modules over group rings.* Quart. Journal Math. **28** (1977), 1–29.

[108] B. Hartley, *Locally finite groups whose irreducible modules are finite dimensional.* Rocky Mountain Math. J. **13** (1983), 255–263.

[109] B. Hartley, T.O. Hawkes, *Rings, modules and linear algebra.* Chapman & Hall, 1974.

[110] B. Hartley, D. McDougal, *Injective modules and soluble groups satisfying the minimal condition for normal subgroups.* Bull. Austral. Math. Soc. **4** (1971), 113–135.

[111] B. Hartley, M.J. Tomkinson, *Splitting over nilpotent and hypercentral residuals.* Math. Proc. Cambridge Phil. Soc. **78** (1975), 215–226.

[112] H. Heineken, I.J. Mohamed, *Groups with normalizer condition.* Math. Annalen **198** (1972), 178–187.

[113] I.N. Herstein, *Conjugates in division rings.* Proc. Amer. Math. Soc. **7** (1956), 1021–1022.

[114] T.W. Hungerford, *Algebra*, 2nd edition. Springer, 1996

[115] B. Huppert, *Endliche Gruppen I.* Springer, 1967.

[116] B. Huppert, N. Blackburn, *Finite groups II.* Springer, 1982.

[117] B. Huppert, N. Blackburn, *Finite groups III.* Springer, 1982.

[118] A.V. Jategaonkar, *Integral group rings of polycyclic-by-finite groups.* J. pure applied Algebra **4** (1974), 337–343.

[119] A.V. Jategaonkar, *Certain injectives are artinian.* Lecture Notes Math. **545** (1976), 128–139.

[120] A.V. Jategaonkar, *Localization in noetherian rings.* Cambridge University Press, 1986.

[121] I. Kaplansky, *Modules over Dedekind and valuation rings.* Trans.Amer. Math. Soc. **72** (1952), 327–340.

[122] I. Kaplansky, *Projective modules.* Annals Math. **68** (1958), 372–377.

[123] I. Kaplansky, *The splitting of modules over integral domains.* Arch. Math. **13** (1962), 341–343

[124] I. Kaplansky, *Infinite abelian groups.* Univ. Michigan Press, 1969.

[125] I. Kaplansky, *Topics in commutative ring theory.* Univ. Chicago Notes from Dept. Math., 1974.

[126] M. Karbe, *Unendliche Gruppen mit schwachen Kettenbedingungen für endlich erzeugte Untergruppen.* Arch. Math. **45** (1985), 97–110.

[127] M. Karbe, L.A. Kurdachenko, *Just infinite modules over locally soluble groups.* Arch. Math. **51** (1988), 401–411.

[128] M.I. Kargapolov, *On soluble groups of finite rank.* Algebra i logika **1** (1962), 37–44.

[129] M.I. Kargapolov, Yu. I. Merzlyakov, *Foundations of the theory of groups.* Springer, 1979.

[130] G. Karpilovsky, *Commutative group algebras.* Marcel Dekker, 1983.

[131] G. Karpilovsky, *Field theory.* Marcel Dekker, 1988.

[132] L.S. Kazarin, L.A. Kurdachenko, *The finiteness conditions and the factorizations in infinite groups.* Russian Math. Surveys **47** (1992), 81–126.

[133] L.S. Kazarin, L.A. Kurdachenko, I.Ya. Subbotin, *On groups saturated with abelian subgroups.* International Journal Algebra and Computation **8** (1998), 443–466.

[134] O.H. Kegel, B.A.F. Wehrfritz, *Locally finite groups.* North Holland, 1973.

[135] Sh.S. Kemhadze, *On stable groups of automorphisms.* Doklady AN USSR **158** (1964), 510–512.

[136] A. Kertesz, *Lectures on artinian rings.* Akademiai Kiado, 1987.

[137] L.G. Kovacs, M.F. Newman, *Direct complementation in group with operators.* Arch. Math. **13** (1963), 427–433.

[138] P.H. Kropholler, *On finitely generated soluble groups with no large wreath product section.* Proc. London Math. Soc. **49** (1984), 155–169.

[139] P.H. Kropholler, P.A. Linnel, J.A. Moody, *Applications of a new K-theoretic theorem to soluble group rings.* Proc. Amer. Math. Soc. **104** (1988), 675–684.

[140] S.A. Kruglyak, *On representations of groups (p, p) over a field of characteristic p.* Doklady AN SSSR **153** (1963), 1253–1256.

[141] L.A. Kurdachenko, *The groups satisfying the weak minimal and maximal conditions for normal subgroups.* Sibir. Math. J. **20** (1979), 1068–1075.

[142] L.A. Kurdachenko, *The groups satisfying the weak minimal and maximal conditions for subnormal subgroups.* Math. Notes **29** (1981), 19–30.

[143] L.A. Kurdachenko, *On some embeddings conditions of FC-groups in the direct product of finite groups and torsion-free abelian groups.* Math. USSR Sbornik **42** (1982), 499–514.

[144] L.A. Kurdachenko, *Locally nilpotent groups with the weak minimal condition for normal subgroups.* Sibir. Math. J. **25** (1984), 589–594.

[145] L.A. Kurdachenko, *Locally nilpotent groups with the weak minimal and maximal comditions for normal subgroups.* Doklady AN Ukrain. SSR, **8** (1985), 9–12.

[146] L.A. Kurdachenko, *The locally nilpotent groups with condition Min-oo-n.* Ukrain. Math. J. **42** (1990), 303–307.

[147] L.A. Kurdachenko, *On some classes of groups with the weak minimal and maximal conditions for normal subgroups.* Ukrain. Math. J. **42** (1990), 1050–1056,

[148] L.A. Kurdachenko, *Artinian modules over groups of finite rank and some finiteness conditions.* "Infinite groups and adjoining algebraic structures", Naukova Dumka Kiev, 1993, 144–159.

[149] L.A. Kurdachenko, *On groups with minimax conjugacy classes.* "Infinite groups and adjoining algebraic structures", Naukova Dumka Kiev, 1993, 160–177.

[150] L.A. Kurdachenko, *Artinian modules over groups of finite rank and the weak minimal condition for normal subgroups.* Ricerche Mat. **44** (1995), 303–335.

[151] L.A. Kurdachenko, *On normal closures of elements in generalized FC-groups.* "Infinite groups 1994" (Ravello 1994). Walter de Gruyter, 1996, 141–151.

[152] L.A. Kurdachenko, *Modules over the group rings with some finiteness conditions.* Ukrain. Math. J. **54** (2002), 931–940.

[153] L.A. Kurdachenko, V.E. Goretsky, *On groups with the weak minimal and maximal conditions for non-normal subgroups.* Ukrain. Math. J. **41** (1989), 1705–1709.

[154] L.A. Kurdachenko, J. Otal, *Some noetherian modules and non-monolithic just non-CC-groups.* J. Group Th. **2** (1999), 53–64.

[155] L.A. Kurdachenko, J. Otal, *Simple modules over CC-groups and monolithic just non-CC-groups.* Boll. Unione Mat. Italiana **8** (2001), 381–390.

[156] L.A. Kurdachenko, J. Otal, *Frattini properties of groups with minimax conjugacy classes.* Quaderni di Matematica **8** (2001), 221–237.

[157] L.A. Kurdachenko, J. Otal, I.Ya. Subbotin, *Groups with prescribed quotient groups and associated module theory.* World Sci. Publ., 2002.

[158] L.A. Kurdachenko, J. Otal, I.Ya. Subbotin, *Baer like decomposition of modules.* Illinois Journal Math. **47** (2003), 329–343.

[159] L.A. Kurdachenko, B.V. Petrenko, I.Ya. Subbotin, *On generalized hypercenters in artinian modules.* Comm. Algebra **25** (1997), 1023–1046.

[160] L.A. Kurdachenko, B.V. Petrenko, I.Ya. Subbotin, *Direct decompositions in artinian modules over FC-hypercentral groups.* Matematica contemporanea **14** (1998), 89–99.

[161] L.A. Kurdachenko, N.V. Polyakov, I.Ya. Subbotin, *On some generalization of a theorem of B. H. Neumann.* Matematica contemporanea **21** (2001), 131–145.

[162] L.A. Kurdachenko, V.V. Pylaev, *On groups with the minimax factor-groups.* Ukrain. Math. J. **42** (1990), 550–553.

[163] L.A. Kurdachenko, V.V. Pylaev, V.E. Goretsky, *The groups with the same systems of minimax factor-groups.* Doklady AN USSR **3A** (1988), 17–20.

[164] L.A. Kurdachenko, N.N. Semko, I.Ya. Subbotin, *Artinian modules over FC-hypercentral groups.* Proceedings of the National Academy of Sciences of Ukraine, 2005, no 2, 29–32.

[165] L.A. Kurdachenko, H. Smith, *Groups with the maximal condition on non-subnormal subgroups.* Boll. Unione Mat. Italiana **10B** (1996), 441–460.

[166] L.A. Kurdachenko, H. Smith, *Groups with the weak minimal condition for non-subnormal subgroups.* Annali Mat. **173** (1997), 299–312

[167] L.A. Kurdachenko, H. Smith, *The nilpotency of some groups with all subgroups subnormal.* Pub. Matematiques **42** (1998), 411–421.

[168] L.A. Kurdachenko, H. Smith, *Groups with the weak maximal condition for non-subnormal subgroups.* Ricerche Math. **47** (1998), 29–49.

[169] L.A. Kurdachenko, I.Ya. Subbotin, *Groups with restrictions for cocentralizers of elements.* Comm. Algebra **24** (1996), 1173–1187.

[170] L.A. Kurdachenko, I.Ya. Subbotin, *Groups whose proper quotients are hypercentral.* J. Austral. Math. Soc., series A **65** (1998), 224–237.

[171] L.A. Kurdachenko, I.Ya. Subbotin, *Groups with many periodic factor-groups.* Comm. Algebra **28** (2000), 1593–1602.

[172] L.A. Kurdachenko, I.Ya. Subbotin, *On minimal artinian modules and minimal artinian linear groups.* International Journal of Mathematics and mathematical Sciences **27** (2001), 704–714.

[173] L.A. Kurdachenko, I.Ya. Subbotin, *On some infinite dimensional linear groups.* South. Asian Bulletin of Math. **26** (2002), 773–787

[174] L.A. Kurdachenko, I.Ya. Subbotin, *On some types of Noetherian modules.* J. Algebra and Its Applications **3** (2004), 169–179.

[175] L.A. Kurdachenko, A.V. Tushev, D.I. Zaitsev, *Noetherian modules over nilpotent groups of finite rank.* Arch. Math. **56** (1991), 433–436.

[176] A.G. Kurosh, *The theory of groups.* NAUKA Moskow, 1967.

[177] A.G. Kurosh, *Lectures on general algebra.* NAUKA Moskow, 1973.

[178] T.Y. Lam, *A first course in noncommutative rings.* Springer, 1991.

[179] S. Lang, Algebra. Addison–Wesley, 1965.

[180] M.D. Larsen, P.J. McCarthy, *Multiplicative theory of ideals*. Academic Press, 1971.

[181] A.I. Maltsev, *On groups of finite rank*. Mat. Sbornik **22** (1948), 351–352.

[182] A.I. Maltsev, *On certain classes of infinite soluble groups*. Mat. Sbornik **28** (1951), 567–588. English translation: Amer. Math. Soc. Translations **2** (1956), 1–21.

[183] E. Matlis, *Injective modules over noetherian rings*. Pacific J. Math. **8** (1958), 511–528.

[184] E. Matlis, *Injective modules over Prüfer rings*. Nagoya Math. J. **15** (1959), 57–69.

[185] E. Matlis, *Divisible modules*. Proc. Amer. Math. Soc. **11**(1960), 385–391.

[186] E. Matlis, *Cotorsion modules*. Mem. Amer. Math. Soc. **49** (1964).

[187] E. Matlis, *Torsion-free modules*. The University Chicago Press, 1972.

[188] E. Matlis, *Generalizations of divisible abelian groups to the theory of modules*. Symp. Mat.Ist. Naz. Alta Mat. **23** (1979), 241–250

[189] D.J. McCaughan, *Subnormality in soluble minimax groups*. J. Austral. Math. Soc.(series A) **17** (1974), 113–128.

[190] J.C. McConnel, J.C. Robson, *Noncommutative noetherian rings*. John Wiley, 1987.

[191] D. McDougall, *Soluble minimax groups with the subnormal intersection property*. Math. Z. **114** (1970), 241–244.

[192] D. McDougall, *Soluble groups with the minimum condition for normal subgroups*. Math. Z. **118** (1970), 157–167.

[193] G.O. Michler, O.E. Villamayor, *On rings whose simple modules are infective*. J. Algebra **25** (1973), 185–201.

[194] I. Musson, *Injective modules for group algebras of locally finite groups*. Math. Proc. Cambridge Phil. Soc. **84** (1978), 247–262.

[195] I. Musson, *Injective modules for group rings of polycyclic groups*. Quart. J. Math. **31** (1980), 429–466.

[196] I. Musson, *Irreducible modules for polycyclic group algebras*. Canad. J.Math. **33** (1981), 901–914.

[197] I. Musson, *Some examples of modules over noetherian rings*. Glasgow Math. J. **23** (1982), 9–13.

[198] I. Musson, *Representations of infinite soluble groups.* Glasgow Math. J. **24** (1983), 43–52.

[199] I. Musson, *On the structure of certain injective modules over group algebras of soluble groups of finite rank.* J. Algebra **85** (1983), 51–75.

[200] I. Musson, *Injective modules, localization and completion in group algebras.* J. pure applied Algebra **57** (1989), 265–279.

[201] W. Narkiewicz, *Elementary and analithic theory of algebraic numbers.* Springer, 1989.

[202] M.L. Newell, *On soluble Min-by-Max groups.* Math. Annalen **186** (1970), 282–286.

[203] M.L. Newell, *A subgroup characterization of soluble Min-by-Max groups.* Arch. Math. **21** (1970), 128–131.

[204] M.L. Newell, *Supplement in abelian-by-nilpotent groups.* J. London Math. Soc. **11** (1975), 74–80.

[205] M.L. Newell, *Some splitting theorem for infinite supersoluble groups.* Math. Z. **144** (1975), 265–275.

[206] D.G. Northcott, *Ideal theory.* Cambridge Univ. Press, 1960.

[207] D.G. Northcott, *Lessons on rings, modules and multiplicities.* Cambridge Univ. Press, 1968.

[208] A.Yu. Ol'shanskii, *Geometry of defining relations in groups.* Kluwer, 1991.

[209] J. Otal, J.M. Peña, *Minimal non-CC-groups.* Comm. Algebra **16** (1988), 1231–1242.

[210] J. Otal, J.M. Peña, *Groups in which every proper subgroups is Chernikov-by-nilpotent or nilpotent-by-Chernikov.* Arch. Math. **51** (1988), 193–197.

[211] J. Otal, J.M. Peña, *Groups with minimal condition related to finiteness properties of conjugate classes.* Rend. Semin. Mat. Univ.Padova **81** (1989), 79–84.

[212] J. Otal, J.M. Peña, *Characterizations of the conjugacy of Sylow p-subgroups of CC-groups.* Proc. Amer. Math. Soc. **106** (1989), 605–610.

[213] J. Otal, J.M. Peña, Sylow theory of *CC*-groups: a survey. London Math. Soc. Lecture Notes Ser. **160** (1991), 400–407.

[214] J. Otal, J.M. Peña, M.J. Tomkinson, *Locally inner automorphisms of CC-groups.* J. Algebra **141** (1991), 382–398.

[215] D.S. Passman, *Group rings satisfying a polynomial identity II.* Pacific J. Math. **39** (1971), 425–438.

[216] D.S. Passman, *Infinite group rings.* Marcel Dekker, 1971.

[217] D.S. Passman, *The algebraic structure of group rings.* John Wiley, 1977.

[218] D.S. Passman, *Group rings of polycyclic groups.* "Group theory: essays for Philip Hall", Academic Press, 1984, 207–256.

[219] D.S. Passman, *A course in ring theory.* Wadsworth & Brooks, 1991.

[220] R.S. Pierce, *Associative Algebras.* Springer, 1982.

[221] B.I. Plotkin, *Automorphism groups of the algebraic systems.* Nauka Moskow, 1966.

[222] C. Polcino-Milies, *Group rings whose units form an FC-groups.* Arch. Math. **30** (1978), 380–384; corrigendum: Arch. Math. **31** (1978), 528.

[223] C. Polcino-Milies, *Units in group rings: A short survey.* London Math. Soc. Lecture Notes Ser. **71** (1982), 281–297.

[224] C. Polcino-Milies, S.K. Sehgal, *FC-elements in a group ring.* Comm. Algebra **9** (1981), 1285–1293.

[225] Ya.D. Polovicky, *Groups with extremal classes of conjugate elements.* Sibir. Math. J. **5** (1964), 891–895.

[226] R. Remak, *Über minimale invariante Untergruppen in der Theorie der endlichen Gruppen.* J. reine angew. Math. **192** (1930), 1–16.

[227] D.J.S. Robinson, *Groups in which normality is a transitive relation.* Proc. Cambridge Philos. Soc. **60** (1964), 21–38.

[228] D.J.S. Robinson, *On finitely generated soluble groups.* Proc. London Math. Soc. **18** (1965), 508–516.

[229] D.J.S. Robinson, *On soluble minimax groups.* Math. Z. **101** (1967), 13–40.

[230] D.J.S. Robinson, *Infinite soluble and nilpotent groups.* Queen Mary College Mathematics Notes, 1968.

[231] D.J.S. Robinson, *Residual properties of some classes of infinite soluble groups.* Proc. London Math. Soc. **18** (1968), 495–520.

[232] D.J.S. Robinson, *A note on groups of finite rank.* Compositio Math. **31** (1969), 240–246.

[233] D.J.S. Robinson, *A theorem of finitely generated hyperabelian groups.* Invent. Math. **10** (1970), 38–43.

[234] D.J.S. Robinson, *Finiteness conditions and generalized soluble groups, part 1.* Springer, 1972.

[235] D.J.S. Robinson, *Finiteness conditions and generalized soluble groups, part 2.* Springer, 1972.

[236] D.J.S. Robinson, *Hypercentral ideals, noetherian modules and theorem of Stroud.* J. Algebra **32** (1974), 234–239.

[237] D.J.S. Robinson, *Splitting theorems for infinite groups.* Sympos. Mat. Ist. Naz. alta Mat. **17** (1975), 441–470.

[238] D.J.S. Robinson, *On the cohomology of soluble groups of finite rank.* J. pure applied Algebra **6** (1975), 155–164.

[239] D.J.S. Robinson, *A new treatment of soluble groups with finiteness conditions on their abelian subgroups.* Bull. London Math. Soc. **8** (1976), 113–129.

[240] D.J.S. Robinson, *The vanishing of certain homology and cohomology group.* J. pure applied Algebra **7** (1976), 145–167.

[241] D.J.S. Robinson, *On the homology of hypercentral groups.* Arch. Math. **32** (1979), 223–226.

[242] D.J.S. Robinson, *A course in the theory of groups.* Springer, 1982.

[243] D.J.S. Robinson, *Applications of cohomology to the theory of groups.* London Math. Soc. Lecture Notes Ser. **71** (1982), 46–80.

[244] D.J.S. Robinson, *Finiteness, solubility and nilpotence.* "Group theory: essays for Philip Hall", Academic Press, 1984, 159–206.

[245] D.J.S. Robinson, *Soluble products of nilpotent groups.* J. Algebra **98** (1986), 183–196.

[246] D.J.S. Robinson, *Cohomology of locally nilpotent groups.* J. pure applied Algebra **48** (1987), 281–300.

[247] D.J.S. Robinson, *Homology and cohomology of locally supersoluble groups.* Math. Proc. Cambridge Philos. Soc. **102** (1987), 233–250.

[248] J.E. Roseblade, *Group rings of polycyclic groups.* J. pure applied Algebra **3** (1973), 307–328.

[249] J.E. Roseblade, *Polycyclic group rings and the Nullstellesatz.* Lecture Notes Math. **319** (1973), 156–167.

[250] J.E. Roseblade, *Application of the Artin–Rees lemma to group rings.* Sympos. Mat. **17** (1976), 471–478.

[251] J.E. Roseblade, *Prime ideals in group rings of polycyclic groups.* Proc. London Math. Soc. **36** (1978), 385–447; Corrigendum: Proc. London Math. Soc. **39** (1979), 216–218.

[252] J.E. Roseblade, *Five lectures on group rings.* London Math. Soc. Lecture Notes Ser. **121** (1985), 93–109.

[253] E. Schenkman, *The splitting of certain solvable groups.* Proc. Amer. Math. Soc. **6** (1935), 286–290.

[254] O.F.G. Schilling, *The theory of valuations.* Amer Math. Soc. Math. Surveys **Vol. 4**, 1950.

[255] D. Segal, *Polycyclic groups.* Cambridge Univ. Press, 1983.

[256] D. Segal, *On the group rings of abelian minimax groups.* J. Algebra **237** (2001), 64–94.

[257] D. Segal, *On modules of finite upper rank.* Trans. Amer. Math. Soc. **353** (2001), 391–410.

[258] S.K. Sehgal, *Topics in group rings.* Marcel Dekker, 1978.

[259] S.K. Sehgal, *Units in integral group rings; a survey.* Proc. First. Symposium in Algebra and Number Theory, Worls Sci. Publ., 1990, 255–268.

[260] S.K. Sehgal, H.J. Zassenhaus, *Group rings whose units form an FC-group.* Math. Z. **153** (1977), 29–35.

[261] S.K. Sehgal, H.J. Zassenhaus, *On the supercenter of a group and its ring theoretic generalizations.* Lecture Notes. Math. **882** (1981), 117–144.

[262] H.H.G. Seneviratne, *On permutable subgroups of soluble minimax groups.* Arch. Math. **44** (1985), 481–484.

[263] R. Sharp, *Steps in commutative algebra.* Cambridge Univ. Press, 1990.

[264] R. Sharp, *Artinian modules over commutative rings.* Math. Proc. Cambridge Phil. Soc. **111** (1992), 25–33.

[265] D.W. Sharpe, P. Vamos, *Injective modules.* Cambridge Univ. Press, 1972.

[266] H. Smith, *Groups with few non-nilpotent subgroups.* Glasgow Math.J. **39** (1997), 141–151.

[267] R.L. Snider, *Group algebras whose simple modules are finite dimensional over their commutative rings.* Comm. Algebra **2** (1974), 15–25.

[268] R.L. Snider, *Primitive ideal in group rings of polycyclic groups.* Proc. Amer. Math. Soc. **51** (1976), 8–10.

[269] R.L. Snider, *Injective hulls of simple modules over group rings.* "Ring theory", Proc. Ohio conf., 1976, 223–226.

[270] R.L. Snider, *Solvable groups whose irreducible modules are finite dimensional.* Comm. Algebra **10** (1982), 1477–1485.

[271] M.J. Tomkinson, *Splitting theorems in abelian-by-hypercentral groups.* J. Austral. Math.Soc.(series A) **25** (1978), 71–91.

[272] M.J. Tomkinson, *FC-groups.* Pitman, 1984.

[273] M.J. Tomkinson, *Abelian-by-FC-hypercentral groups*. Ukrain Math. J. **43** (1991), 1038–1042

[274] M.J. Tomkinson, *FC-groups: recent progress*. "Infinite groups 1994" (Ravello 1994), Walter de Gruyter, 1996, 271–285.

[275] M.J. Tomkinson, *A splitting theorem for abelian-by-FC-hypercentral groups*. Glasgow Math. J. **42** (2000), 55–65.

[276] A.V. Tushev, *On the primitivity of group algebras of certain classes of soluble groups of finite rank*. Sbornik Math. **186** (1994), 447–463.

[277] A.V. Tushev, *Spectra of conjugated ideals in group algebras of abelian groups of finite rank and control theorems*. Glasgow Math. J. **38** (1996), 309–320.

[278] O.E. Villamayor, *On weak dimension of algebras*. Pacific J. Math. **9** (1959), 941–951.

[279] J.M. Wedderburn, *On hypercomplex numbers*. Proc. London Math. Soc. **6** (1908), 77–118.

[280] B.A.F. Wehrfritz, *Infinite linear groups*. Springer, 1973.

[281] B.A.F. Wehrfritz, *Groups whose irreducible representations have finite degree*. Math. Proc. Cambridge Phil.Soc. **90** (1981), 411–421.

[282] B.A.F. Wehrfritz, *Groups whose irreducible representations have finite degree*. Proc. Edinburgh Math. Soc. **25** (1982), 237–243.

[283] B.A.F. Wehrfritz, *Groups whose irreducible representations have finite degree*. Math. Proc. Cambridge Phil. Soc. **91** (1982), 397–406.

[284] J.S. Wilson, *Some properties of groups inherited by normal subgroups of finite index*. Math.Z. **114** (1970), 19–21.

[285] J.S. Wilson *Locally soluble groups satisfying the minimal condition for normal subgroups*. Lecture Notes Math. **573** (1977),130–142.

[286] J.S. Wilson, *Soluble products of minimax groups and nearly surjective derivations*. J. pure applied Algebra **53** (1988), 297–318.

[287] J.S. Wilson, *Soluble groups which are products of minimax groups*. Arch. Math. **50** (1988), 19–21.

[288] A.V. Yakovlev, *To the problem of classification of torsion-free abelian groups of finite rank*. Zapiski nauchn. semin. Lomi **57** (1976), 171–175.

[289] D.I. Zaitsev, *Groups satisfying the weak minimal condition*. Ukrain. Math. J. **20** (1968), 472–482.

[290] D.I. Zaitsev, *On groups satisfying the weak minimal condition*. Mat. Sb. **78** (1969), 323–331.

[291] D.I. Zaitsev, *To the theory of minimax groups.* Ukrain. Math. J. **23** (1971), 652–660.

[292] D.I. Zaitsev, *Groups satisfying the weak minimal condition for non-abelian subgroups.* Ukrain. Math. J. **23** (1971), 661–665.

[293] D.I. Zaitsev, *On soluble groups of finite rank.* "Groups with restrictions on subgroups", Naukova Dumka Kiev 1971, 115–130.

[294] D.I. Zaitsev, *On complementations of subgroups in extremal groups.* "Investigations of groups with prescribed properties of subgroups", Math. Inst. Kiev, 1974, 72–130.

[295] D.I. Zaitsev, *On the index of minimality of groups.* "Investigations of groups with prescribed properties of subgroups", Math. Inst. Kiev, 1974, 72–130.

[296] D.I. Zaitsev, *Groups with complemented normal subgroups.* "Some problems in Group Theory", Math. Inst. Kiev, 1975, 30–74.

[297] D.I. Zaitsev, *On existence of direct complements in groups with operators.* "Investigations in Group Theory", Math. Inst. Kiev, 1976, 26–44.

[298] D.I. Zaitsev, *On soluble groups of finite rank.* Algebra i logika **16** (1977), 300–312.

[299] D.I. Zaitsev, *Infinitely irreducible normal subgroups.* "Structure of groups and properties of its subgroups", Math. Inst. Kiev, 1978, 17–38.

[300] D.I. Zaitsev, *On locally soluble groups of finite rank.* Doklady AN SSSR **240** (1978), 257–259.

[301] D.I. Zaitsev, *Hypercyclic extensions of abelian groups.* "Groups defined by the properties of systems of subgroups", Math. Inst. Kiev, 1979, 16–37.

[302] D.I. Zaitsev, *On extensions of abelian groups.* "A constructive description of groups with prescribed properties of subgroups", Math. Inst. Kiev, 1980, 16–40.

[303] D.I. Zaitsev, *Products of abelian groups.* Algebra i logika **19** (1980), 94–106.

[304] D.I. Zaitsev, *On splitting of extensions of abelian groups.* "Investigations of groups with the prescribed properties of systems of subgroups", Math. Inst. Kiev, 1981, 14–25.

[305] D.I. Zaitsev, *The residual nilpotence of metabelian groups.* Algebra i logika **20** (1981), 638–653.

[306] D.I. Zaitsev, *On properties of groups inherited by its normal subgroups.* Ukrain. Math. J. **38** (1986), 707–713.

[307] D.I. Zaitsev, *Splitting extensions of abelian groups.* "The structure of groups and properties of its subgroups", Math. Inst. Kiev, 1986, 22–31.

[308] D.I. Zaitsev, *Hyperfinite extensions of abelian groups.* "Investigations of groups with restrictions on subgroups", Math. Inst. Kiev 1988, 17–26.

[309] D.I. Zaitsev, *On direct decompositions of abelian groups with operators.* Ukrain. Math. J. **40** (1988), 303–309.

[310] D.I. Zaitsev, *On locally supersoluble extensions of abelian groups.* Ukrain. Math. J. **42** (1990), 908–912.

[311] D.I. Zaitsev, M.I. Kargapolov, V.S. Charin, *Infinite groups with prescribed properties of systems of subgroups.* Ukrain. Math. J. **24** (1972), 618–633.

[312] D.I. Zaitsev, L.A. Kurdachenko, A.V. Tushev, *Modules over nilpotent groups of finite rank.* Algebra and logic **24** (1985), 412–436.

[313] D.I. Zaitsev, V.A. Maznichenko, *On the direct decomposition of artinian modules over hypercyclic groups.* Ukrain. Math. J. **43** (1991), 930–934.

Index

$\Pi(G)$, 17
$s(G)$, 19
$t(G)$, 18

algebra
    locally Wedderburn algebra, 95
annihilator
    of a module, 3
    of a subset of a module, 2
    of a subset of a ring, 59
augmentation
    ideal augmentation, 51
    unit augmentation, 51

chain condition
    ascending, 1
    descending, 1
chain conditions
    Max-Σ, 8
    Max-G, 8
    Max-n, 8
    Min-Σ, 8
    Min-G, 8
    Min-n, 8
cocentralizer, 25
conjugacy classes, 25

formation, 26
    infinitely hereditary, 126
    overfinite, 114

group
    CC-group, 26
    FC-group, 25
    Σ-invariant, 7
    Σ-operator group, 7

α^{th}-$\mathfrak{X}C$-hypercenter, 27
$\mathfrak{X}$-group, 25
$\mathfrak{X}C$-group, 27
$\mathfrak{X}C$-hypercenter, 27, 112
$\mathfrak{X}C$-hypercentral, 27
$\mathfrak{X}C$-hyperecccenter, 112
$\mathfrak{X}C$-nilpotent, 27
abelian $\hat{\mathfrak{U}}$-group, 15
abelian $\mathfrak{A}_1$-group, 16
abelian A_0-group, 15
artinian, 8
central direct product, 26
Charin group, 44
Chernikov, 8
complement, 213
factor
    $\mathfrak{X}$-central, 112
    $\mathfrak{X}$-eccentric, 112
generalized radical, 22
hyperfinite radical, 206
just infinite, 9
layer of a group, 9
layers of a p-group, 11
lower layer of a p-group, 11
minimax, 16
noetherian, 8
Prüfer, 8
quasifinite, 9
restricted, 96
soluble $\hat{\mathfrak{U}}$-group, 15
soluble $\mathfrak{S}_1$-group, 16
soluble A_0-group, 15
soluble A_1-group, 14
soluble A_2-group, 16
soluble A_3-group, 16

Tarsky Monster, 9
upper CC-hypercenter, 27
upper FC-hypercenter, 27
upper hypercenter, 27

idempotent
  orthogonal, 84
  primitive, 84

maximal condition, 1
minimal condition, 1
module
  FC-center, 113
  I-module, 60
  RG-center, 113
  $\mathcal{H}$-module, 164
  $\mathcal{M}_c$-module, 89
  $\mathfrak{X}$-RG-hypereccentric, 113
  $\mathfrak{X}C$-RG-hypercentral, 113
  $\mathfrak{X}C$-RG-nilpotent, 113
  $\mathfrak{X}C$-RG-center of a module, 113
  $\mathfrak{X}C$-RG-hypercenters, 113
  $\mathfrak{X}C$-center of a module, 113
  artinian, 2
  assasinator of a module, 60
  complemented submodule, 65
  component of a module, 60
  composition length of a module,
    39
  conjugate, 98
  decomposable, 6
  decompositions
    Z-decomposition, 114
    $\mathfrak{F}$-decomposition, 114
    Baer $\mathfrak{X}$-decomposition, 113
  divisible
    R-divisible, 66
    x-divisible, 66
    divisible part of a module, 70
    envelope divisible, 69
    minimal divisible, 69
  element-module, 156
  essential extension of a submod-
    ule, 67

    maximal, 67
factor
  $\mathfrak{X}$-central, 112
  $\mathfrak{X}$-eccentric, 112
Frattini submodule of a module,
  90
hereditary, 194
indecomposable, 6
injective, 66
injective envelope of a module,
  66
monolith of a module, 6
monolithic, 6
nearly injective, 164
noetherian, 2
periodic, 59
periodic part of a module, 59
Prüfer P-module, 72
primary, 60
primary decomposition, 62
quasifinite, 192
semisimple, 38
series
  $\mathfrak{X}C$-RG-central series, 113
  lower DG-central series, 130
simple, 6
socle of a module, 37
stabilizer, 98
strong locally semisimple, 91
the upper $\mathfrak{X}C$-RG-hypercenter of
  a module, 113
the upper $\mathfrak{X}C$-hypercenter of a
  module, 113
torsion-free, 59
uniserial, 43
  length, 43
upper FC-hypercenter, 113
upper RG-hypercenter, 113

rank
  P-rank of a module, 74
  torsion-free rank, 74
rank of a group
  0-rank, 13

p-rank, 15
p-rank of an abelian group , 14
Hirsch number, 13
index of minimality, 16
minimax rank, 16
Prüfer–Maltsev rank, 15
reduced rank, 15
section rank, 15
special rank, 15
torsion-free rank, 13
total rank, 16
rationally irreducible, 20
ring
  artinian, 2
  Dedekind
   Dedekind *Z*-domain, 156
  Dedekind domain, 54
  element integral over a ring, 54
  hereditary, 194
  ideal
   divisor, 55
   greatest common divisor, 55
   least common multiple, 56
   relatively prime, 56
  ideal fractional, 54
   invertible, 54
   product, 54
  integral closure, 54
  integrally closed, 54
  Jacobson radical of a ring, 84
  noetherian, 2
  prime spectrum of a ring, 54

series, 87
  composition series, 87
  factor, 87
  refinement, 87
  term, 87
  upper $\mathfrak{X}C$-central series, 27
series of a module
  ascending Loewy series, 40
  composition series, 39
  upper socular series, 40
   socular height, 40

splitting, 213
  conjugately splitting, 213

Frontiers in Mathematics

Your Specialized Publisher in Mathematics

Birkhäuser

Available Titles

■ **Bouchut, F.**, CNRS & Ecole Normale Sup., Paris, France

Nonlinear Stability of Finite Volume Methods for Hyperbolic Conservation Laws and Well-Balanced Schemes for Sources

2004. 142 pages. Softcover. ISBN 3-7643-6665-6

■ **Clark, J.**, Otago Univ., New Zealand / **Lomp, C.**, Univ. di Porto, Portugal / **Vanaja, N.**, Mumbai Univ., India / **Wisbauer, R.**, Univ. Düsseldorf, Germany

Lifting Modules

2006. 408 pages. Softcover. ISBN 3-7643-7572-8

■ **De Bruyn, B.**, Ghent University, Ghent, Belgium

Near Polygons

2006. 276 pages. Softcover. ISBN 3-7643-7552-3

■ **Henrot, A.**, Université Henri Poincaré, Vandoeuvre-les-Nancy, France

Extremum Problems for Eigenvalues of Elliptic Operators

2006. 216 pages. Softcover. ISBN 3-7643-7705-4

■ **Kasch, F.**, Universität München, Germany / **Mader, A.**, Hawaii University

Rings, Modules, and the Total

2004. 148 pages. Softcover. ISBN 3-7643-7125-0

■ **Krausshar, R.S.**, Ghent University, Ghent, Belgium

Generalized Analytic Automorphic Forms in Hypercomplex Spaces

2004. 182 pages. Softcover. ISBN 3-7643-7059-9

■ **Kurdachenko, L.**, Dnipropetrovsk National University, Ukraine / **Otal, J.**, University of Zaragoza, Spain / **Subbotion, I.Ya.**, National University, Los Angeles, USA

Artinian Modules over Group Rings

2006. 264 pages. Softcover. ISBN 3-7643-7764-X

■ **Lindner, M.**, University of Reading, UK

Infinite Matrices and their Finite Sections. An Introduction to the Limit Operator Method

2006. 208 pages. Softcover. ISBN 3-7643-7766-6

■ **Perthame, B.**, Ecole Normale Supérieure, Paris

Transport Equations in Biology

2006. 208 pages. Softcover. ISBN 3-7643-7841-7

■ **Thas, K.**, Ghent University, Ghent, Belgium

Symmetry in Finite Generalized Quadrangles

2004. 240 pages. Softcover. ISBN 3-7643-6158-1

■ **Xiao, J.**, Memorial University of Newfoundland, St. John's, Canada

Geometric Q_p Functions

2006. 252 pages. Softcover. ISBN 3-7643-7762-3

■ **Zaharopol, R.**, Math. Reviews, Ann Arbor, USA

Invariant Probabilities of Markov-Feller Operators and Their Supports

2005. 120 pages. Softcover. ISBN 3-7643-7134-X